Antimicrobials and Antimicrobial Resistance in the Environment

Antimicrobials and Antimicrobial Resistance in the Environment

Special Issue Editors

Ashok J. Tamhankar
Cecilia Stålsby Lundborg

MDPI • Basel • Beijing • Wuhan • Barcelona • Belgrade

Special Issue Editors
Ashok J. Tamhankar
Indian Initiative for the Management of
Antibiotic Resistance (IIMAR)
KIIT University
RD Gardi Medical College
India

Cecilia Stålsby Lundborg
Karolinska Institutet
Sweden

Editorial Office
MDPI
St. Alban-Anlage 66
4052 Basel, Switzerland

This is a reprint of articles from the Special Issue published online in the open access journal *International Journal of Environmental Research and Public Health* (ISSN 1660-4601) from 2017 to 2019 (available at: https://www.mdpi.com/journal/ijerph/special_issues/Antimicrobial).

For citation purposes, cite each article independently as indicated on the article page online and as indicated below:

LastName, A.A.; LastName, B.B.; LastName, C.C. Article Title. *Journal Name* **Year**, *Article Number*, Page Range.

ISBN 978-3-03928-030-8 (Hbk)
ISBN 978-3-03928-031-5 (PDF)

Cover image courtesy of Cecilia Stålsby Lundborg.

Contents

About the Special Issue Editors

Ashok J. Tamhankar, Professor and Scientific Advisor, has been working in the field of integrated management of environmental pathogens and agricultural pests for the last 50 years. At various times, he has been associated with eminent academic and research institutes such as Bhabha Atomic Research Centre, India, Cornell University, USA, Karolinska Institutet, Sweden, R. D. Gardi Medical College, and KIIT University, India. He is a founding member and the National Coordinator of the Indian Initiative for Management of Antibiotic Resistance (IIMAR), a voluntary organization devoted to spreading awareness about prudent antibiotic use and the threat of antibiotic resistance. With more than 200 publications to his credit, his current research interests include the detection of antibiotic-resistant pathogens and antibiotic residues in the environment and their disinfection and decontamination, newer approaches to disinfection of pathogens, One health, etc. He has been a WHO panelist in the area of antibiotic residues and resistance in aquatic environments.

Cecilia Stålsby Lundborg, Professor, Research Group Leader 'Global Health—Health Systems and Policy', Department of Global Public Health, Karolinska Institutet, Sweden, was previously the Scientific Advisor for Development Research at the Swedish Research Council and professor at the Nordic School of Public Health. She is a member of the Swedish JPIAMR advisory group and is a WHO consultant regarding antibiotics in the environment. She has been active in Swedish Strama and ReAct – Action on Antibiotic resistance for 15 years. Her research concerns antibiotic use and resistance and prevention of infections and spread of resistance, often with a One health perspective. She is involved in vast international collaborations in Africa, Asia, and Europe, including in India, China, Laos, South Africa, and Uganda. She has published about 200 international peer-reviewed papers, supervised about 30 doctoral and post-doctoral students from a range of European, Asian, and African countries and led international courses on antibiotic stewardship and antimicrobial resistance.

International Journal of
*Environmental Research
and Public Health*

Editorial

Antimicrobials and Antimicrobial Resistance in the Environment and Its Remediation: A Global One Health Perspective

Ashok J. Tamhankar [1,2,3,4,]* and **Cecilia Stålsby Lundborg** [1]

1   Global Health—Health Systems and Policy: Medicines, Focusing Antibiotics, Department of Public Health Sciences, Karolinska Institutet, 171 77 Stockholm, Sweden; Cecilia.Stalsby.Lundborg@ki.se
2   Indian Initiative for Management of Antibiotic Resistance, 302, Aryans, Deonar farm road, Deonar, Mumbai 400088, India
3   Department of Environmental Medicine, R.D. Gardi Medical College, Ujjain 456006, India
4   School of Biotechnology, Kalinga Institute of Industrial Technology, Bhubaneswar 751024, India
*   Correspondence: ashok.tamhankar@ki.se

Received: 30 October 2019; Accepted: 31 October 2019; Published: 20 November 2019

1. Introduction

The awareness about pollution of the environment by antimicrobials/antibiotics is increasing globally. So is the literature, which is predominantly on antibiotic resistant bacteria, antibiotic resistance genes and antibiotic residues in the environment. The main concern about this, is the fear that resistance in the environment will get transferred to the clinical pathogens (for example, through horizontal gene transfer) leading to untreatable infectious diseases. It is estimated that antibiotic resistance may result in deaths of several million per year, if suitable measures are not taken up to mitigate the resistance problem [1]. The World Health Organization and the United Nations General Assembly have therefore called antimicrobial resistance a global threat that needs to be resolved with top priority [2,3].

The resistance generating sources in the environment are mainly human waste, animal waste and manufacturing waste. Both humans and animals (agriculture, poultry, aquaculture etc.), release large amounts of antimicrobials/antibiotics, which are consumed by them for therapeutic and prophylactic use, in the environment through excretions and improper disposal, and also the resistant bacteria in their systems, and make the environment prone to multiplication of resistant bacteria and abundance of resistance genes. An additional issue in this is the inappropriate use of antibiotics by humans for themselves and for their animals, because of lack of awareness regarding appropriate use of antibiotics. Interventions in the form of increasing public awareness and knowledge are the most commonly used strategies for effecting appropriate antimicrobial use and reducing antimicrobial resistance [4]. For example, in a survey in China it was found that the pig farmers' knowledge regarding antibiotic use for their pigs was very poor and it was accompanied with improper behaviour. The survey results further showed that the probability of improper antibiotic use decreased with the increase in farmers' knowledge about appropriate antibiotic use, and about the hazards of antibiotic residues in the environment [5]. The drug manufacturing units also, through their effluents, pollute the environment by antimicrobials. The available treatments/treatment plants for treating wastewater/effluents not being efficient to neutralize these pollutants, there is an abundance of antimicrobials/antibiotics, resistant microbes/bacteria and resistance genes in the environment. The share of literature is higher for antibiotic resistant bacteria compared to antibiotic residues and resistance genes as the detection of the latter two is relatively more expensive and also requires a little higher level of technology. In this article, we will mainly deal with antibiotic resistant bacteria, resistance genes and antibiotic residues in the environment.

2. Non-Aquatic Environment

Studies from several parts of the world have reported the presence of antibiotic residues, antibiotic resistant bacteria, as well as resistance genes in various non-aquatic environmental compartments such as soil, manure, animal meat, plants etc. [6–10]. In the context of resistance, the literature is more abundant for poultry, it being known as an extensive user of antimicrobials for growth promotion/prophylaxis besides the therapeutic use, and there are several reports of antibiotic residues, resistant bacteria and resistance genes being detected in poultry environment [11–13].

3. Aquatic Environment

Most of the antimicrobials/antibiotics used for various purposes and that from manufacturing plant effluents end up in the aquatic systems of the world environment, as well as the resistant bacteria and resistance genes generated by them. Thus, there are reports of their occurrence in hospital wastewater [14–18], rivers [19–21], rainwater-harvesting tanks [22], canal waters [23], recreational waters [24], municipal/community wastewater [25], and pharmaceutical plant effluent [26]. These wastewater discharges further have impact on various water bodies and contaminate them [27].

4. Non-Aquatic and Aquatic Environment Combined

While there are studies which look into only one of the many non-aquatic or aquatic compartments of the environment, there are many studies that cover both these types encompassing a composite environment. For example, water and plants [28], water and sand [24], wastewaters, natural and drinking waters and solid matrices such as sludge, sediment, and soil [8,29].

5. Resistance Built up in Bacteria after Exposure to Antibiotics in Environment

While in vitro studies show a link between antibiotic exposure and antibiotic resistance, experiments are also needed to be done in actual environmental niches to see whether resistance gets built up in the presence of antibiotics in an environmental compartment and whether antibiotic exposure causes any adverse effects on the environmental system. Two such experiments are cited here. In one experiment, in a turkey farm, it was found that resistance to enrofloxacin was detected at a very high frequency after treatments with enrofloxacin via drinking water, a representation of poultry drinking water from natural sources contaminated with antibiotic residues [30]. In another hydroponic experiment, representing plants growing in antibiotic contaminated waters, exposing pakchoi (*Brassica chinensis* L.) to antibiotic contaminated waters, resulted in detection of target antibiotics at concentrations ranging from 6.9 to 48.1 $\mu g \cdot kg^{-1}$ in the vegetable grown in contaminated water, and the rates of antibiotic-resistant endophytic bacteria as well as the resistance genes significantly increased in the plants [31].

6. Environmental Contamination by Antibiotics and One Health

We define here 'One Health' in the context of environment and antimicrobial resistance as, One Health is a study and interpretation of an integrated paradigm of antimicrobials and antimicrobial resistance dynamics and epidemiology, that encompasses human health, biodiversity health and ecosystem health including socio-behavioural aspects, that informs on the processes leading to the occurrence and recurrence of infectious agents and resistance and their dissemination and extinction in organic and inorganic habitats/environments, for the purpose of development of antimicrobial resistance management strategies. Few studies, projects or literature reviews encompassing all these dimensions for an organism or an antimicrobial in a particular niche/geographical area/ecosystem are evident in literature (e.g., [27,29]). Studies mostly occur in separate events and not as a conscious integrated event. In our project in India entitled 'APRIAM-Studies on Antibiotic Use, Antibiotic Resistance and Antibiotic Residues in the Environment of India with a Context of Antibiotic Resistance Management in a One Health Approach', we kept in focus a One Health approach while using varying study dimensions and while creating certain protocols [32,33]. Although the project is still ongoing, a mention of some of its results is worthwhile here to create a context between environmental antibiotic residues, antibiotic resistance, resistance genes and One Health. We found that, in people's and healthcare worker's perception, environment was intimately connected to occurrence of infectious diseases, antibiotic use and resistance development [34,35]; a time-series analysis study also showed that climatic factors influenced occurrence of Methicillin-Resistant *Staphylococcus aureus* (MRSA) skin and soft-tissue infections [36], and further we found that, resistance patterns were shared for *Escherichia coli* from humans, animal (cow) and their associated water when from an inland area, whereas, when located in the proximity of sea, resistance of *E. coli* from humans, animals and water had a shared pattern but it was different from the inland one [37]. We also found that in a niche area in a village, there was not only commonality of a resistance pattern of *E. coli* in humans, animals and the water in their environment but the commonality also extended to resistance genes [38]. In further exploration, we found that antibiotic residues, antibiotic resistance and resistance genes in water and sediments of a nearby river share some commonality [21]. As socio-behavioural and anthropogenic aspects also have an impact on the generation of resistance in the environment [39]. We also conducted studies on the same river about the impact of a special anthropogenic activity particular to India, holy dip and congregative holy dip of millions of persons in a holy river (Kumbh Mela) on antibiotic residues, antibiotic resistance and resistance genes (to be published). When our studies are complete, all these will be mapped from a One Health perspective.

7. Current Wastewater Treatment Failure

Wastewater is produced daily from various sectors and segments of society. Worldwide, 113 countries have data available on wastewater production, 103 countries on wastewater treatment, and 62 countries on wastewater use [40] Even after treatment, antibiotic residues, antibiotic resistant bacteria and resistance genes are still present in the wastewater, and the wastewater treatment plants (WWTP) are considered 'hot spots' of resistance multiplication [41] The wastewater from households, animal rearing facilities and WWTP effluents mostly get released into nearby waterways, wherefrom it might be used for irrigation purposes and studies have shown that some antibiotics have very long half-lives in agricultural soils: 55 to 578 days for tetracycline and 120 to 2310 days for ciprofloxacin [42–44]. Conventional wastewater treatment facilities typically have biological degradation, for example using the activated sludge process, whereas advanced facilities have tertiary treatment processes, such as reverse osmosis, ozonation, sonolysis and advanced oxidation technologies like fenton oxidation, heterogenous-photocatalysis with TiO2 etc. These treatments do not necessarily fully remove antibiotic residues, antibiotic resistant bacteria and resistance genes from the wastewater For example, there are reports that antibiotic residues, antibiotic resistant bacteria and resistance genes still remain even after the conventional treatment [16,45]. Additionally, even after the advanced treatment processes currently in use, the problem is not fully resolved, for example, a study showed that even after ozonation treatment about 20% of sulfonamides, trimethoprim and macrolides still remained in the effluent [46].

8. Complete Remediation of Environmental Antibiotic Residues, Resistant Bacteria and Resistance Genes

Considering these issues and also that the normal photocatalysts used for disinfection are expensive materials like silver (Ag), titanium (Ti) etc., there was a need to develop photocatalysis based on inexpensive resources. Our research group has developed a technique using cheap resources like iron (Fe) or kaolinite nanoparticles and sunlight or visible light, that results in complete disinfection of multi-drug resistant pathogenic enteric bacteria and *Salmonella* from natural waters such as from ponds, rivers, lakes, tap water etc. [47–49]. The same technique using the cheap resource of Fe and sunlight is also successful in 100% decontamination of environmentally highly stable antibiotics like ciprofloxacin from natural waters [50]. This technique using sunlight is also useful with the conventional expensive photocatalysts [51]. Further, for this technique, we have been able to show that the genetic resistance material gets completely degraded by this technique and in the process, we have also developed an insight into how the resistance gets broken down [49,52].

9. Conclusions

There is a need for regulations to be established and implemented in many areas related to antimicrobials in the environment. The areas to focus are the pharmaceutical industry, hospitals, wastewater treatment plants, aquaculture farms, poultry farms, pig farms, and households. Other key areas to focus are strengthening and persevering awareness and education, antimicrobial stewardship strategies inclusive of environmental risk sensitization and management, pharmaceutical take-back programs, designing greener antimicrobials with better degradability in the environment, implementing environmental risk assessment prior to the launch of new drugs, monitoring release of antimicrobials into the environment, and eco-pharmacovigilance. The risk of using sewage/wastewater for irrigation needs to be carefully evaluated. Toxicological effects of antimicrobial use on non-target organisms and the environment should be addressed and informed to practitioners. There is a need to use less costly methods for antimicrobial residue measurements. Additionally, there should be methods of monitoring progress of correctives.

The whole gamut of antimicrobial/antibiotic use, antimicrobial/antibiotic residues, antimicrobial/antibiotic resistance, resistance genes, and horizontal gene transfer is interconnected, one leading to another and finally resulting in increased antimicrobial/antibiotic use, which further leads to the same consequences. Therefore, there is a need to develop and implement instruments to carefully monitor antimicrobial/antibiotic use in community, animals, and hospitals, as well as residues, resistant microbes/bacteria and resistance genes in all compartments of the environment, and to update this information on a continuous basis. The crisis of antimicrobial/antibiotic resistance is reaching unmanageable proportions and if immediate measures are not taken to resolve the problem, simple infections may become life threatening.

Funding: This research received no external funding.

Conflicts of Interest: The authors declare no conflicts of interest.

References

1. O'Neill, J. Antimicrobial Resistance: Tackling A Crisis for the Health and Wealth of Nations. Available online: https://amr-review.org/sites/default/files/AMR%20Review%20Paper%20-%20Tackling%20a%20crisis%20for%20the%20health%20and%20wealth%20of%20nations_1.pdf (accessed on 9 November 2019).
2. Global Action Plan on Antimicrobial Resistance. I. World Health Organization. ISBN 978 92 4 150976 3. Available online: http://www.wpro.who.int/entity/drug_resistance/resources/global_action_plan_eng.pdf (accessed on 25 October 2019).
3. Draft Political Declaration of the High-Level Meeting of the General Assembly on Antimicrobial Resistance. Available online: https://www.un.org/pga/71/wp-content/uploads/sites/40/2016/09/DGACM_GAEAD_ESCAB-AMR-Draft-Political-Declaration-1616108E.pdf (accessed on 25 October 2019).
4. Van Katwyk, S.R.; Grimshaw, J.M.; Nkangu, M.; Nagi, R.; Mendelson, M.; Taljaard, M.; Hoffman, S.J. Government policy interventions to reduce human antimicrobial use: A systematic review and evidence map. *PLoS Med.* **2019**, *16*, e1002819. [CrossRef] [PubMed]
5. Chen, X.; Wu, L.; Xie, X. Assessing the Linkages between Knowledge and Use of Veterinary Antibiotics by Pig Farmers in Rural China. *Int. J. Environ. Res. Public Health* **2018**, *15*, 1126. [CrossRef] [PubMed]
6. Pikkemaat, M.G.; Yassin, H.; van der Fels-Klerx, H.J.; Berendsen, B.J.A. *Antibiotic Residues and Resistance in the Environment*; RIKILT Report; RIKILT Wageningen UR (University & Research centre): Wageningen, The Netherlands, 2016; Volume 9, p. 32.

7. Oloso, N.O.; Fagbo, S.; Garbati, M.; Olonitol, S.O.; Awosany, E.J.; Aworh, M.K.; Adamu, H.; Odetokun, A.I.; Fasina, F.O. Antimicrobial Resistance in Food Animals and the Environment in Nigeria: A Review. *Int. J. Environ. Res. Public Health* **2018**, *15*, 1284. [CrossRef] [PubMed]

8. Hanna, N.; Pan, S.; Sun, Q.; Li, X.W.; Yang, X.; Xiang, J.; Zou, H.; Ottoson, J.; Nilsson, L.E.; Berglund, B.; et al. Presence of antibiotic residues in various environmental compartments of Shandong province in eastern China: Its potential for resistance development and ecological and human risk. *Environ. Int.* **2018**, *114*, 131–142. [CrossRef]

9. Pérez-Etayo, L.; Berzosa, M.; González, D.; Vitas, A.I. Prevalence of Integrons and Insertion Sequences in ESBL-Producing E. coli Isolated from Different Sources in Navarra, Spain. *Int. J. Environ. Res. Public Health* **2018**, *15*, 2308. [CrossRef]

10. Berendonk, T.U.; Manaia, C.M.; Merlin, C.; Fatta-Kassinos, D.; Cytryn, E.; Walsh, F.; Bürgmann, H.; Sørum, H.; Norström, M.; Pons, M.N.; et al. Tackling antibiotic resistance: The environmental framework. *Nat. Rev. Microbiol.* **2015**, *13*, 310–317. [CrossRef]

11. Muaz, K.; Riaz, M.; Akhtar, S.; Park, S.; Ismail, A. Antibiotic Residues in Chicken Meat: Global Prevalence, Threats, and Decontamination Strategies: A Review. *J. Food Prot.* **2018**, *81*, 619–627. [CrossRef]

12. Lee, J.; Jeong, J.; Lee, H.; Ha, J.; Kim, J.; Choi, Y.; Oh, H.; Seo, K.; Yoon, Y.; Lee, S. Antibiotic Susceptibility, Genetic Diversity, and the Presence of Toxin Producing Genes in Campylobacter Isolates from Poultry. *Int. J. Environ. Res. Public Health* **2017**, *14*, 1400. [CrossRef]

13. Odoch, T.; Sekse, C.; L'abee-Lund, T.; Hansen, H.C.H.; Kankya, C.; Wasteson, Y. Diversity and Antimicrobial Resistance Genotypes in Non-Typhoidal Salmonella Isolates from Poultry Farms in Uganda. *Int. J. Environ. Res. Public Health* **2018**, *15*, 324. [CrossRef]

14. Diwan, V.; Tamhankar, A.J.; Khandal, R.K.; Sen, S.; Aggarwal, M.; Marothi, Y.; Iyer, R.V.; Sundblad-Tonderski, K.; Stålsby-Lundborg, C. Antibiotics and antibiotic-resistant bacteria in waters associated with a hospital in Ujjain, India. *BMC Public Health* **2010**, *3*, 414. [CrossRef]

15. Diwan, V.; Stålsby Lundborg, C.; Tamhankar, A.J. Seasonal and Temporal Variation in Release of Antibiotics in Hospital Wastewater: Estimation Using Continuous and Grab Sampling. *PLoS ONE* **2013**, *8*, e68715. [CrossRef] [PubMed]

16. Lien, T.Q.; Hoa, N.Q.; Chuc, N.T.; Thoa, N.T.; Phuc, H.D.; Diwan, V.; Dat, N.T.; Tamhankar, A.J.; Stålsby Lundborg, C. Antibiotics in Wastewater of a Rural and an Urban Hospital before and after Wastewater Treatment, and the Relationship with Antibiotic Use-A One Year Study from Vietnam. *Int. J. Environ. Res. Public Health.* **2016**, *13*, 588. [CrossRef] [PubMed]

17. Lien, L.T.Q.; Lan, P.T.; Chuc, N.T.K.; Hoa, N.Q.; Nhung, P.H.; Thoa, N.T.M.; Diwan, V.; Tamhankar, A.J.; Stålsby Lundborg, C. Antibiotic resistance and antibiotic resistance genes in Escherichia coli isolates from hospital wastewater in Vietnam. *Int. J. Environ. Res. Public Health* **2017**, *14*, 699. [CrossRef] [PubMed]

18. Martins, A.F.; Vasconcelos, T.G.; Henriques, D.M.; da S. Frank, C.; König, A.; Kümmerer, K. Concentration of Ciprofloxacin in Brazilian Hospital Effluent and Preliminary Risk Assessment: A Case Study. *Clean Soil Air Water* **2008**, *36*, 264–269. [CrossRef]

19. Hladicz, A.; Kittinger, C.; Zarfel, G. Tigecycline Resistant Klebsiella pneumoniae Isolated from Austrian River Water. *Int. J. Environ. Res. Public Health* **2017**, *14*, 1169. [CrossRef] [PubMed]

20. Kittinger, C.; Kirschner, A.; Lipp, M.; Baumert, R.; Mascher, M.; Farnleitner, A.H.; Zarfel, G. Antibiotic Resistance of Acinetobacter spp. Isolates from the River Danube: Susceptibility Stays High. *Int. J. Environ. Res. Public Health* **2018**, *15*, 52. [CrossRef] [PubMed]

21. Diwan, V.; Hanna, N.; Purohit, M.; Chandran, S.; Riggi, E.; Parashar, V.; Tamhankar, A.J.; Stålsby Lundborg, C. Seasonal Variations in Water-Quality, Antibiotic Residues, Resistant Bacteria and Antibiotic Resistance Genes of Escherichia coli Isolates from Water and Sediments of the Kshipra River in Central India. *Int.J. Environ. Res. Public Health* **2018**, *15*, 1281. [CrossRef]

22. Malema, M.S.; Abia, A.L.K.; Tandlich, R.; Zuma, B.; Kahinda, J.M.M. Ubomba-Jaswa E Antibiotic-Resistant Pathogenic Escherichia Coli Isolated from Rooftop Rainwater-Harvesting Tanks in the Eastern Cape, South Africa. *Int. J. Environ. Res. Public Health* **2018**, *15*, 892. [CrossRef]

23. Ebomah, K.E.; Adefisoye, M.A.; Okoh, A.I. Pathogenic Escherichia coli Strains Recovered from Selected Aquatic Resources in the Eastern Cape, South Africa, and Its Significance to Public Health. *Int. J. Environ. Res. Public Health* **2018**, *15*, 1506. [CrossRef]

24. Akanbi, O.E.; Njom, H.A.; Fri, J.; Otigbu, A.C.; Clarke, A.M. Antimicrobial Susceptibility of Staphylococcus aureus Isolated from Recreational Waters and Beach Sand in Eastern Cape Province of South Africa. *Int. J. Environ. Res. Public Health* **2017**, *14*, 1001. [CrossRef]

25. Seifrtová, M.; Pena, A.; Lino, C.M.; Solich, P. Determination of fluoroquinolone antibiotics in hospital and municipal or wastewaters in Coimbra by liquid chromatography with a monolithic column and fluorescence detection. *Anal. Bioanal. Chem.* **2008**, *391*, 799–805. [CrossRef]

26. Larsson, D.G.J.; de Pedro, C.; Paxeus, N. Effluent from drug manufactures contains extremely high levels of pharmaceuticals. *J. Hazard. Mater.* **2007**, *148*, 751–755. [CrossRef] [PubMed]

27. Stålsby Lundborg, C.; Tamhankar, A.J. Antibiotic residues in the environment of South East Asia. *BMJ* **2017**, *358*, 2440. [CrossRef] [PubMed]

28. Hashemi, M.M.; Mmuoegbulam, A.O.; Holden, B.S.; Coburn, J.; Wilson, J.; Taylor, M.F.; Reiley, J.; Baradaran, D.; Stenquist, T.; Deng, S.; et al. Susceptibility of Multidrug-Resistant Bacteria, Isolated from Water and Plants in Nigeria, to Ceragenins. *Int. J. Environ. Res. Public Health* **2018**, *15*, 2758. [CrossRef] [PubMed]

29. Wang, J.; He, B.; Hu, X. Human-use antibacterial residues in the natural environment of China: Implication for ecopharmacovigilance. *Environ. Monit. Assess.* **2015**, *187*, 331. [CrossRef] [PubMed]

30. Chuppava, B.; Keller, B.; El-Wahab, A.A.; Meißner, J.; Kietzmann, M.; Visscher, C. Resistance of *Escherichia coli* in Turkeys after Therapeutic or Environmental Exposition with Enrofloxacin Depending on Flooring. *Int. J. Environ. Res. Public Health* **2018**, *15*, 1993. [CrossRef] [PubMed]

31. Zhang, H.; Li, X.; Yang, Q.; Sun, L.; Yang, X.; Zhou, M.; Deng, R.; Bi, L. Plant Growth, Antibiotic Uptake, and Prevalence of Antibiotic Resistance in an Endophytic System of Pakchoi under Antibiotic Exposure. *Int. J. Environ. Res. Public Health* **2017**, *14*, 1336. [CrossRef]

32. Stålsby Lundborg, C.; Diwan, V.; Pathak, A.; Purohit, M.R.; Shah, H.; Sharma, M.; Mahadik, V.K.; Tamhankar, A.J. Protocol: A 'One health' two year follow-up, mixed methods study on antibiotic resistance, focusing children under 5 and their environment in rural India. *BMC Public Health.* **2015**, *15*, 1321. [CrossRef]

33. Diwan, V.; Purohit, M.; Chandran, S.; Parashar, V.; Shah, H.; Mahadik, V.K.; Lundborg, C.S.; Tamhankar, A.J. A three-year follow-up study of antibiotic and metal residues, antibiotic resistance and resistance genes, focusing on Kshipra-a river associated with holy religious mass-bathing in India: Protocol paper. *Int. J. Environ. Res. Public Health* **2017**, *14*, 574. [CrossRef]

34. Sahoo, K.C.; Tamhankar, A.J.; Johansson, E.; Stålsby Lundborg, C. Community perceptions of infectious diseases, antibiotic use and antibiotic resistance in context of environmental changes: A study in Orissa, India. *Health Expect.* **2014**, *5*, 651–663. [CrossRef]

35. Sahoo, K.C.; Tamhankar, A.J.; Johansson, E.; Stalsby Lundborg, C. Antibiotic use, resistance development and environmental factors: A qualitative study among healthcare professionals in Orissa, India. *BMC Public Health* **2010**, *10*, 629. [CrossRef] [PubMed]

36. Sahoo, K.C.; Sahoo, S.; Marrone, G.; Pathak, A.; StålsbyLundborg, C.; Tamhankar, A.J. Climatic Factors and Community—Associated Methicillin-Resistant Staphylococcus aureus Skin and Soft-Tissue Infections—A Time-Series Analysis Study. *Int. J. Environ. Res. Public Health.* **2014**, *29*, 8996–9007. [CrossRef] [PubMed]

37. Sahoo, K.C.; Tamhankar, A.J.; Sahoo, S.; Sahu, P.S.; Klintz, S.R.; Lundborg, C.S. Geographical Variation in Antibiotic-Resistant *Escherichia coli* Isolates from Stool, Cow-Dung and Drinking Water. *Int. J. Environ. Res. Public Health* **2012**, *9*, 746–759. [CrossRef] [PubMed]

38. Purohit, M.R.; Chandran, S.; Shah, H.; Diwan, V.; Tamhankar, A.J.; Stålsby Lundborg, C. Antibiotic resistance in an Indian rural community: A 'One-health' observational study on commensal coliform from humans, animals and water. *Int. J. Environ. Res. Public Health* **2017**, *14*, 386. [CrossRef] [PubMed]

39. Stålsby Lundborg, C.; Tamhankar, A.J. Understanding and changing human behavior—Antibiotic mainstreaming as an approach to facilitate modification of provider and consumer behavior. *Upps. J. Med. Sci.* **2014**, *119*, 125–133. [CrossRef]

40. Sato, T.; Qadir, M.; Yamamoto, S.; Endo, T.; Zahoor, A. Global, regional, and country level need for data on wastewater generation, treatment, and use. *Agric. Water Manag.* **2013**, *130*, 1–13. [CrossRef]

41. Michael, I.; Rizzo, L.; McArdell, C.S.; Manaia, C.M.; Merlin, C.; Schwartz, T.; Dagot, C.; Fatta-Kassinos, D. Urban wastewater treatment plants as hotspots for the release of antibiotics in the environment: A review. *Water Res.* **2013**, *47*, 957–995. [CrossRef]

42. Bound, J.P.; Kitsou, K.; Voulvoulis, N. Household disposal of pharmaceuticals and perception of risk to the environment. *Environ. Toxicol. Pharmacol.* **2006**, *3*, 301–307. [CrossRef]

43. Campagnolo, E.R.; Johnson, K.R.; Karpati, A.; Rubin, C.S.; Kolpin, D.W.; Meyer, M.T.; Esteban, J.E.; Currier, R.W.; Smith, K.; Thu, K.M.; et al. Antimicrobial residues in animal waste and water resources proximal to large-scale swine and poultry feeding operations. *Sci. Total Environ.* **2002**, *299*, 89–95. [CrossRef]

44. Walters, E.; McClellan, K.; Halden, R.U. Occurrence and loss over three years of 72 pharmaceuticals and personal care products from biosolids-soil mixtures in outdoor mesocosms. *Water Res.* **2010**, *44*, 6011–6020. [CrossRef]

45. Galler, H.; Feierl, G.; Petternel, C.; Reinthaler, F.; Haas, D.; Habib, J.; Kittinger, C.; Luxner, J.; Zarfel, G. Multiresistant Bacteria Isolated from Activated Sludge in Austria. *Int. J. Environ. Res. Public Health* **2018**, *15*, 479. [CrossRef] [PubMed]

46. Hollender, J.; Zimmermann, S.G.; Koepke, S.; Krauss, M.; McArdell, C.S.; Ort, C.; Singer, H.; von Gunten, U.; Siegrist, H. Elimination of organic micropollutants in a municipal wastewater treatment plant upgraded with a full-scale postozonation followed by sand filtration. *Environ. Sci. Technol.* **2009**, *43*, 7862–7869. [CrossRef] [PubMed]

47. Das, S.; Sinha, S.; Das, B.; Jayabalan, R.; Suar, M.; Mishra, A.; Tamhankar, A.J.; Stålsby Lundborg, C.; Tripathy, S.K. Disinfection of Multidrug Resistant Escherichia coli by Solar-Photocatalysis using Fe-doped ZnO Nanoparticles. *Sci. Rep.* **2017**, *7*, 104. [CrossRef] [PubMed]

48. Misra, A.J.; Das, S.; Rahman, A.H.; Das, B.; Jayabalan, R.; Behera, S.K.; Suar, M.; Tamhankar, A.J.; Mishra, A.; Lundborg, C.S.; et al. Doped ZnO nanoparticles impregnated on Kaolinite (Clay): A reusable nanocomposite for photocatalytic disinfection of multidrug resistant *Enterobacter* sp. under visible light. *J. Colloid Interface Sci.* **2018**, *530*, 610–623. [CrossRef] [PubMed]

49. Habeeb Rahman, A.P.; Misra, A.J.; Das, S.; Das, B.; Jayabalan, R.; Suar, M.; Mishra, A.; Tamhankar, A.J.; Stålsby Lundborg, C.; Tripathy, S.K. Mechanistic insight into the disinfection of *Salmonella* spp. by sun-light assisted sonophotocatalysis using doped ZnO nanoparticles. *Chem. Eng. J.* **2018**, *336*, 476–488. [CrossRef]

50. Das, S.; Ghosh, S.; Misra, A.J.; Mishra, A.; Tamhankar, A.J.; Stålsby Lundborg, C.; Tripathy, S.K. Sunlight assisted Photocatalytic Degradation of Ciprofloxacin in Water using Fe doped ZnO Nanoparticles for Potential Public Health Applications. *Int. J. Environ. Res. Public Health* **2018**, *15*, 2440. [CrossRef] [PubMed]

51. Das, S.; Ranjana, N.; Misra, A.J.; Suar, M.; Mishra, A.; Tamhankar, A.J.; Stålsby Lundborg, C.; Tripathy, S.K. Dsinfection of the Water Borne Pathogens *Escherichia coli* and *Staphylococcus aureus* by Solar Photocatalysis Using Sonochemically Synthesized Reusable Ag@ZnO Core-Shell Nanoparticles. *Int. J. Environ. Res. Public Health.* **2017**, *14*, 747. [CrossRef]

52. Das, S.; Habeeb Rahman, A.P.; Misra, A.J.; Das, B.; Jayabalan, R.; Suar, M.; Tamhankar, A.J.; Mishra, A.; Stålsby Lundborg, C.; Tripathy, S.K. Ag@SnO$_2$@ZnO core-shell nanocomposites assisted solar-photocatalysis downregulates multidrug resistance in *Bacillus* sp: A catalytic approach to impede antibiotic resistance. *Appl. Catal. B Environ.* **2019**, *259*, 118065. [CrossRef]

International Journal of
Environmental Research and Public Health

Article

Assessing the Linkages between Knowledge and Use of Veterinary Antibiotics by Pig Farmers in Rural China

Xiujuan Chen [ID], Linhai Wu * and Xuyan Xie

Institute for Food Safety Risk Management, School of Business, Jiangnan University, Wuxi 214122, China; xjchen@jiangnan.edu.cn (X.C.); xiexuyan1@yahoo.com (X.X.)
* Correspondence: wlh6799@jiangnan.edu.cn; Tel.: +86-135-0617-9899

Received: 29 March 2018; Accepted: 29 May 2018; Published: 31 May 2018

Abstract: Improper use of veterinary antibiotics (VAs) has led to antibiotic resistance and food safety issues that are harmful for sustainable development and public health. In this study, farmers' knowledge influencing their usage of veterinary antibiotics was analyzed based on a survey of 654 pig farmers in Funing County, China. A behavior probability model was constructed, and a Matlab simulation was used to evaluate the dynamic changes in farmers' behavioral choice regarding VAs use. The survey results showed that the 654 pig farmers' knowledge of VAs were relatively poor, along with a high occurrence of improper behavior. Specifically, 68.35% of the 654 surveyed pig farmers admitted their violation of VAs use regulations, while 55.50% among them overused and 24.31% among them misused VAs. The simulation results showed that the probability of improper VA use decreased with the increase in farmers' knowledge about VA use specification, and when farmers' knowledge about the hazards of VA residues increased. However, when farmers had a high level of knowledge about relevant laws and their penalties, there was still a high probability of improper VA use.

Keywords: farmer; veterinary antibiotics use; knowledge; behavior probability model; China

1. Introduction

Antibiotics are the most important finding in the 20th century for controlling bacterial infections and protecting health [1]. In addition to human treatment, antibiotics have been widely used in agriculture, the food industry, and aquaculture [2]. Veterinary antibiotics (VAs) are widely used in the treatment and prophylaxis of diseases in food-producing animals and in non-therapeutic applications [3,4]. However, the misuse or overuse of VAs is the culprit for increasing antibiotic resistance and food chain contamination [5,6]. Improper use of VAs, on one hand, leads to a high proportion of VA residues that pollute the ecological environment and exacerbate antibiotic resistance. On the other hand, VA residues may accumulate in animals and enter the food chain in the form of chemical hazards, thus causing food safety risks that endanger the health of consumers (i.e., public health) [7–9]. It is noteworthy that the development of antibiotic resistance has exacerbated the overuse of antibiotics in veterinary drugs [10], while the release of antibiotics into the environment has accelerated the development of antibiotic-resistant bacteria. This results in a vicious cycle that poses a tremendous threat to the ecological environment and public health.

Since antibiotic resistance has become a common global problem, [11,12], there is increasing concern regarding VAs in developing countries. China is not only the largest producer and user of antibiotics in the world [1,13], but also the largest pig producer and consumer [14]. In 2013, antibiotic consumption in animals accounted for approximately 52% of the total antibiotic consumption of approximately 162,000 tons in China [15]. The negative effects of improper use of VAs by pig farmers are evident to varying degrees in the vast rural areas of China [16]. In many countries around the world, including China, there are a large number of small-scale farmers of meat-producing animals who are the direct users of VAs. In China, farmers tend to overuse VAs, use human antibiotics, or do not follow the withdrawal time recommendations due to their poor knowledge of VAs and the pursuit of economic benefits from meat-producing animals [14].

Minimizing antibiotic resistance should be the responsibility of all members of society [17]. The crucial role of farmers in shaping and preserving multifunctional agro-ecosystems, has been highlighted by agricultural scientists over the past decades [18]. It has been pointed out in some studies that improper antibiotic use by farmers is closely related to their knowledge of antibiotics [19,20]. Kuipers et al. (2016) [21] found that professionally trained farmers (i.e., farmers with higher knowledge level) tend to use less VAs in dairy herds. However, the possible use of VAs by farmers with different knowledge levels and under different regulatory policies (e.g., in China) has been rarely reported. Therefore, this study empirically investigated the knowledge and use of VAs by pig farmers in rural areas in China. A behavior probability model was constructed based on the knowledge of pig farmers that affected their VA use. The dynamic changes in farmers' behavioral choice regarding the use of veterinary antibiotics, was then observed by Matlab simulation, when considering their knowledge regarding VAs and the different government regulation environments. Based on the findings, policy recommendations were made to regulate improper VA use by farmers.

2. Materials and Methods

2.1. Sample Site

Funing County (located in Jiangsu Province) was selected as it is a famous pig farming base in China (Figure 1), known as "the hometown of piglets". Pig farming is an important source of family income for farmers in Funing, and more than 50% of the pigs were produced by small-scale household farming. As small-scale household pig farming where VAs are directly used by farmers will persist over a long period of time in China, investigating the use of VAs in Funing has important practical significance.

Figure 1. Location of the survey area in China. Note: This is merely a schematic diagram and does not cover the issue of territorial sovereignty.

2.2. Study Design

Prior to the formal survey, a preliminary survey was conducted among pig farmers in Xinlian Village, Sanzao Town, Wangji Village, Longwo Village, and Shuanglian Village in Funing County. A final questionnaire was developed after problems were identified and solved based on the findings of the preliminary survey. The formal survey was conducted by random sampling and home visits in all 13 towns/villages in Funing County. Since face-to-face interviews can effectively avoid the respondents' possible misunderstanding of survey questions and improve the response rate, the survey was performed by properly trained investigators (postgraduate and doctoral students) who were familiar with the questionnaire and interview process. A total of 654 valid questionnaires were collected for the final analysis.

2.3. Instruments

The questionnaire was developed based on the literature review and the authors' field observations [14,16,22]. The questionnaire was divided into three parts. The first part was designed to collect the demographic characteristics of respondents including gender, age, education, annual production, farming income, and years of farming experience. In the second part, the use and knowledge of VAs were assessed. In view of the diversity and complexity of improper use of VAs, the three most common types: overdose (addition of VAs at a higher than specified concentration), use of human antibiotics instead of VAs, and non-compliance with withdrawal time requirements were investigated. To assess the respondents' knowledge about VAs, their knowledge about VA use specifications, hazards of VA residues, and relevant laws and their penalties were examined. Theirs level of knowledge was scored on five-point Likert scale where 1 = no knowledge, 2 = little knowledge, 3 = moderate knowledge, 4 = good knowledge, and 5 = complete knowledge. The third part examined the effect of government regulation the use of VAs by farmers. The respondents were asked about the frequency of spot checks of pig farmers' VA use by local government regulators, how and to what extent government spot checks affected pig farmers' VA use, whether farmers were punished for improper use of VAs according to law, and what was the effect of punishment. Note that, strictly in China's newest Regulations on Administration of Veterinary Drugs, there is no permitted use of human drugs on animals

3. Model Approach and Simulation Scenarios

This paper referred to a literature for the modeling and scenario methods [22].

3.1. Model Construction for Farmers' Behavior Choice

3.1.1. Basic Model Assumptions

Due to the fact that farmers operate in a complex environment, the simulation could not take into account all the factors that may affect their VA use. Therefore, this study focused on how the differences in farmers' knowledge affected their VA use during pig farming. The following assumptions were made:

(1) There are only two choices—either proper or improper—for pig farmers regarding the use of VAs. Proper use refers to the use of VAs in a correct and reasonable way according to requirements. Improper use comprises of one or more behavior of VA overdose, use of human antibiotics, and non-compliance with withdrawal time requirements.

(2) Pig farmers are economically rational. Their use of VAs follows the cost–benefit approach.

(3) The government makes spot checks of farmers' VA use during pig farming. Farmers will be subject to financial penalties, pressure of public opinion, and moral pressure, if improper use is discovered.

(4) Pig farmers' choice regarding VA use is a dynamic process affected by the behaviors of peers in real-world situations.

3.1.2. Farmers' Knowledge

In the simulation experiments, pig farmers were the primary actor in economic activity. Their knowledge and cognitive capacity were the main factors affecting their estimation of expected return [23], thus influencing their use of VAs. According to the literature research and the author's field observations, the farmers' knowledge was summarized into three categories: knowledge of VA use specification, knowledge of hazards of VA residues, and knowledge of relevant laws and their penalties, represented by φ_{i1}, φ_{i2}, and φ_{i3}, respectively. As the five-point Likert scale was used in the measurement of the farmers' knowledge level, it was assumed that φ_{i1}, φ_{i2}, and φ_{i3} take a value in [1,5], respectively, where 1 means no knowledge and 5 means complete knowledge. Since the knowledge level of each farmer is not exactly the same in reality, the values were given for φ_{i1}, φ_{i2}, and φ_{i3}, respectively, in the simulation.

3.1.3. Farmers' Expected Returns

Based on the above basic assumptions, the farmers' expected returns were related to government regulation in their behavioral decision regarding the use of VAs. To regulate the use of VAs, government regulators make spot checks to monitor pig farmers' VA use, and punish the improper use of VAs in accordance with laws and regulations. Government regulation and punishment of pig farmers for improper VA use have an impact on their use of VAs. Therefore, the farmers' expected returns can be described as follows.

Farmers' expected return from proper VA use is:

$$W_1 = G \tag{1}$$

Farmers' expected return from improper VA use is:

$$W_2 = (1 - q) \times (\Delta G + G) + q \times (\Delta G + G - C_1 - C_2) \tag{2}$$

where $\Delta G = \theta \times G$, where G is the farmers' return from proper VA use; ΔG is the farmers' extra return from improper VA use; C_1 are the financial penalties imposed by government regulators on farmers for improper VA use; C_2 are the social costs of discovered improper VA use for farmers including pressure from public opinion and moral pressure, etc.; θ is the ratio of farmers' increased return from improper VA use to that from proper use; q is the probability of the farmers' improper VA use to be checked by government regulators.

3.1.4. Behavior Probability Model

As pig farmers' VA use is affected by multiple factors, the choice probability for VA use varies among farmers. Sun et al. [23] developed a mathematical model of behavior probabilities to assess the probability of choosing a certain behavior under the general reward expectation on that behavior. For individual pig farmers, behavior probability is a description of behavioral uncertainty, that is, the probability of a farmer choosing a certain use of VAs in the "behavior set". Correspondingly, for the pig farmer group, behavior probability is the proportion of individual farmers who choose a certain use of VAs in the group. If all individuals in the group have the same return expectation on each use of VAs, they will all choose the same use of VAs, and there is no need to discuss behavior probability. However, in fact, there is a big difference in farmers' return expectation on each use of VAs. The differences in cognitive capacity and bias regarding VA use specification, hazards of VA residues, and relevant laws and regulations among each individual actor in the group lead to different probabilities for each farmer in choosing the use of VAs. Based on the literature [23] and the knowledge of farmers, a behavior probability model was developed in this study to simulate the farmers' VA use during pig farming under different return expectations.

According to the assumptions, the farmers' VA use was simplified into two categories: either proper use a_+ or improper use a_-. The behavior set was $A = \{a_+, a_-\}$. The following behavior probability model was developed:

$$\begin{cases} p_i(a_+) = \dfrac{e^{\{\varphi_{i0}+(\varphi_{i1}+\varphi_{i2}+\varphi_{i3})w_i(a_+)-(\varphi_{i4}+\varphi_{i5}+\varphi_{i6})w_i(a_-)\}}}{1+e^{\{\varphi_{i0}+(\varphi_{i1}+\varphi_{i2}+\varphi_{i3})w_i(a_+)-(\varphi_{i4}+\varphi_{i5}+\varphi_{i6})w_i(a_-)\}}} \\[2em] p_i(a_-) = 1 - p_i(a_+) = 1 - \dfrac{e^{\{\varphi_{i0}+(\varphi_{i1}+\varphi_{i2}+\varphi_{i3})w_i(a_+)-(\varphi_{i4}+\varphi_{i5}+\varphi_{i6})w_i(a_-)\}}}{1+e^{\{\varphi_{i0}+(\varphi_{i1}+\varphi_{i2}+\varphi_{i3})w_i(a_+)-(\varphi_{i4}+\varphi_{i5}+\varphi_{i6})w_i(a_-)\}}} \\[2em] \quad = \dfrac{1}{1+e^{\{\varphi_{i0}+(\varphi_{i1}+\varphi_{i2}+\varphi_{i3})w_i(a_+)-(\varphi_{i4}+\varphi_{i5}+\varphi_{i6})w_i(a_-)\}}} \\[2em] \dfrac{p_i(a_+)}{p_i(a_-)} = e^{\{\varphi_{i0}+(\varphi_{i1}+\varphi_{i2}+\varphi_{i3})w_i(a_+)-(\varphi_{i4}+\varphi_{i5}+\varphi_{i6})w_i(a_-)\}} \end{cases} \quad (3)$$

where φ_{ij} is the regression coefficient, $i \in [1,2, \ldots ,N]$, $j \in [1,2, \ldots ,6]$, and $\varphi_{ij} > 0$. It should be noted that when $j = 0$, $\varphi_{i0} \in (-\infty,+\infty)$. When φ_{i0} determines that the expected returns from the two different behavioral choices, i.e., proper and improper VA use, are both 0, that is, $w_i(a_+) = w_i(a_-) = 0$, the i-th actor's behavior occurs without a driving force. The behavior probability in this case is called spontaneous probability. In fact, the farmer's choice regarding VA use is influenced by their judgment of the expected return. Based on the behavior probability model, the probabilities of proper use a_+ and improper use a_-, $p_i(a_+)$ and $p_i(a_-)$, were simulated under the influence of farmers' knowledge and return expectations. It is assumed that when $p_i(a_+) \geq p_i(a_-)$, the i-th actor chooses proper use; otherwise, they choose improper use. The group behavior probability was obtained by the observation of a total of N actors.

3.2. Simulation Experiment Description

In this study, the independence and interaction of individual pig farmers as an actor were simulated in a computer-generated environment when considering the influences of their knowledge and actor-to-actor information exchange on their VA use. The simulation was performed using Matlab based on Wu's and Zhou's research [22,24], and is described as follows:

(1) The simulation area is a 20 × 20 square area. At the start of the simulation, 100 farmers were randomly distributed in this area. Specific parameters are listed in Table 1 below.

Table 1. Settings of experimental parameters.

Model Parameters	Parameter Value (Symbol)
Area	20 × 20
Total number of farmers, N	100
Farmers: proper use	A
Farmers: improper use	B
Vacancy	O

(2) Vision values of farmers. Farmers' VA use is closely related to the behavior of their peers [25]. "Vision value" was used to indicate the ability of farmers to collect surrounding information in the model. The larger the value, the higher the ability to collect surrounding information. At the start of the simulation, 100 vision values were randomly generated and assigned to each farmer. A vision value of two means that a farmer can observe the behaviors of other farmers in 2 × 4 grids surrounding them. It was assumed that: (a) If a farmer's behavior is A, and the number of A within their range of vision ≥ the number of B, they will maintain their own behavior; otherwise, their behavior will change to B; and (b) if a farmer's behavior is B, they will maintain their own behavior if the number of B within their range of vision ≥ the number of A; otherwise, their behavior will change to A.

(3) Knowledge of farmers. As set forth, φ_{i1} (the farmers' knowledge of VA use specification), φ_{i2} (knowledge of hazards of VA residues), and φ_{i3} (knowledge of relevant laws and their

penalties), take a value in $[1,5]$ in the simulation, respectively, where 1 means no knowledge and 5 means complete knowledge. Based on the behavior probability model, φ_{i1}, φ_{i1}, and φ_{i3} are the coefficient part of proper use, and φ_{i4}, φ_{i5}, and φ_{i6} are the coefficient part of improper use. As proper and improper VA uses are two opposite behaviors, when a farmer has a high willingness to perform one behavior, the willingness to perform the other behavior will be relatively low. Therefore, it is assumed that the relationship between the two sets of coefficients is as follows:

$$\begin{cases} \varphi_{i1} + \varphi_{i4} = 5 \\ \varphi_{i2} + \varphi_{i5} = 5 \\ \varphi_{i3} + \varphi_{i6} = 5 \end{cases} \tag{4}$$

To ensure scientific rigor and practical relevance of the simulation, the 100 farmers were assumed to have a lower-middle level of knowledge at the start of the simulation. It was assumed that φ_{i0} = 2 and $\varphi_{i1} = \varphi_{i2} = \varphi_{i3} = 3$, that is, the farmers' three categories of knowledge fluctuated in the range of $[1,7]$.

(4) Farmers' expected return. The farmers' expected return can be calculated by Equations (1) and (2). Farmers' normal return, G, follows uniform distribution in $[5,9]$ (in ten thousand yuan). θ is the ratio of farmers' increased return from improper VA use to that from proper usage. In general, the higher the knowledge level regarding VA use specification, the lower the probability of an improper return. Therefore, θ is correlated with φ_{i1}. To ensure that θ is nonnegative, it was assumed that $\theta + \varphi_{i1} = 5$. Based on the finding of field interview regarding spot checks for pig farmers that were conducted by government regulators each year, the initial value of q was set to 0.3. According to the Regulations on Administration of Veterinary Drugs in China, the penalty for improper VA use was set to 30,000 yuan considering the various forms of improper use. Hence, $C_1 = 3$. The higher the farming return, the higher the pressure from public opinion and moral pressure when the misconduct is disclosed and sanctioned. Hence, it is assumed that $C_2 = 2 \times G$.

4. Results and Discussion

4.1. Sample Characteristics

Statistical analysis was performed by SPSS 20.0 (SPSS Inc., Chicago, IL, USA). Demographic characteristics of respondents are shown in Table 2. According to our investigation, of the 654 pig farmers surveyed, 59.2% were male and 40.8% were female. The average age was 56.2 years. Generally, economic development in rural agricultural zones in China is lower than urban areas. The fact makes most younger generation workforce leave the rural agricultural zones to seek jobs in urban ones. This explains well why the average age of our sample is relatively high. Future studies can focus on the age factor and see if this variable can impact the relationship between knowledge and VA use. Furthermore, 58.7% of the respondents had an education level of primary school or below, and 28.4% had junior high school. A large proportion of respondents (51.4%) had a family size of five or more. The majority of respondents (66.1%) had a pig farming income accounting for 30% or less of total household income, and 78.9% of them had over ten years of pig farming experience.

4.2. Behaviors and Knowledge of Farmers Regarding Veterinary Antibiotics (VAs) Use

In terms of behaviors regarding VA use, 68.3% of the 654 pig farmers surveyed reported non-compliance with withdrawal time requirements, 55.5% overdosed VAs, and 24.3% used human antibiotics instead of VAs. Some farmers reported two or more types of improper VA use.

With regard to knowledge regarding VA use (Table 3), 78.0% and 19.3% of the respondents had no and little knowledge of VA use specification, respectively (97.3% altogether); 66.2% and 22.5% had no and little knowledge that antibiotics customized for humans cannot be used in pig farming (88.7% altogether); 48.2% and 28.9% had no and little knowledge of hazards of VA residues, respectively (77.1% altogether); and 64.7% and 22.0% had no and little knowledge of punishment for violating VA use regulations, respectively.

Table 2. Demographic characteristics of the surveyed farmers.

Characteristics	Categories	Frequency (*n*)	Percentage (%)
Gender	Male	387	59.2
	Female	267	40.8
Education Attainment	Primary school and lower	384	58.7
	Middle school	186	28.4
	High School and Above	84	12.9
Number of household members	1	12	1.8
	2	57	8.7
	3	93	14.2
	4	156	23.9
	5 or more	336	51.4
Proportion of pig production to family income	30% or less	432	66.1
	31–50%	78	11.9
	51–80%	54	8.3
	81–90%	33	5.0
	91% or more	57	8.7
Years of farming	1–3 years	45	6.9
	4–6 years	42	6.4
	7–10 years	51	7.8
	Over 10 years	516	78.9
Slaughter amount	1–30 pigs	417	63.8
	31–100 pigs	135	20.6
	Over 100 pigs	102	15.6

In terms of effect of government regulation, the majority of respondents (68.81%) believed that government supervision and inspection had no effect on their daily farming behaviors. Only 3.21% reported a great effect or a very great effect. Moreover, the vast majority of respondents (91.74%) were not penalized for violating VA use regulations. Only 7.34% and 0.92% were occasionally and frequently penalized for violations, respectively.

Table 3. Farmers' knowledge about VAs (in %).

Knowledge	1 = No Knowledge	2 = Little Knowledge	3 = Moderate Knowledge	4 = Good Knowledge	5 = Complete Knowledge
VAs should be used as directed by a veterinarian in strict accordance with the manufacturer's instructions	78.03	19.27	0.30	1.22	1.22
Antibiotics customized for human cannot be used in pig farming	66.21	22.48	1.22	7.34	2.75
VA residues can cause antibiotic resistance and endanger human health	48.17	28.90	7.80	12.84	2.29
Farmers will be punished by the government for improper VA use	64.68	22.02	3.67	8.26	1.37

4.3. Simulation Experiment Results

The effects of each knowledge category on pig farmers' VA use were simulated using Matlab. The effectiveness of government regulation on preventing and controlling pig farmers' improper VA use was also analyzed. In figures regarding the simulation experiments, the black and gray curves indicate the probabilities of proper and improper uses in the farmer group, respectively.

4.3.1. Influence of Knowledge about VA Use Specification on Farmer's Behavioral Choices

The farmers were randomly distributed in the simulation area at the start of the simulation and then interacted with each other over time. Repeated experiments revealed relatively obvious curve changes when the value of knowledge of VA use specification, φ_{i1}, was set to 1, 2, 3, and 5. Figure 2 reflects the co-variation between knowledge about VA use specification and behavioral choices of pig farmers. As shown in Figure 2, the behavioral choices of farmers appeared to have some regularity under the four different parameter settings of VA use knowledge of farmers—likelihood of good VA use behavior increases with increasing knowledge about VS use specification. When the value of φ_{i1} was 1, that is, the farmers generally have a low level of knowledge of VA use specification, there was a high probability of improper VA use, fluctuating between 95% and 100%, in the farmer group, as shown in Figure 2a. The probability of improper VA use decreased gradually when the value of φ_{i1} changed from 1 to 2 and 3. When $\varphi_{i1} = 3$, the probabilities of proper and improper uses fluctuated around 50%. When φ_{i1} further increased to 5, the probability of improper VA use was significantly lower than that of proper use. The above findings indicated that the probability of improper VA use decreased with an increase in the farmers' knowledge of VA use specification. This was consistent with the conclusion of Wu [26]. However, the probability of improper VA use was still higher than that of proper use. Only when the level of knowledge was sufficiently high were farmers inclined to use VAs properly. Also, such result echoes to the finding of Pham and colleagues [27] that the farmers seldom know the real and specific purpose of using VA. Therefore, persistent improvement of pig farmers' knowledge about VA use specification plays a fundamental role in promoting proper VA use. Note that in the model, the x-axis represents a parameter of time, but we did not specifically assign a time unit for that parameter. By not specifying time period can extend the flexibility and generalizability of the models and results [24].

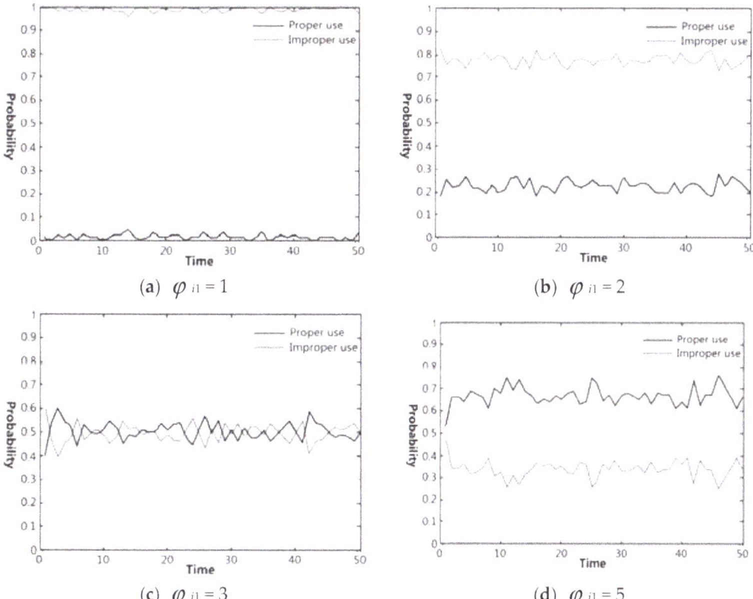

Figure 2. Simulation results of the changes of farmers' behavioral choices regarding VA use under the variation of their knowledge about VA use specification: (**a**) the value of φ_{i1} (the farmers' knowledge of VA use specification) was set to 1; (**b**) the value of φ_{i1} was set to 2; (**c**) the value of φ_{i1} was set to 3; (**d**) the value of φ_{i1} was set to 5.

4.3.2. Influence of Knowledge about the Hazards of VA Residues on Farmer's Behavioral Choices

Figure 3 demonstrates the relationship between farmers' knowledge about the hazards of VA residues on the behavioral choices of them. As can be seen from Figure 3a ($\varphi_{i2} = 1$), when farmers had no knowledge about the hazards of VA residues, there was a high probability of improper VA use, fluctuating around 90%, in the farmer group. This result was consistent with the survey finding that respondents with improper VA use had a poor knowledge about the hazards of VA residues. Moreover, the probability of proper VA use increased significantly when the whole group's knowledge about the hazards of VA residues increased to a certain level, as shown in Figure 3b. When $\varphi_{i2} = 3$, the probabilities of improper and proper VA use fluctuated between 40% and 60%. A comparison of Figure 3b ($\varphi_{i2} = 3$) and 3c ($\varphi_{i2} = 4$) indicated that the probability of improper VA use did not significantly decrease with the further increase in knowledge about the hazards of VA residues. One possible reason is the difficulties in government regulation due to decentralized farming. Moreover, the economic benefits from improper VA use in pig farming are attractive enough for most farmers due to the general absence of strict supervision and punishment by the government [26]. Therefore, it is necessary to educate farmers about the hazards of improper VA use, and at the same time impose financial penalties for improper VA use to reduce willful misconduct.

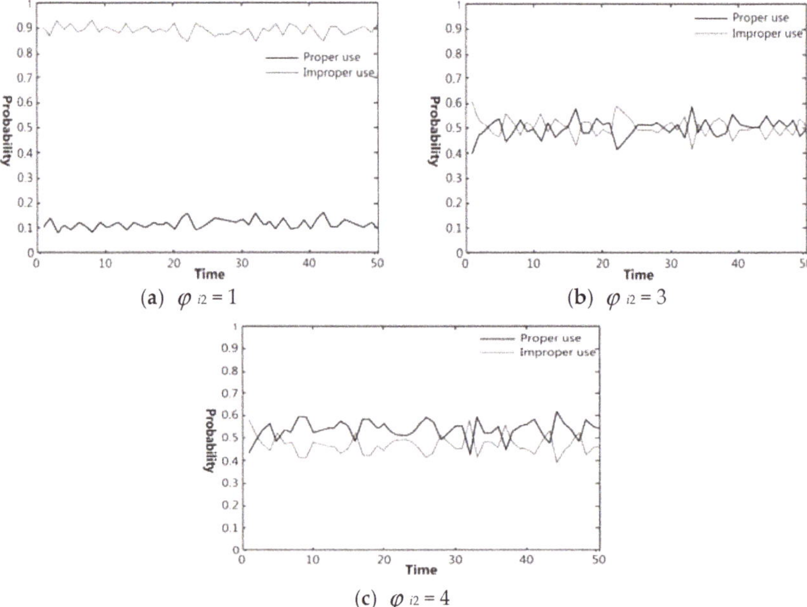

Figure 3. Simulation results of the changes of farmers' behavioral choices regarding VA use under the variation of their knowledge about hazards of VA residues: (**a**) the value of φ_{i2} (knowledge of hazards of VA residues) was set to 1; (**b**) the value of φ_{i2} was set to 3; (**c**) the value of φ_{i2} was set to 4.

4.3.3. Influence of Knowledge about the Relevant Laws and Their Penalties on Farmer's Behavioral Choices

Figure 4 demonstrates the influences of knowledge about the relevant laws and penalties on farmers' behavioral choice. Farmers' VA use is closely related to their knowledge about relevant laws and their penalties. When farmers' knowledge about relevant laws and their penalties $\varphi_{i3} = 1$, there was a relatively high probability of improper VA use, fluctuating between 70% and 80%, as shown in Figure 4a,b. When the value of φ_{i3} changed from 1 to 3, the probability of proper VA use in the farmer group did not increase substantially, while the probability of improper use decreased by 5–10%. As can be seen from Figure 4c, when farmers had a relatively high level of knowledge about the relevant laws and their penalties, improper VA use still occurred at a probability of around 50%, which was similar to the probability of proper use. In fact, current pre- and post-slaughter pig quarantine in China only focuses on foot-and-mouth disease, swine fever, swine vesicular disease, and other diseases. VA residues in live pigs are not strictly monitored. The testing of antibiotic residues only includes several common types of VAs. This has resulted in a low probability of discovering improper VA use by farmers, and consequently, there has been insufficient punishment. From the perspective of policy regulation, pig farmers in China is allowed to execute routine treatments by themselves, just like some advanced nations including The Netherlands (e.g., Kuipers et al., 2016 [21]). This and other similar permissions have allowed farmers in China more autonomy in medical related behaviors. Therefore, it is possible that farmers, driven by economic interests and endorsed with higher behavioral autonomy, deliberately choose improper VA use, despite knowing the penalties.

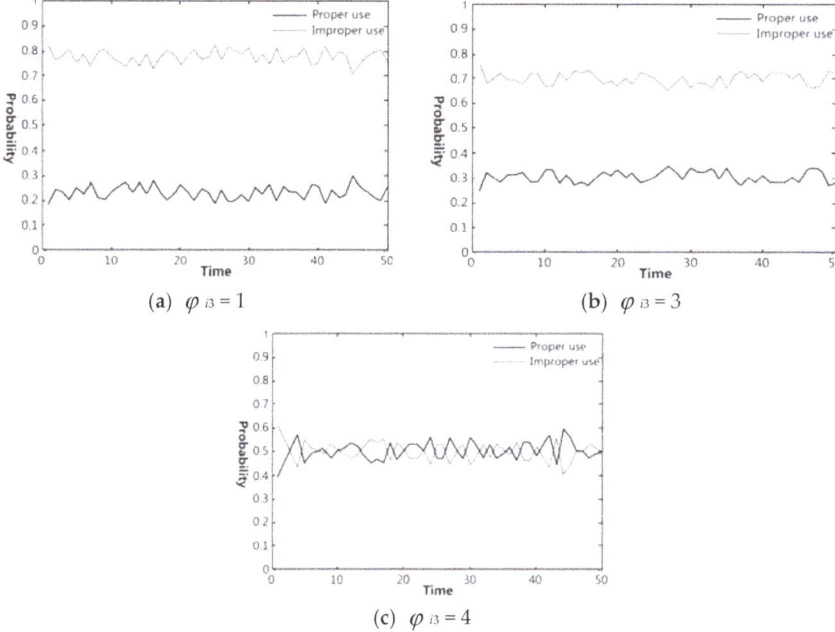

(a) $\varphi_{i3} = 1$ (b) $\varphi_{i3} = 3$

(c) $\varphi_{i3} = 4$

Figure 4. Simulation results of the changes of farmers' behavioral choices regarding VA use under the variation of their knowledge about relevant laws and their penalties: (**a**) the value of φ_{i3} (knowledge of relevant laws and their penalties) was set to 1; (**b**) the value of φ_{i3} was set to 3; (**c**) the value of φ_{i3} was set to 4.

4.4. Influence of Government Regulation on Farmer's Behavioral Choices

Regulative tactics can influence antibiotic use in different ways [28]. Hence, in our simulation experiments, government regulation of pig farmers' VA use was reflected by spot checks and penalties for improper use. The experimental results are shown in Figure 5, which illustrates the relationships between government regulation (in terms of different numbers of random checking and amount of penalty) and farmers' VAs use. When the sampling rate in spot checks and the penalties were both low, the proportion of farmers with improper VA use (approximately 80%) was much larger than that with proper use (approximately 20%). When the sampling rate in spot checks increased, the proportion of farmers with proper VA use (fluctuating between 50% and 60%) was slightly higher than that with improper use. Furthermore, when the penalties were increased, the number of farmers with proper VA use was significantly higher than that with improper use. This was consistent with the findings of Chen et al. [29] on the behaviors of pig farmers.

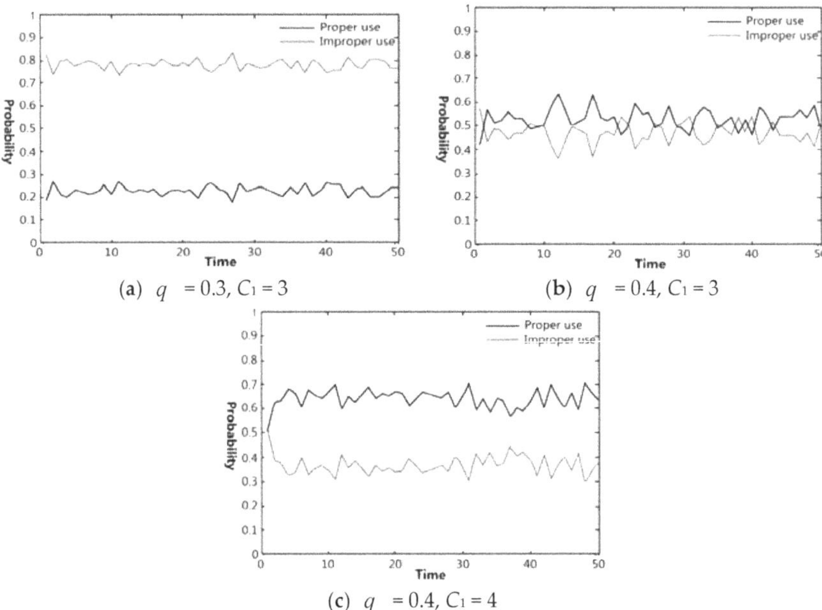

Figure 5. Simulation results of the changes of farmers' behavioral choices regarding VA use under the variation of government regulation: (**a**) the probability of the farmers' improper VA use checked by government regulators was set to 0.3 (q = 0.3), and the penalty for improper VA use was set to 30,000 yuan (C_1 = 3); (**b**) q = 0.4, C_1 = 3; (**c**) q = 0.4, C_1 = 4.

5. Conclusions and Policy Implications

In this study, the dynamic changes in farmers' behavioral choice regarding VA use were observed by simulation when considering their knowledge regarding VAs and farmer-to-farmer interaction. First, the simulation results showed that the probability of improper VA use decreased with the increase in farmers' knowledge about VA use specification. When the level of this knowledge was high enough, farmers were inclined to make proper use of VAs. In short, their use of VAs was significantly affected by their knowledge about VA use specification. Second, the probability of improper VA use decreased at a decreasing rate as farmers' knowledge about the hazards of VA residues increased. In general, the farmers' use of VAs is related to their knowledge about the hazards of VA residues increased. In general, the farmers' use of VAs is related to their knowledge about the hazards of VA residues. Third, when farmers had a high level of knowledge about the relevant laws and their penalties, there was still a high probability of improper VA use, which was similar to that of proper use. The farmers' choice regarding the use of VAs was not significantly affected by their knowledge about the relevant laws and their penalties.

These important findings call for the improvement of VA management policies and the development of sustainable interventions in China to prevent the improper use of VAs by pig farmers, in order to reduce antibiotic resistance and improve pork safety for the protection of public health. Considering the fact that improper VA use is common among pig farmers in China, the following policy recommendations are offered based on the above conclusions. First, support should be provided to help farmers, the end-user of VAs, to improve their knowledge about VA use specification and to keep records of VA use. Changes of management practices (e.g., veterinary professionals' involvement and professionals-farmers communications) may help increase the level of farmers' awareness [21]. Second, support should be provided to help farmers understand the hazards of VAs and thus make proper use of them. Nonetheless, such provision of supportive resources and information should be highly relevant to farmers' special situations (e.g., Garforth et al., 2013 [30]), so as to be highly appreciated and adopted by farmers. For farmers with different levels of knowledge, specific and different resources and information should be endorsed in different ways. For example, for farmers with lower knowledge level of VA, more visual (non-text) and life-related case stories should be told, while for farmers with higher-level knowledge, more systematic information and resource packages should be supplied. Third, as the food safety regulator, the government should improve and publicize relevant laws and regulations to enhance the legal awareness of farmers [28]. Moreover, the government should enhance supervision and inspection, increase the sampling rate in spot checks, and impose harsher penalties for improper VA use.

Author Contributions: X.C. conceived and designed the experiments and wrote the paper; X.X. analyzed the data; L.W. checked the paper and provided important suggestions for the paper.

Funding: This research was funded by a key project of the National Natural Science Foundations of China (Grant No. 71633002) and Jiangsu Province Social Science Foundation (Grant No. 15JD003).

Acknowledgments: We thank all the field workers who supported data collection and subjects who participated in this study.

Conflicts of Interest: The authors declare no conflict of interest.

References

1. Yang, Q.; Zhang, H.; Guo, Y.; Tian, T. Influence of Chicken Manure Fertilization on Antibiotic-Resistant Bacteria in Soil and the Endophytic Bacteria of Pakchoi. *Int. J. Environ. Res. Public Health* **2016**, *13*, 662. [CrossRef] [PubMed]

2. Van Boeckel, T.P.; Brower, C.; Gilbert, M.; Grenfell, B.T.; Levin, S.A.; Robinson, T.P.; Teillant, A.; Laxminarayan, R. Global trends in antimicrobial use in food animals. *Proc. Natl. Acad. Sci. USA* **2015**, *112*, 5649–5654. [CrossRef] [PubMed]

3. Center for Veterinary Medicine. CVM Updates—FDA Annual Summary Report on Antimicrobials Sold or Distributed in 2015 for Use in Food-Producing Animals. Available online: https://www.fda.gov/AnimalVeterinary/NewsEvents/CVMUpdates/ucm534244.htm (accessed on 30 May 2018).

4. Sarmah, A.K.; Meyer, M.T.; Boxall, A.B.A. A global perspective on the use, sales, exposure pathways, occurrence, fate and effects of veterinary antibiotics (VAs) in the environment. *Chemosphere* **2006**, *65*, 725–759. [CrossRef] [PubMed]

5. Ghosh, S.; LaPara, T.M. The Effects of Subtherapeutic Antibiotic Use in Farm Animals on the Proliferation and Persistence of Antibiotic Resistance among Soil Bacteria. *ISME J.* **2007**, *1*, 191–203. [CrossRef] [PubMed]

6. Capita, R.; Alonso-Calleja, C. Antibiotic-Resistant Bacteria: A Challenge for the Food Industry. *Crit. Rev. Food Sci.* **2013**, *53*, 11–48. [CrossRef] [PubMed]

7. Kummerer, K. Drugs in the Environment: Emission of Drugs Diagnostic Aids and Disinfectants into Wastewater by Hospitals in Relation to Other Sources—A Review. *Chemosphere* **2001**, *45*, 957–969. [CrossRef]

8. Pan, M.; Wong, C.K.C.; Chu, L.M. Distribution of antibiotics in wastewater-irrigated soils and their accumulation in vegetable crops in the Pearl River Delta southern China. *J. Agric. Food Chem.* **2014**, *62*, 11062–11069. [CrossRef] [PubMed]

9. Fahrion, A.S.; Jamir, L.; Richa, K.; Begum, S.; Rutsa, V.; Ao, S.; Padmakumar, V.P.; Deka, R.P.; Grace, D. Food-Safety Hazards in the Pork Chain in Nagaland, North East India: Implications for Human Health. *Int. J. Environ. Res. Public Health* **2014**, *11*, 403–417. [CrossRef] [PubMed]

10. Neu, H.C. The crisis in antibiotic resistance. *Science* **1992**, *257*, 1064–1073. [CrossRef] [PubMed]

11. Ying, G.; He, L.; Ying, A.; Zhang, Q.; Liu, Y.; Zhao, J. China Must Reduce Its Antibiotic Use. *Environ. Sci. Technol.* **2017**, *51*, 1072–1073. [CrossRef] [PubMed]

12. Wilson, C. Battling resistance to antibiotics and pesticides: An economic approach. *Ecol. Econ.* **2004**, *50*, 159–160. [CrossRef]

13. Hvistendahl, M. China takes aim at rampant antibiotic resistance. *Science* **2012**, *336*, 795. [CrossRef] [PubMed]

14. Chen, X.; Wu, L.; Xie, X.; Zhu, D.; Wang, J.; Zhang, X. Influence of pig farmer characteristics on improper use of veterinary drugs. *Trop. Anim. Health Prod.* **2016**, *48*, 1395–1400. [CrossRef] [PubMed]

15. Zhang, Q.Q.; Ying, G.G.; Pan, C.G.; Liu, Y.S.; Zhao, J.L. Comprehensive evaluation of antibiotics emission andfate in the river basins of China: Source analysis, multimedia modeling, and linkage to bacterial resistance. *Environ. Sci. Technol.* **2015**, *49*, 6772–6782. [CrossRef] [PubMed]

16. Gao, L.; Tan, Y.; Zhang, X.; Hu, J.; Miao, Z.; Wei, L.; Chai, T. Emissions of *Escherichia coli* Carrying Extended-Spectrum β-Lactamase Resistance from Pig Farms to the Surrounding Environment. *Int. J. Environ. Res. Public Health* **2015**, *12*, 4203–4213. [CrossRef] [PubMed]

17. Lee, C.-R.; Cho, I.H.; Jeong, B.C.; Lee, S.H. Strategies to Minimize Antibiotic Resistance. *Int. J. Environ. Res. Public Health* **2013**, *10*, 4274–4305. [CrossRef] [PubMed]

18. Walder, P.; Kantelhardt, K. The Environmental Behaviour of Farmers—Capturing the Diversity of Perspectives with a Q Methodological Approach. *Ecol. Econ.* **2018**, *143*, 55–63. [CrossRef]

19. Visschers, V.H.M.; Iten, D.M.; Riklin, A.; Hartmann, S.; Sidler, X.; Siegrist, M. Swiss pig farmers' perception and usage of antibiotics during the fattening period. *Livest. Sci.* **2014**, *162*, 223–232. [CrossRef]

20. Huang, J.H. Survey of the Status of Veterinary Use in Pig Farming. *Heilongjiang Anim. Sci. Vet. Med.* **2010**, *9*, 117–119.

21. Kuipers, A.; Koops, W.J.; Wemmenhove, H. Antibiotic use in dairy herds in the Netherlands from 2005 to 2012. *J. Dairy Sci.* **2016**, *99*, 1632–1648. [CrossRef] [PubMed]

22. Wu, L.; Xu, G.; Li, Q.; Hou, B.; Hu, W.; Wang, J. Investigation of the disposal of dead pigs by pig farmersin mainland China by simulation experiment. *Environ. Sci. Pollut. Res.* **2017**, *24*, 1469–1483. [CrossRef] [PubMed]

23. Sun, S.; Jiao, Y.; Liu, C. The Mathematical Model of Behavior Probability. *Syst. Eng. Theory Pract.* **2007**, *27*, 79–86. [CrossRef]

24. Zhou, R.; Yu, D.; Tu, G. System Model and Simulation for Diffusion of Agricultural Technology based on Farmer Knowledge Behavior. *J. Syst. Manag.* **2017**, *26*, 701–712.

25. Chen, Y.; Fang, R. An Empirical Analysis of Influence Factors of Farmers' Behavior of Admindtering Fishery Drugs. *Chin. Rural Econ.* **2011**, *8*, 72–80.

26. Wu, X.; Mao, Y.; Zhan, S.; Yu, X.; Zhang, Y. Behavior of Pig Farmers to Use Veterinary Drugs and Antibiotic Analysis Based on Questionnaire Investigation into 964 Pig Farmers. *Chin. J. Anim. Sci.* **2013**, *49*, 19–23.

27. Pham, D.K.; Chu, J.; Do, N.T.; Brose, F.; Degand, G.; Delahaut, P.; De Pauw, E.; Douny, C.; Van Nguyen, K.; Vu, T.D.; et al. Monitoring antibiotic use and residue in freshwater aquaculture for domestic use in Vietnam. *EcoHealth* **2015**, *12*, 480–489. [CrossRef] [PubMed]

28. Poizat, A.; Bonnet-Beaugrand, F.; Rault, A.; Fourichon, C.; Bareille, N. Antibiotic use by farmers to control mastitis as influenced by health advice and dairy farming systems. *Prev. Vet. Med.* **2017**, *146*, 61–72. [CrossRef] [PubMed]

29. Chen, B.; Zheng, Q. Study on Influence Factors on Intention of Pollution-Free Products Supply and Empirical Research. *World Agric.* **2010**, *2*, 21–25.

30. Garforth, C.J.; Bailey, A.P.; Tranter, R.B. Farmers'attitudes to diseaserisk management in England: A comparative analysis of sheep andpig farmers. *Prev. Vet. Med.* **2013**, *110*, 456–466. [CrossRef] [PubMed]

International Journal of
Environmental Research and Public Health

Article

Antimicrobial Resistance in Food Animals and the Environment in Nigeria: A Review

Nurudeen Olalekan Oloso [1,*] , Shamsudeen Fagbo [2], Musa Garbati [3], Steve O. Olonitola [4], Emmanuel Jolaoluwa Awosanya [5], Mabel Kamweli Aworh [6], Helen Adamu [7], Ismail Ayoade Odetokun [8] and Folorunso Oludayo Fasina [1,9,*]

1   Department of Production Animal Studies (Epidemiology section), Faculty of Veterinary Science, Onderstepoort Campus 0110, University of Pretoria, 0110, South Africa
2   Public Health Agency, Ministry of Health, Riyadh, 11176, Saudi Arabia; oloungbo@yahoo.com
3   Department of Medicine, Infectious Diseases and Immunology Unit, University of Maiduguri, PMB 1069, Maiduguri 600230, Borno State, Nigeria; musagarbati@gmail.com
4   Department of Microbiology, Faculty of Life Sciences, Ahmadu Bello University, Zaria 810241, Nigeria; olonisteve@yahoo.com
5   Department of Veterinary Public Health and Preventive Medicine, University of Ibadan, Ibadan 200284, Nigeria; emmafisayo@yahoo.com
6   Veterinary Drugs/Animal Welfare Branch, Quality Assurance and Standards Division, Department of Veterinary & Pests Control Services, Federal Min. of Agric. & Rural Dev. F.C.D.A, Area 11, Garki, Abuja 900001, Nigeria; mabelaworh@yahoo.com
7   Center for Clinical Care and Clinical Research, Plot 784, By Glimor Engineering, Off Life camp, Gwarimpa Express Way, Jabi, Abuja 240102, Nigeria; hadamu@cccr-nigeria.org
8   Department of Veterinary Public Health and Preventive Medicine, Faculty of Veterinary Medicine, University of Ilorin, Ilorin 240272, Kwara State, Nigeria; odetokun.ia@unilorin.edu.ng
9   Emergency Centre for Transboundary Diseases (ECTAD-FAO), Food and Agricultural Organization of the United Nation, Dar es Salaam 0701072, Tanzania
*   Correspondence: nurudeenoloso@gmail.com (N.O.O.); daydupe2003@yahoo.co.uk (F.O.F.); Tel.: +234-80-5525-4702 or +27-83-529-2326 (N.O.O.); +255-686-132-852 (F.O.F.)

Received: 4 April 2018; Accepted: 14 June 2018; Published: 17 June 2018

Abstract: Antimicrobial resistance (AMR) has emerged as a global health threat, which has elicited a high-level political declaration at the United Nations General Assembly, 2016. In response, member countries agreed to pay greater attention to the surveillance and implementation of antimicrobial stewardship. The Nigeria Centre for Disease Control called for a review of AMR in Nigeria using a "One Health approach". As anecdotal evidence suggests that food animal health and production rely heavily on antimicrobials, it becomes imperative to understand AMR trends in food animals and the environment. We reviewed previous studies to curate data and evaluate the contributions of food animals and the environment (2000–2016) to the AMR burden in Nigeria using a Preferred Reporting Items for Systematic Reviews and Meta-Analyses (PRISMA) flowchart focused on three areas: Antimicrobial resistance, residues, and antiseptics studies. Only one of the 48 antimicrobial studies did not report multidrug resistance. At least 18 bacterial spp. were found to be resistant to various locally available antimicrobials. All 16 residue studies reported high levels of drug residues either in the form of prevalence or concentration above the recommended international limit. Fourteen different "resistotypes" were found in some commonly used antiseptics. High levels of residues and AMR were found in food animals destined for the human food chain. High levels of residues and antimicrobials discharged into environments sustain the AMR pool. These had evolved into potential public health challenges that need attention. These findings constitute public health threats for Nigeria's teeming population and require attention.

Keywords: antimicrobial resistance; antibiotics residue; food animals; environment; bacteria; Nigeria

1. Introduction

The reliance of public health and animal health on antimicrobials since the last century is well known and undisputable [1]. Paradoxically, this reliance (sometimes, over-reliance) and its attendant successes have evolved to become a threat to global animal and human health through the phenomenon of antimicrobial resistance (AMR) [2]. Following the development and use of an antimicrobial, various pathogens, in their attempt to survive or evade current and new antimicrobials, undergo evolutionary processes, which results in a short to long term resistance [3]. AMR is the ability of a microorganism (bacteria, viruses, and certain parasites) to prevent an antimicrobial (antibiotics, antivirals, and antimalarials) from working against it [4]. This may lead to resultant ineffectiveness of standard treatments and the infections may persist, with a higher likelihood of spread [5]. The World Health Organization (WHO) presented the level of exposure of the challenges of AMR through the report of the general worldwide situation analysis [4]. This magnitude of threat associated with AMR then received the highest level of political commitment from world leaders and was discussed at the United Nations General Assembly in 2016, where a political declaration on AMR was issued [6]. Hitherto, WHO and the Food and Agriculture Organization of the United Nations (FAO) produced some fundamental documents toward curbing the threat of AMR. These include the WHO Global action plan on antimicrobial resistance and the FAO action plan on antimicrobial resistance 2016–2020, respectively [7,8]. The report from the monitoring of the global action plan by FAO has suggested and recommended the need for situation analysis and production of action plans for individual countries [9].

Food producing animals are linked to humans via the food chain and shared environment [10]. Thus, a One Health approach is necessary to study and understand how to control burdens of AMR, including those presented through foodborne transmission routes [11,12], as well as create a sound and broad-based antimicrobial stewardship program worldwide [12].

Nigeria is also confronted with the burdens of AMR. The Nigerian Centre for Disease Control (NCDC), in collaboration with other institutions, has made efforts to develop an approach to combat AMR using an evidence-based method. Meanwhile, NCDC (2017) reported that Nigeria has experienced huge resistance to antimicrobials in humans, especially in sepsis, respiratory, and diarrheal infections. These include childhood-related life-threatening diseases and are supported by empirical evidence, which are replete and scattered in peer-reviewed and grey literature, as well as commissioned reports [13]. In addition, the situation analysis and recommendations on AMR and drug use in Nigeria has recently been documented [13]. This document still requires detailed information about several sources of AMR, creating a gap in the trend, status, and situation of AMR arising from food animals and the environment. This study fills that gap through a systematic review of published studies and available reports. Specifically, the study collates, curates, and analyzes data on AMR in Nigeria related to food producing animals and the environment, and the immediate human link as contributors to the burdens of AMR in Nigeria. This study is required as a reference source towards the development of a good antimicrobial stewardship program by stakeholders through the "One Health Platform" for Nigeria.

2. Materials and Methods

2.1. Research Question(s)

We developed some research questions that were used as guides during the study to pursue the attainment of our objectives towards establishing the situation analysis of AMR in the Nigerian environment from food animals. What was the status of antimicrobial resistance in the food producing animals and the environment in Nigeria in the previous studies? What was the pattern of resistance among the classes of antimicrobials tested? What was the status of resistance among the common Nigerian antiseptics and disinfectants that sought to control pathogens at the environmental interface? What were the common organisms and their AMR resistance patterns studied in Nigeria to date?

2.2. Search Design

We searched specific databases (Pub Med-NCBI, Google Scholar, Cabdirect, Medline, Embase, Cochrane, and African Journals Online) and various institutional repository of Nigeria using broad terms, "antimicrobial, resistance, and Nigeria". Where necessary, search terms were stated as strings: Antimicrobial resistance OR Antibiotic resistance OR Antibiotic residue OR Antimicrobial susceptibility OR Antibiotic abuse OR Antibiotic misuse AND Nigeria AND animals; "animals" was substituted with environment and different animal names (poultry, goat, sheep, cattle, camel, pig, etc.). References in the identified materials were also searched and contacted. This effort yielded a broad list of 2393 studies from all sources by the contributors. After removing duplicates, we obtained 435 studies, which were screened to 235 studies by excluding studies conducted prior to the year, 2000, and those with Nigerian authors or affiliations, but focused on samples from outside Nigeria. Upon assessment, we obtained 139 publications and a further 80 were excluded to give 59 publications included in the review and analysis. Each publication was treated as a study, which contains single or multiple reports. The 80 studies excluded did not directly relate to the objectives or yielded information that could be subjected to organized peer review and data analysis. The 59 included studies were sorted into three categories of 42 antimicrobial resistance studies [14–55], 16 antimicrobial residue studies [56–71], and 1 antiseptic or disinfectants study [72]. The PRISMA-style flowchart was modified and used for this analytical review (Figure S1) [73].

2.3. Analysis

The number of publications (Table 1a), diversity of methods of data reporting, multiple appearances of study populations reported (Table 1b) in each study, and the objectives of the various studies of the 59 publications we reviewed made it expedient to find a system of accommodating the information through a uniform standard for data harmonization and interpretation in line with the objectives of this study. The various methods of data analysis in all the studies were reviewed to form a unified scale as presented in Table 2. This scale was developed to harmonize the diverse data for analyzing the situation of AMR in Nigeria within the 42 antimicrobial resistance studies (AMRS) and 16 antimicrobial residue studies (ARS). Therefore, the data of reported resistance and residue in the studies were categorized and interpreted according to the standard developed (Table 2). Percentage in Table 2 referred to the percentage (portion) of resistant microbe populations (species) per study. The methods used in most studies were descriptive statistics simple percentages. Some ARS reports were presented in relation to the FAO or WHO standard of maximum residue limit (MRL) at the time of publication. In such studies, the report where no residue was found is categorized as "No residue", the report where there was residue below standards is categorized as "Low residue", and the report where the mean residue level was above the MRL is categorized as "Very high residue". Analysis of the data was then done with MS Excel using simple descriptive statistical analysis, pivot tables, and charts.

Table 1. Rate of publication per year (**a**) and population groups identified in the studies (**b**).

a. Rate of Publication per Year				
Publication Year	**AMRS**	**ARS**	**SDA**	**Total Reports**
2001		1		1
2002		2		2
2003			1	1
2005		1		1
2007	2			2
2008	1			1
2009	4			4
2010	4	1		5
2011	2	1		3
2012	6	7		13

Table 1. *Cont.*

a. Rate of Publication per Year				
Publication Year	**AMRS**	**ARS**	**SDA**	**Total Reports**
2013	7	1		8
2014	4	1		5
2015	5			5
2016	7	1		8
Total	42	16	1	59

b. Population Groups Identified in the Studies				
Sample Population	**AMRS**	**ARS**	**SDA**	**Total Reports**
Environment	45	-	1	46
Cattle	28	6	-	34
Poultry	26	6	-	32
Pig	10	2	-	12
Goat	6	3	-	9
Vegetables	3	-	-	3
Human	3	-	-	3
Bats	2	-	-	2
Camel	2	-	-	2
Sheep	2	-	-	2
Fish	1	1	-	1
Total	128	18	1	146

AMRS: Antimicrobial resistance studies. ARS: Antimicrobial residue studies. SDA: Surface disinfectants and antiseptics. Table 1: This is a table to show the number of studies for different measurement parameters: (**a**) Showed the number of studies on each measured parameter for each year; and (**b**) showed the total number of reports of appearance of each population group for each measurement parameter.

Table 2. Categorization for the measure of resistance or residue level.

Group Scale	Categorization	Antimicrobial Resistance Studies	Antimicrobial Residue Studies
1	$\leq 1\%$	Sensitive or No resistance	No residue
2	$>1 \leq 24\%$	Moderately sensitive or very low resistance	Low residue
3	$>24 \leq 50\%$	Weakly sensitive or Low resistance	Slightly high residue
4	$>50 \leq 74\%$	Low sensitive or High resistance	High residue
5	$>74\%$	Very low (no) sensitive or Very high resistance	Very high residue

Table 2: This is a table showing the scale developed to measure the level of resistance or residue in a harmonized form from different diverse measurements from the several studies. Percentage referred to the proportion of resistant microbe populations (species) per study.

3. Results

We observed that few studies were undertaken before 2009, with no AMRS, but only four ARS, after which there was an increase in AMRS research from 2009 until recently (Table 1a). The study population involved were environment, cattle, poultry, pig, goat, vegetables, human, bats, camel, sheep, and fish listed in descending order of the number of reports and the type of resistance reported (Table 1b). The study populations appeared singly or in multiple in a study (Table 1b). Also, each study reported from one zone or several geopolitical zones of Nigeria (Figures S2 and S3). Our review revealed that these studies on samples from animals and the environment carried out between 2000 and 2016 fell into three categories (Table 1b).

3.1. Antimicrobial Resistance Studies (AMRS)

This category included 42 studies, with the inclusive eligibility criteria in which diverse phenotypic or genotypic methods were utilized ([14–55], Tables S1–S3). These studies sought to detect the presence and extent of AMR in collected samples with a selected panel of antibiotics.

Cumulatively, these 42 studies tested 68 antimicrobials (Table 3) belonging to different classes and generations of antibiotics from the first to fourth generation of antibiotics, including others that cannot be classified based on generations that were placed on "no generational classification" (NGC) in the course of the analysis (Table 3, Figure S4a,b, and Figure S5a). These resulted in the report of 1139 antimicrobial resistance findings. Out of the 42 studies, only one study on camel samples [45] did not report multidrug resistance (MDR). Two studies [30,38] reported low MDR in cattle and camel samples, and the remaining 39 studies confirmed various patterns of MDR. The AMRS were based on 18 organisms (genus) with species or serovars appearing at least once (Figure 1). The five most important pathogens in which AMR testing was carried out were *E. coli*, *Salmonella* serovars, *Staphylococcus aureus*, *Pseudomonas* spp., and *Klebsiella* spp. *Enterococcus* spp., *Vibrio* spp., *Proteus* spp., and *Listeria* spp. are other microbes used by researchers in AMRS (Figure 1).The nationwide geographical distribution pattern based on geopolitical zones demonstrated that the highest number of reports were from South West Nigeria (44 studies) and, in descending order, from South South (28), North West (16), North Central (10), North East (4), and the lowest was South East (1), which showed poor distribution of studies at the North East and South East (Figure 2, Figure S2).

Table 3. List of antibiotics used and the number of reports of each antimicrobial resistance.

Antibiotics in Peer-Reviewed Studies (n)	Class	Generation	Number of Reports & Category of Resistance Level					
			Very High	High	Low	Very Low	No	Total
Amikacin (AMK) (5)	Aminoglycoside	NGC	1	0	3	6	7	17
Amoxicillin (AMX) (10)	β-lactam	3	17	4	10	8	1	40
Amoxycillin-clavunanic acid (AMC) (23)	β-lactam +	4	18	7	9	4	7	45
Ampicillin (AMP) (20)	β-lactam	3	22	5	5	7	8	47
Ampicillin-cloxacillin (APX) (3)	β-lactam	4	4	2	5	4	3	18
Ampicillin-sulbactam (AMS) (1)	β-lactam +	4	0	0	0	3	0	3
Apramycin (APR) (5)	Aminoglycoside	NGC	0	0	0	0	5	5
Aztreonam (AZT) (5)	β-lactam	1	2	0	2	7	4	15
Cabenicillin (CBN) (3)	β-lactam	4	2	1	1	3	1	8
Cefalexin (CLX) (1)	β-lactam	2	0	0	0	0	1	1
Cefalotin (CLT) (3)	β-lactam	1	0	0	0	4	1	5
Cefazoline (CFZ) (1)	β-lactam	1	0	0	0	2	1	3
Cefepime (CFP) (3)	β-lactam	4	1	0	0	0	6	7
Cefixime (CFX) (1)	β-lactam	3	0	0	1	0	0	1
Cefoperazone (CPZ) (1)	β-lactam	3	1	1	3	0	1	6
Cefotaxime (CTX) (10)	β-lactam	3	2	1	3	2	10	18
Cefoxitin (CXT) (4)	β-lactam	2	1	0	1	0	4	6
Cefpodoxime (CPM) (2)	β-lactam	3	0	1	0	2	1	4
Ceftazidime (CAZ) (6)	β-lactam	3	5	1	2	3	7	18
Ceftiofur (XNL) (6)	β-lactam	3	0	0	0	0	6	6
Ceftriaxone (CRO) (8)	β-lactam	3	2	2	13	13	3	33
Cefuroxime (CXM) (6)	β-lactam	2	4	0	3	3	7	17
Chloramphenicol (CHL) (21)	Phenicol	NGC	16	3	11	12	6	48
Ciprofloxacin (CIP) (30)	Quinolone	2	8	6	8	28	23	73
Clindamycin (CLI) (5)	Macrolide	NGC	0	3	1	1	0	5
Cloxacillin (CXL) (4)	β-lactam	2	2	0	2	0	0	4
Colistin (COL/CT) (7)	Polypeptide	1	0	0	0	0	7	7
Enrofloxacin (ENR) (3)	Quinolone	2	1	0	1	1	0	3
Ertapenem (ETP) (1)	β-lactam	NGC	0	0	0	0	3	3
Erythromycin (E) (17)	Macrolide	NGC	18	2	12	3	4	39
Florfenicol (FFC) (6)	Phenicol	NGC	0	0	0	2	4	6
Fosfomycin (FFM) (1)	Organophosphate	NGC	2	1	2	1	0	6
Fusidic acid (FUA) (3)	Steroid	NGC	2	0	0	1	0	3
Gentamycin (CN/GEN) (33)	Aminoglycoside	NGC	7	2	17	29	26	81
Imipenem (IMP) (4)	β-lactam	NGC	1	0	0	2	6	9
Kanamycin (K) (2)	Aminoglycoside	NGC	0	0	1	1	0	2
Levofloxacin (LVF) (1)	Quinolone	3	0	1	3	1	1	6
Linezolid (LIZ) (2)	Oxazolidinone	NGC	0	0	0	1	1	2
Lomeofloxacin (LMF) (1)	Quinolone	2	2	2	2	0	0	6
Nalidixic acid (NAL) (16)	Quinolone	1	7	5	9	9	6	36
Neomycin (N) (8)	Aminoglycoside	NGC	1	1	4	6	5	17
Nitrofuran (NIT) (8)	Furan	NGC	5	6	4	5	2	22
Norfloxacin (NOR) (3)	Quinolone	2	0	0	2	3	1	6
Meropenem (MPM) (2)	β-lactam	NGC	1	0	0	0	3	4
Mezlocillin (MZC) (1)	β-lactam	4	2	1	2	0	1	6
Mupirocin (MP) (2)	Carbolic acid	NGC	0	0	0	0	2	2
Ofloxacin (OFX) (11)	Quinolone	2	5	1	9	17	8	40
Oxacillin (OX) (7)	β-lactam	2	3	1	1	1	3	9
Penicillin (P) (6)	β-lactam	1	7	0	1	1	1	10

Table 3. *Cont.*

Antibiotics in Peer-Reviewed Studies (n)	Class	Generation	Number of Reports & Category of Resistance Level					
			Very High	High	Low	Very Low	No	Total
Pefloxacin (PEF) (9)	Quinolone	2	10	1	5	13	15	44
Piperacillin (PPC) (1)	β-lactam	4	0	1	0	0	0	1
Piperacillin-tazobactam (PTB) (4)	β-lactam + β-LI	4	1	0	2	1	9	13
Quinupristin (QUI) (1)	Streptogramins	2	0	0	1	0	0	1
Sparfloxacin (SPF) (4)	Quinolone	3	6	0	7	12	1	26
Rifampicin (RIF) (1)	Ansamycin	NGC	0	0	0	1	0	1
Spectinomycin (SPE) (6)	Aminoglycoside	NGC	0	0	1	2	4	7
Streptomycin (S) (22)	Aminoglycoside	NGC	17	10	16	13	9	65
Sulphadimidine (SDN) (1)	Sulfonamides	NGC	8	0	0	0	1	9
Sulfamethoxazole (SMX) (10)	Sulfonamides	NGC	7	2	2	4	1	16
Triple sulphur (TS) (1)	Sulfonamides	NGC	1	0	0	0	0	1
Co-trimoxazole (COT) (17)	Sulfonamides + DI	NGC	22	9	14	9	3	57
Teicoplan (TCP) (1)	Glycopeptide	NGC	6	0	0	0	0	6
Tetracycline (T) (30)	Tetracycline	NGC	32	10	15	8	7	72
Ticarcillin (TCC) (2)	β-lactam	4	1	1	2	2	3	9
Tigecycline (TGC) (1)	Tetracycline	NGC	0	0	0	0	3	3
Tobramycin (TMN) (3)	Aminoglycoside	NGC	1	0	2	3	4	10
Trimethoprim (TMP) (10)	DI	NGC	4	4	3	1	2	14
Vancomycin (V) (4)	Glycopeptide	NGC	1	2	0	1	2	6
Total (42)			289	100	223	266	261	1139

NGC: No generation classification. 1,2,3 and 4: First and second generation antibiotics, respectively. β-lactam + β-LI : β-lactam + β-lactamase inhibitor. β-lactam + means β-lactam combined with another antibiotics; DI: Diaminopyrimidine inhibitor. Sulfonamides + DI: Sulfonamides + Diaminopyrimidine inhibitor. (n): Number of peer reviewed studies for each antibiotic are placed in bracket after each antibiotic.

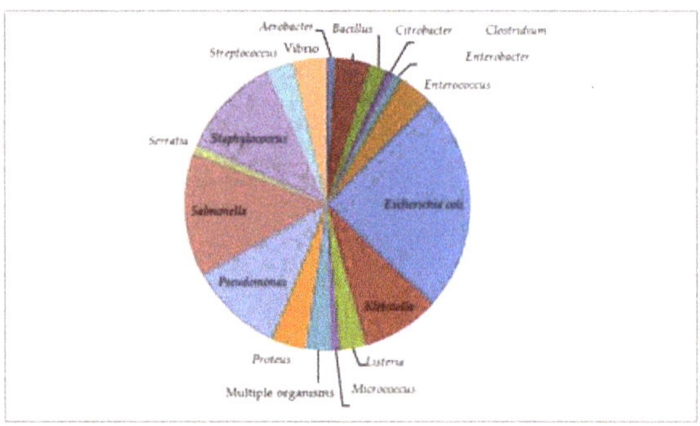

Figure 1. Distribution of organisms studied in the antimicrobial resistance studies based on reports.

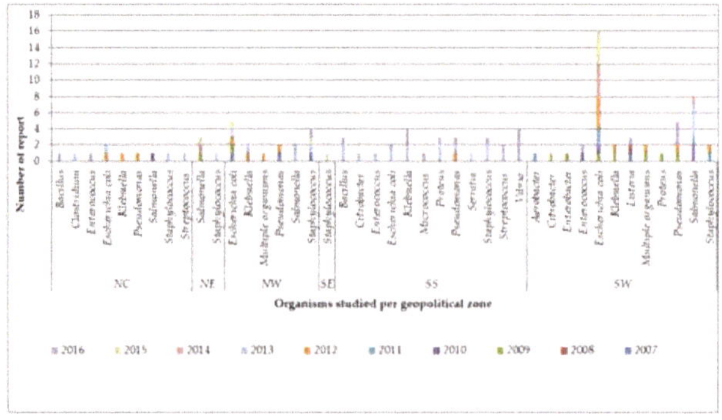

Figure 2. Number of reports yearly per organism for the geopolitical zones of Nigeria. NC = North central, NE = North east, NW = North West, SE = South East, SS = South South, SW = South West.

3.1.1. Antimicrobial Resistance According to Generation of Antibiotics

Antimicrobial resistance within the generational classification of antibiotics used in AMRS (Table 3) revealed that the 68 antibiotics used in all 42 studies involved first, second, third, and fourth generations, and NGC. The generational classification (Table 3) was done using the WHO and the World Organization for Animal Health (OIE) lists of critically important antimicrobial in humans and animals [74–76]. This classification is, essentially, based on the spectrum of activity, which increased from first to fourth generation, implying narrow to broad coverage of antibiotics' action [74]. Cumulatively, of the 1139 antimicrobial report findings, the NGC had the highest number of reports of 537 in the studies of different resistance levels, followed by second and third generation at 210 and 205 reports, respectively; then, fourth generation at 100 and first generation at 86 reports of the resistance findings (Table 3, Figure S4a). The pattern of resistance (Figure S4b) based on proportional percentages of reports showed about 30% of reports on third and fourth generation, and NGC antimicrobials; 20% of first and second generation had very high levels of resistance. It was only 30% of the reports on first, second, and fourth generation, then 20% of third and NGC antimicrobials that had no resistance (Table 3, Figure S4b).

3.1.2. Resistance Level within the Classes of Antibiotics

The 1139 antimicrobial report findings from the 68 antimicrobials included in the panels of all the studies (AMRS) belonged to 19 classes of antibiotics: Aminoglycoside, Ansamycin, Carbolic acid, Diaminopyrimidine inhibitor (DPI), Furan, Glycopeptide, Macrolide, Organophosphate, Oxazolidinone, Phenicol, Polypeptide, Quinolone, Steroid, Streptogramins, Sulfonamides, Sulfonamides + Diaminopyrimidine combinations (SDPI), Tetracycline, β-lactam, and β-lactam + β-lactamase inhibitor combination (Tables 3 and 4). The number of appearances along the resistance level of these classes (Table 4) revealed β-lactam, Quinolone, and Aminoglycoside as the predominant classes studied. The distribution of these classes along the generation showed that β-lactam derivatives, Quinolone, polypeptide, and streptogramins were the antibiotics with generational classification, while others fall in NGC (Tables 3 and 4, Figure S5a,b). Therefore, the distribution of resistance within them have great connected implications in human health as they are mostly used in treating disease conditions in hospitals [76].

Using the developed standard (Table 2), we observed the distribution pattern of resistance levels within classes (Table 4) demonstrated that polypeptides and carbolic acids were the only classes where organisms studied had all the reports to be the "no resistance" category (Table 4). Oxazolidinone, Ansamycim, streptogramins, and Aminoglycosides antibiotics were, at best, categorized as "very low resistance". Meanwhile, phenicol, β-lactam DPI, SDPI, furan, glycopeptides, macrolides, organophosphate, and tetracycline were, at best, of the "very high resistance" category. The highest level of resistance within the resistance pattern distributions among the antibiotic classes were in steroids and sulfonamides, with 70% of the reports on them having "high resistance" to "very high resistance" (Table 4, Figure S5a,b). Each class had peculiar patterns of resistance among the antibiotics belonging to them, which is important for further exposure of the situation of AMR.

Table 4. Number of reports of each resistance level category within the classes of antimicrobial in the Antimicrobial resistance studies.

Class of Antimicrobials	Number of Reports of Each Resistance Level Category					
	Very High	High	Low	Very Low	No	Total n (%)
Aminoglycoside	27	13	44	60	60	204 (17.9%)
Ansamycin	0	0	0	1	0	1 (0.09%)
Carbolic acid	0	0	0	0	2	2 (0.18)
DPI	4	4	3	1	2	14 (1.2)
Furan	5	6	4	5	2	22 (1.9)
Glycopeptide	7	2	0	1	2	12 (1.1%)
Macrolide	18	5	13	4	4	44 (3.9)
Organophosphate	2	1	2	1	0	6 (0.5%)

Table 4. *Cont.*

Class of Antimicrobials	Number of Reports of Each Resistance Level Category					
	Very High	High	Low	Very Low	No	Total n (%)
Oxazolidinone	0	0	0	1	1	2 (0.16)
Phenicol	16	3	11	14	10	54 (4.8%)
Polypeptide	0	0	0	0	7	7 (0.6%)
Quinolone	39	16	46	84	55	240 (21.1%)
Steroid	2	0	0	1	0	3 (0.2%)
Streptogramins	0	0	1	0	0	1 (0.08)
Sulfonamides	16	2	2	4	2	26 (2.3%)
Sulfonamides + DI	22	9	14	9	3	57 (5.0%)
Tetracycline	32	10	15	8	10	75 (6.6%)
β-lactam	80	22	57	64	85	308 (27.0%)
β-lactam + β-LI	19	7	11	8	16	61 (5.4%)
Total	289	100	223	266	261	1139 (100%)

DI = Diaminopyrimidine inhibitor β-LI = β-lactamase inhibitor.

β-lactam Derivatives

These were the most tested, constituting 32.4% of all classes of antimicrobials in this study (Table 4). The β-lactam combinations consisted of β-lactam 27% and β-lactam combinations (β-lactam and β-lactamase inhibitors) at 5.4%. The combinations were supposed to improve the sensitivity of the antibiotics against resistant organisms. However, in this study, the organism tested demonstrated higher levels of resistance to β-lactam combinations (19/61) over β-lactam (80/308), which reported very high resistance levels (Table 4, Figure 3a, Figure S6). We observed Amoxycillin-clavunalic acid as one of the most studied β-lactam derivatives, with organisms showing the highest resistance levels to it among the β-lactam combinations, while Piperacillin-tazobactam was the most sensitive, with a lesser proportion of reports of resistance among β-lactam combinations (Tables 3 and 4, Figure 3a). Among the β-lactams, the third generation antibiotics were the most researched, with Ampicillin and Amoxycillin highest in study rate and also with the highest number of reported resistance, with above 50% of reports on them having very high resistance (Figure 3a, Figure S6). Among all β-lactam derivatives, cefalexin in second generation, Ceftioufur in third generation, and ertapenem in NGC were the only antimicrobials that had all reports on them to be "no resistance" (Figure 3a, Figure S6). All other β-lactams had various patterns of resistance level.

Quinolones

This was the second most studied (21.1%) class of antibiotics (Table 4). It comprised nine antimicrobials, with Ciprofloxacin as the most studied. Lomeofloxacin, of the second generation antibiotics, had the highest resistance level, with over 65% of its reports being "high resistance" to "very high resistance" (Figure 3b, Figure S7). The pattern of resistance had little difference along the generation within this class.

Aminoglycosides

These constituted 17.77% of the studied antibiotics (Table 4), with gentamycin and streptomycin dominating the antibiotics researched in this group. Streptomycin had the highest level of resistance from organisms tested, with a proportion of 40% of its report to be "high to very high resistance" (Figure 3c, Tables 3 and 4, Figure S8). Apramycin was the only antibiotic that was not resisted; all reports on it had "no resistance", while spectinomycin had 80% of its reports with no resistance. The antibiotics in this class demonstrated various patterns of resistance levels (Figure 3c, Figure S8).

Macrolide, Phenicol, and Tetracycline

All these three classes belonged to the NGC. Tetracycline, chloramphenicol, and erythromycin dominated, in descending order, respectively. Tetracycline had the highest level of resistance, with 58%

of its report to be "high" to "very high resistance" from the organisms researched. It was followed by erythromycin (50%) and chloramphenicol (40%) had "high" to "very high resistance", then clindamycin, with 60% of reports on it being "high resistance". Tigercycline was the only one that had all the reports on it as "no resistance" and florfenicol, with 65% as no resistance (Figure 3d, Figure S9).

Sulfonamides Derivatives

All the sulfonamides studied belonged to NGC. The three classes and antibiotics studied were Diaminopyrimidine inhibitor (Trimethroprim), Sulfonamides (sulfamethoxazole, sulphadimidine, and triple sulphur), and sulfonamides-diaminopyrimidine inhibitor combination (co-trimoxazole). The reported proportional resistance level in these classes of antibiotics was the most heightened. The combination (co-trimoxazole) was the most studied and 55% of the studies on it reported "high" to "very high resistance level" from organisms studied. The triple sulphur had only one report and the study reported "very high resistance" level to it. Sulphadimidine had eight out of nine reports (90%) to be "very high resistance level", while trimethoprim and sulfamethoxazole both had 55% that reported a "high" to "very high resistance" level (Figure 3e, Table 4, Figure S10).

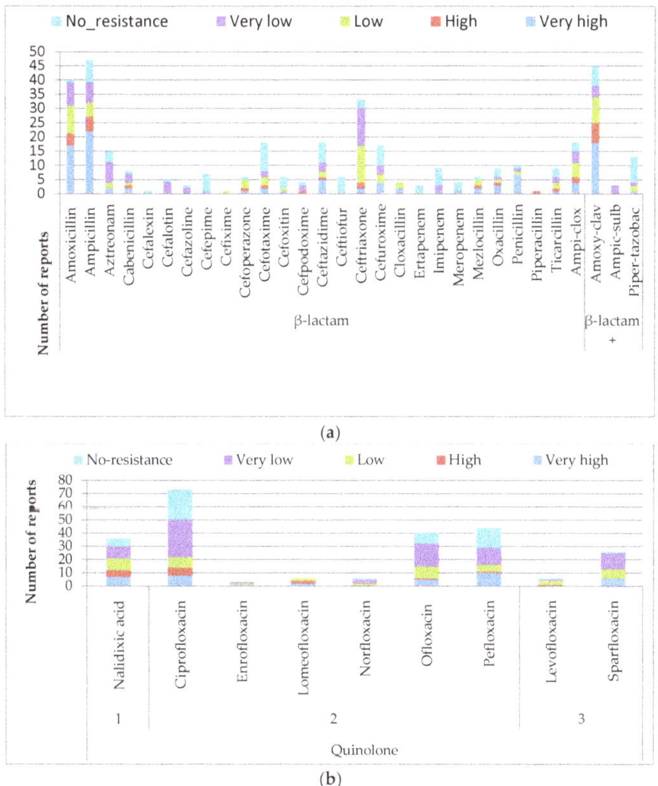

(a)

(b)

Figure 3. *Cont.*

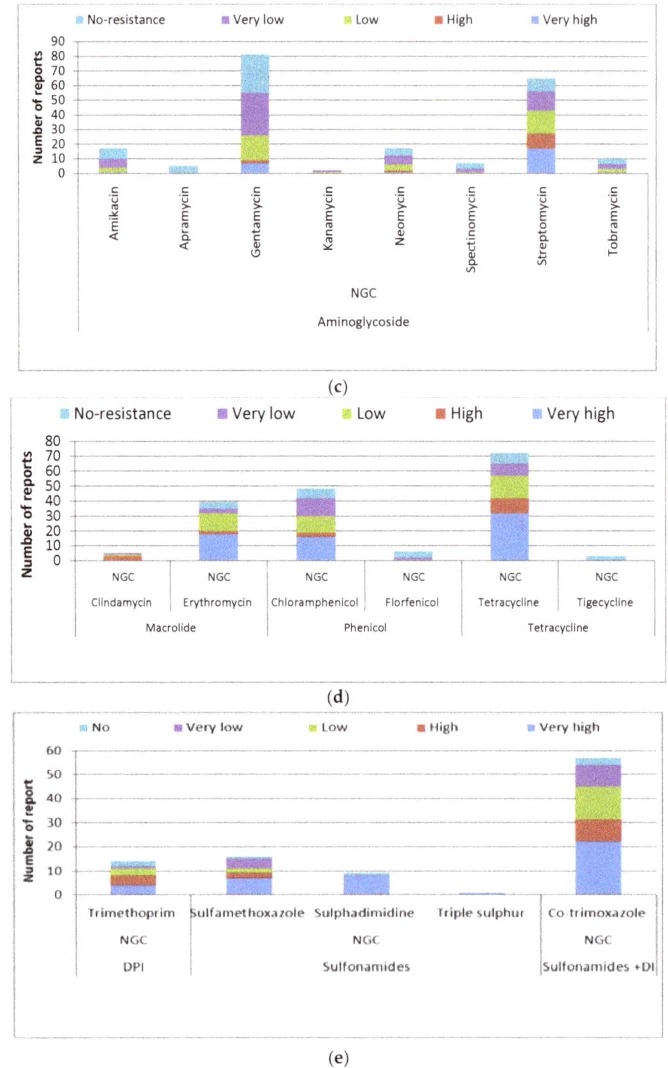

Figure 3. Number of reports of antimicrobial resistance levels of (**a**) β-lactam derivatives, (**b**) Quinolones; (**c**) Aminiglycosides; (**d**) Macrolides, Phenicols, and Tetracyclines; and (**e**) Sulfonamides derivatives antimicrobials.

Other Classes of Antibiotics

The other classes contributed a minute number of report findings, with each class consisting of one antibiotic only; hence, they were pooled together for analysis. Among them, nitrofuran was the most studied and had a high resistance level like vancomycin and fosfomycin, with 50% of the reports having a "high" to "very high resistance" from the organisms studied. In this group, colistin and mupirocin were the most sensitive because they had all reports on them as the "no resistance" level; rifampicin had all its report as "very low resistance", while teicoplan had the highest resistance, with all the reports on it as "very high resistance" from organisms studied. Then fusidic acid responded to the isolates, with about 70% of the reports to be "very high resistance" (Figures S11 and S12).

3.1.3. Resistance along the Organisms Studied

The AMRS were based on 18 organisms (genus), with species or serovars appearing at least once (Figure 1). The organisms' appearance, in descending order, were: *Escherichia coli*, *Salmonella*, *Staphylococcus*, *Pseudomonas*, *Klebsiella*, *Bacillus*, *Enterococcus*, *Proteus*, *Vibrio*, *Listeria*, *streptococcus*, *Citrobacter*, *Aerobacter*, *Clostridium*, *Enterobacter*, *Micrococcus*, and *Serratia* (Figure 1). The distribution of the organisms studied yearly at geopolitical zones demonstrated some organisms were studied more in particular regions or geopolitical zones of Nigeria (Figure 2).

Escherichia coli (E. coli)

It was the most studied organism (25%) in Nigeria, but had a skewed distribution, with a higher concentration of *E. coli* studies in South West Nigeria and none in the North East and South East (Figure 2). The distribution of the studies revealed that 57 antibiotics were used to test AMR in *E. coli* isolates, with gentamycin, tetracycline, ciprofloxacin, cotrimoxazole, ampicillin, streptomycin, amoxicillin-clavulanic acid, ofloxacin, ceftriaxone, nitrofuran, perfloxacin, amoxicillin, nalidixic acid, chloramphenicol, cefuroxime, ceftazidime, neomycin, and sparfloxacin being the most prominent in descending order, respectively (Figure 4a). All the reports on *E. coli* isolates revealed "no resistance" to Apramycin, cefepime, cefoxitin, ceftiofur, colistin, florfenicol, Imipenem, meropenem, vacomycin, cefazoline, ertapenem, and tigecycline in the studies that incorporated into the panel of antimicrobial tested. The *E. coli* isolates researched showed "very-low resistance" to "no-resistance" levels in some antibiotics: Amikacin, aztreonam, norfloxacin, ofloxacin, tobramycin, cefalotin, ticarcillin clavulanate, and cefpodoxime in all reports that used them. However, all reports had a "very high resistance" level to cloxacillin, penicillin, teicoplanin, and sulphadimidine where they were included. We observed other various patterns of resistance levels to the remaining antibiotics studied (Figure 4a, Figure S13).

Salmonella

It was the second most studied organism (14%) in all geopolitical zones, except the South East and South South where there were none (Figures 1 and 2). The distribution of the studies revealed that 27 antibiotics were used to test for AMR in *Salmonella* isolates, with a close distribution in the number of appearances of individual antibiotics (Figure 4b). The pattern of resistance reported showed that *Salmonella*, studied in all the reports, had no resistance to apramycin, aztreonam, cefotaxime, ceftiofur, colistin, pefloxacin, and co-trimoxazole. We observed that florfenicol, neomycin, ofloxacin, and spectinomycin, respectively, had 40%, 50%, 50%, and 30% of the report on them to be "very low resistance", but had the remaining 60%, 50%, 50%, and 70% of their reports as the "no resistance" category. Cefalotin and kanamycin had all their reports as the "very low resistance" category. However, all reports on amoxicillin, enrofloxacin, and triple sulphur had "very high resistance". Other various patterns of resistance were observed in the remaining antibiotics studied (Figure 4b, Figure S14).

Staphylococcus

This genus was the third most studied (12%) pathogen for AMR in Nigeria, with the widest spread across all geopolitical zones (Figures 1 and 2). The distribution of the studies of all antibiotics used revealed that 32 antibiotics were used to test the AMR of *Staphylococcus* isolates (Figure 4c). The pattern of resistance reported for *Staphylococcus* showed that all studies that tested cefuroxime, nitrofuran, mupirocin, and cefalexin revealed "no resistance". All that tested rifampicin and tombromycin reported "very low resistance". Only two studies reported on linezolid, with one each of "very low resistance" and "no resistance", and the only study that tested trimethoprim on *Staphylococcus* showed "low resistance" (Figure 4c). However, the two reports on ampicillin had "very high resistance" for *Staphylococcus* isolates. Other patterns (mixed) for the remaining antibiotics tested were observed (Figure 4c, Figure S15).

Pseudomonas

This represents the fourth most studied organism (11%) for AMR in Nigeria and had a spread similar to *E. coli* research (Figures 1 and 2). The distribution of the studies of all antimicrobials used revealed that 38 antimicrobials were used to test the AMR in *Pseudomonas* isolates (Figure 4d). Unlike other organisms, there were no antibiotics from the 38 tested with *Pseudomonas* without resistance (Figure 4d). There was "very high resistance" by all *Pseudomonas* studied to amoxicillin, amoxicillin-clavunanic acid, ampicillin-cloxacillin, cefuroxime, meropenem, mezlocillin, and teicoplanin and "high resistance" to cefotaxime, erythromycin, nitrofuran, piperacillin, tobramycin, ticarcillin clavulanate, cefoperazone, lomeofloxacin, and fosfomycin (Figure 4d). All reports of studies that tested chloramphenicol with *Pseudomonas* spp. had 75% of them to be "very high resistance" and the remaining 25% of reports were "high resistance". Various resistance patterns were observed in the remaining antibiotics studied (Figure 4d, Figure S16).

Klebsiella

This is the fifth most studied organism and contributed 9% of the overall studies for AMR in Nigeria, with spread across four out of the six geopolitical zones (South West, South South, North West, and North Central) of Nigeria (Figures 1 and 2). The distribution pattern of the appearance of all antimicrobials used revealed that 33 antimicrobials were used to test the AMR of *Klebsiella* isolates (Figure 4e). All the *Klebsiella* spp. studied demonstrated "no resistance" to amikacin, aztreonam, cefotaxime, ceftazidime, piperacillin-texobactam, tobramycin, mezlocillin, ticarcillin clavulanate, and cefoperazone and "low resistance" to cefuroxime and levofloxacin; but, "very high resistance" to ampicillin-cloxacillin, nitrofuran, lomeofloxacin, teicoplanin, fosfomycin, and sulphadimidine (Figure 4a). Meanwhile, it demonstrated a high proportion of "very high resistance" in amoxicillin (60%), amoxicillin-clavunanic acid (75%), ampicillin (75%), chloramphenicol (50%), erythromycin (50%), neomycin (33%), and co-trimoxazole (80%) (Figure 4e, Figure S17).

Other Organisms

All other organisms that made minute contributions were pooled together for analysis. They were spread across the four geopolitical zones of South West, South South, North West, and North Central of Nigeria (Figures 1 and 2). Analysis revealed 43 antimicrobials were used to test for AMR in these organisms (Figure 4f). The organisms were *Proteus, Listeria, Enterococcus, Enterobacter, Citrobacter, Aerobacter, Vibrio, Streptococcus, Serratia, Micrococcus, Bacillus,* and *Clostridium* (Figure 1). All of them had "very high resistance" to fusidic acid and teicoplanin; "high resistance" to clindamycin; but, "very low resistance" to enrofloxacin (Figure 4f). However, they had resistance levels that were "very high resistance" and "high resistance" (combined) to some popular antimicrobials in Nigeria: Amoxicillin (30%), amoxicillin-clavunanic aicd (65%), ampicillin (82%), ampicillin-cloxacillin (20%), aztreonam (15%), cefotaxime (15%), ceftazidime (15%), cefuroxime (35%), chloramphenicol (50%), ciprofloxacin (15%), cloxacillin (30%), erythromycin (55%), nalidixic acid (40%), nitrofuran (30%), ofloxacin (30%), Oxacillin (100(50/50)%), penicillin (75%), perfloxacin (20%), sparfloxacin (25%), streptomycin (50%), sulfamethoxazole (60%), co-trimoxazole (50%), tetracycline (75%), tobramycin (35%), trimethoprim (100(50/50)%), vacomycin (100(50/50)%), carbenicillin (20%), mezlocillin (30%), ticarcillin clavulanate (30%), cefoperazone (30%), lomeofloxacin (30%), and fosfomycin (35%) (Figure 4f, Figure S18).

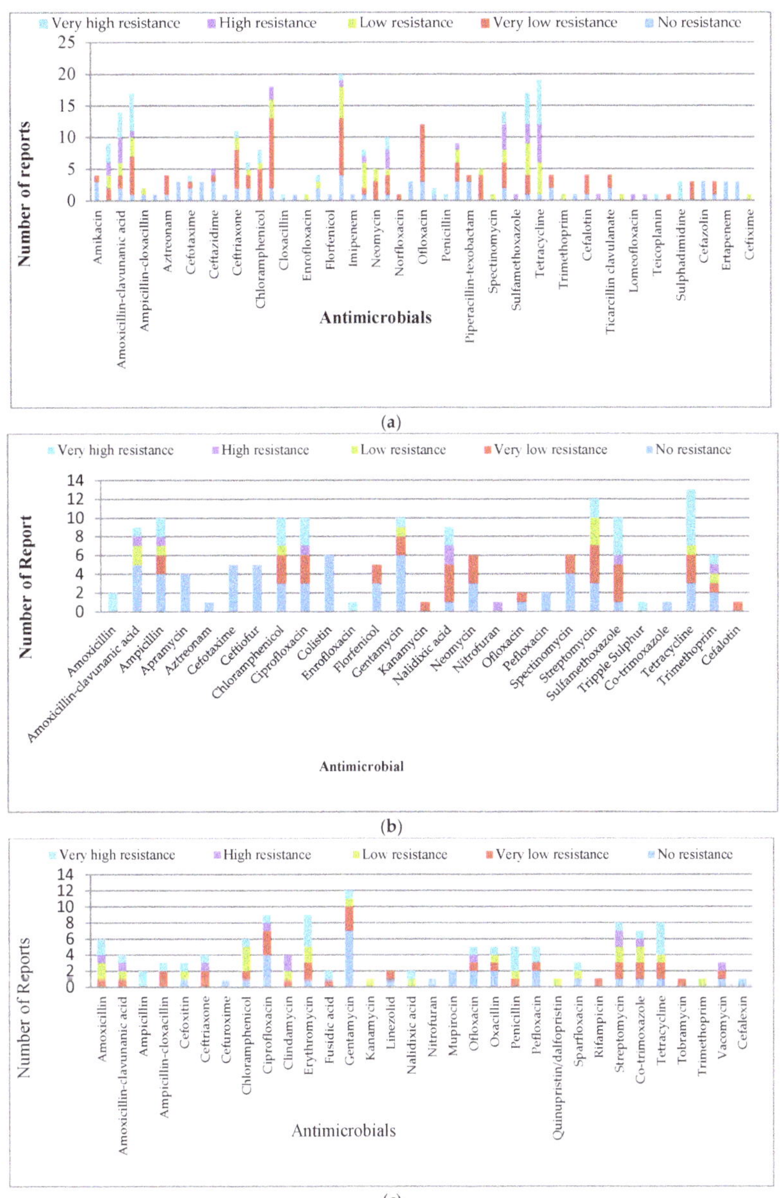

(a)

(b)

(c)

Figure 4. *Cont.*

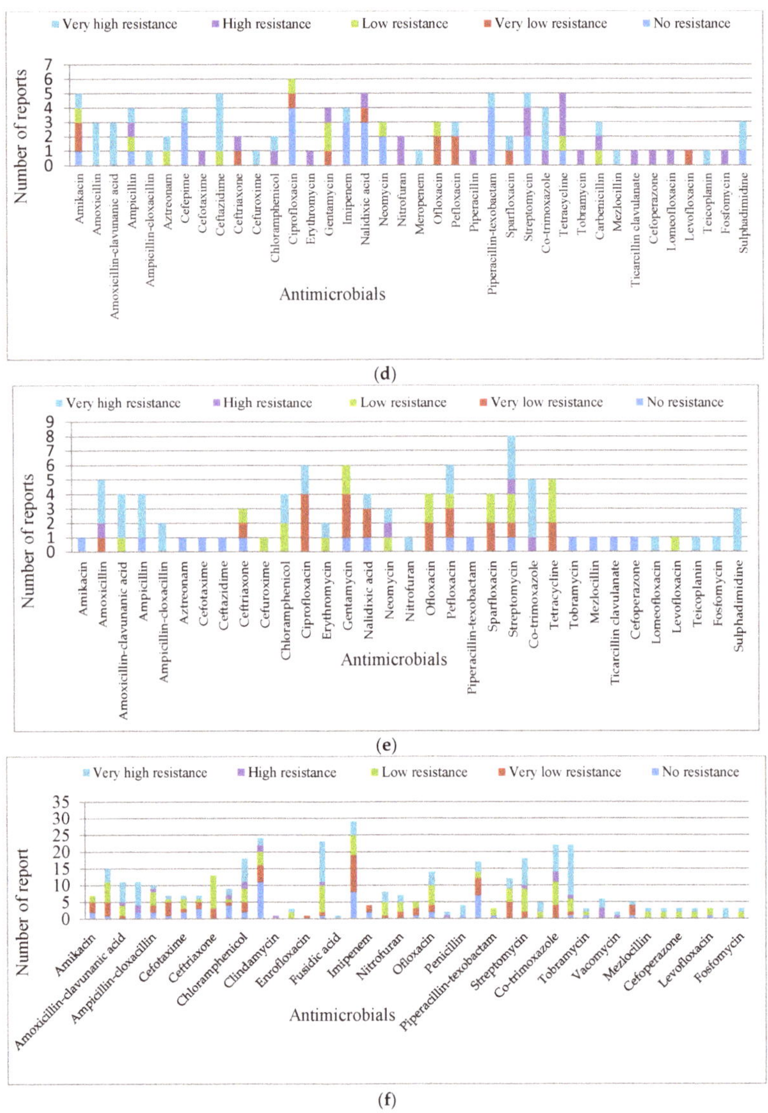

Figure 4. Number of reports of antimicrobial resistance categories for (**a**) *Escherichia coli*; (**b**) *Salmonella*; (**c**) *Staphylococcus*; (**d**) *Pseudomonas*; (**e**) *Klebsiella*; and (**f**) other bacteria.

3.2. Antimicrobial Residue Studies (ARS)

Summarized in Table 5, in this category, 16 studies were identified that dealt with antimicrobial residues in animals and the environment between 2000 and 2016. We considered published research involving qualitative and quantitative assessment of antimicrobial residues in tested samples. We observed the geographical spread of the studies in this category was poor and was skewed to the South West of Nigeria, with few studies in the South East, North Central, and North West, and no studies from the North East and South South (Table 5, Figure S3). The test procedures utilized by the researchers in the studies included microbiological assay (MA), immunological assay, and chromatography. Specifically, the Ridascreen chloramphenicol ELISA kits, Premi test kit (version 0505, Gelen contain *Bacillus stearothermophilus*), MA (seeded with *Bacillus subtilis*), MA (seeded with *Bacillus stearothermophilus*), microbial inhibition test (contain *Micrococcus luteus*), liquid chromatography, High Performance Liquid Chromatography (HPLC), four plate agar diffusion test (FPT), antibody-online ELISA kits, and the agar diffusion method was used (Table 5, Figure 5a). The reference drugs used for the measurement of residue in all studies singly or in pairs were penicillin, amoxicillin, oxytetracycline, and chloramphenicol, and some researchers only measured antimicrobial residue without mentioning a specific drug (Table 5, Figure 5a). Using a unified scale developed (Table 2), no study revealed "No residue"; while they all reported different levels of residue (Table 5, Figure 5a,b). Tetracycline demonstrated to be the most researched (Figure 5a), with reports demonstrating about 40% as a "Very high residue" level (Figure 5b). Other antibiotics demonstrated lower "Very high residue" levels, with the exception of amoxicillin as shown in Figure 5a,b (Table 5, Figure 5b).

Table 5. Summary evaluation of antimicrobial residue studies.

Ref.	Sample				Zone	Test Procedure	Positive Tested Antimicrobial Residue Level				
	Population	Type	Size	Site			TET	CHL	AMX	PEN	AR
56	Cattle	Liver, kidney & muscle	180	Ogun Lagos	SW	Agar diffusion method	Low (16.63%)	-	-	-	-
57	Cattle	Urine	500	Zaria	NW	Microbial Inhibition Test with *Micrococcus luteus*		-	-	-	Low (7.4%)
58	Goat and pig	liver, kidney & muscle	360	Ogun Lagos	SW	Agar diffusion method	Low (15.6%)	-	-	-	-
59	Poultry	Imported layer birds meat	100	Ogun, Lagos, Oyo	SW	Microbiological assay seeded with B.S 1	Low (14%)	-	-	-	-
60	Cattle	Beef	180	Akure	SW	High Performance Liquid Chromatography	High (54.4%)	-	-	-	-
61	Poultry	Eggs	35	Enugu	SE	Microbiological assay seeded with B.S 2	-	-	-	-	Slightly high (30–36%)
62	Goat	Milk	166	Ibadan,	SW	Liquid Chromatography	-	-	Very high (100%)	Very high (100%)	-
63	Poultry	Chicken egg	125	Ibadan	SW	High Performance Liquid Chromatography	Very high >80%	-	-	-	-
64	Goat and pig	Muscle, liver & kidney	240	Nsukka	SE	Four plate agar diffusion test (FPT)	-	-	-	-	Slightly high 25–30%
65	Cattle	Kidney, Liver, Muscle, Urine	448	Abuja	NC	Premi test kit, version 0505, Gelen contain B.S 2	-	-	-	-	Very high 89.3%
66	Poultry	Eggs, muscles, liver, & kidney	168	Ibadan	SW	Ridascreen CHL ELISA kits	-	High	-	-	-
67	Fish	Fresh & frozen fish	60	Ibadan	SW	High Performance Liquid Chromatography	Very high	Very high	-	-	-
68	Poultry	Frozen chicken	100	Lagos & Ibadan	SW	High Performance Liquid Chromatography	Very high	-	-	-	-
69	Cattle	Organs: kidney, liver, muscles	90	South west	SW	High Performance Liquid Chromatography	Low	-	-	Low	-
70	Poultry	Chicken eggs	288	Abuja	NC	Antibody-online ELISA kits	-	Low	-	-	-
71	Cattle	Dairy products	598	Oyo state	SW	High Performance Liquid Chromatography	-	-	-	Slightly high	-

TET: Tetracycline, CHL: Chloramphenicol, AMX: Amoxicillin, PEN: Penicillin B.S 1: *Bacillus subtilis*. B.S 2: *Bacillus stearothermophilus* AR: Antimicrobial residue.

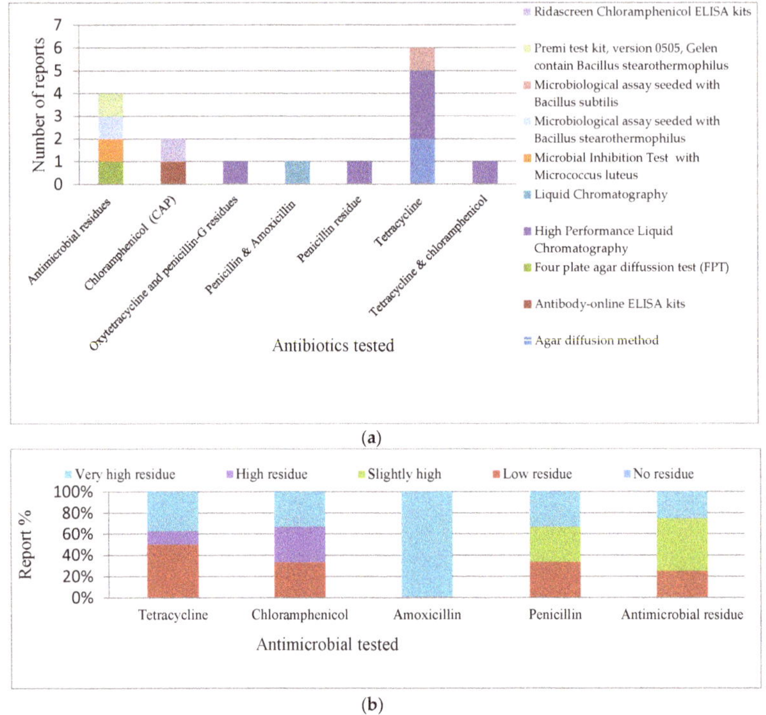

(a)

(b)

Figure 5. (**a**) Test procedure for each antibiotic tested in antimicrobial residue studies; (**b**) Relative level of antimicrobial residue. Tetracycline: Very high residue ($n = 3$), High ($n = 1$), Low ($n = 4$); Chloramphenicol: Very high ($n = 1$), High ($n = 1$), Low ($n = 1$); Amoxicillin No: ($n = 1$); Penicillin: Very high ($n = 1$), Slight high ($n = 1$), Low ($n = 1$); and Antimicrobial residue generally: Very high ($n = 1$), Slightly high ($n = 2$), Low ($n = 1$).

3.3. Antiseptics and Disinfectant Chemicals

Only one study identified human and chicken isolates of *Campylobacter jejuni* to show resistance to at least 19 different commonly used chemicals to control microbes [72].

4. Discussion

We found that several patterns of multidrug resistance were reported in the different studies reviewed and confirmed high levels of resistance to various antimicrobials and common chemical agents [76–79]. mostly used in Nigeria for prophylactic and therapeutic purposes in animals, as well as for the control and management of multiple bacterial pathogens encountered in veterinary and human medical environments [76]. These corroborated the reports of some researchers that antibiotics were readily available over the counter (without prescription) against the existing legislation, prompting a very high level of self-medication [77].

The geographical spread of the reviewed studies showed that the problem of AMR is developing nationwide despite increased awareness demonstrated by the number of studies over time. Few human samples were involved in this study where the researchers collected samples from humans along with other samples without separating the result based on sample population. The results in this study, therefore, reflected an interaction with humans. The overall outcome is an indication towards the situation in humans. However, a similar evaluation in the human health system like the current study had revealed that *Escherichia coli*, *Shigella*, *Salmonella Typhimurium*, and *S. Enteritidis* were more isolated in human diagnostic samples, with evidence of zoonotic infections [78]. Patterns of antimicrobial resistance in humans are similar to what we have also established in animal populations and the environment as indicated in this work. Resistance to penicillin, tetracycline, ampicillin, nalixidic acid, chloramphenicol, and cotrimaxole, among others, has been established in humans [78]. Whether the patterns in humans, animals, and the environment have some association cannot be established in this study, but anecdotal evidence suggests that food animals are often slaughtered and pass into the human food chain without the establishment of residual antimicrobials. We found from observation of study populations that camels were relatively free compared to other animals, but this is only in one study. We are careful to make deductions in this regards as a single study may be tricky to make predictions on the level of antimicrobials in camels, although field situation does not support the widespread use of antibiotics in camels. Also, other studies also demonstrated very low levels of AMR in camels. These studies suggested that the situation of AMR reported may be from the predictor of production management because the herders rarely use antimicrobials in camels in comparison to their use in other food animals [38,44].

Staphylococcus was the only organism included in all studies in all geopolitical zones and had the widest spatial spread. Therefore, the analysis of studies on *Staphylococcus* had the greatest national reflection of the situation of AMR in Nigerian food animals and the environment. All studies on *Staphylococcus* reported very high levels of resistance to ampicillin. This corroborated the report that ampicillin and its combinations were the most consumed over-the-counter (self-medicated) drugs by humans and in animals in Nigeria [77]. This is of great concern because ampicillin is a third generation and ampicillin-cloxacillin is a fourth generation, both of β-lactams. Although, *E. coli* was the most studied, *Pseudomonas* spp. had the highest AMR because this pathogen demonstrated resistance to all antibiotics tested. Also, *Salmonella* demonstrated greater AMR than *E. coli*.

Observations of AMR within the classes of antibiotics along the generation reflected higher percentages of resistance in the antibiotics belonging to the β-lactam derivatives and quinolones of the third and fourth generation, and aminoglycosides. This raised further concerns of the threat posed by AMR. These concerns are heightened as these drugs are listed by WHO as critically important antibiotics required in the management of severe disease conditions. Considering the concerns raised by the drop in global inventions and lack of advances in the production of new antibiotics in the last three decades, which has necessitated monitoring of the circulation of antibiotics worldwide, this current situation is critical. WHO, in response to the above, produced and categorized all antibiotics, which is regularly updated yearly. Therefore, the heightened concerns are necessary to stimulate the Nigerian government and the "One Health Platform", which is under formation, to be proactive towards monitoring, improving, and controlling the current trend.

The reported rate of "high to very high level" of residue levels in the ARS is a confirmation of the demonstration of resistance levels in the AMR studies. All reference drugs tested in the ARS are commonly used in human and food animals in Nigeria [79]. Very high levels of drug residue in goats' milk (100%) is of concern. This portends a problem of AMR of food origins in humans [62]. Meanwhile, the high drug residues in Nigeria food delicacies, including muscle, liver, kidney, and milk, means that human exposure risk is high.

High level of resistance implies that most antibiotics are insensitive to most pathogens in the Nigerian environment. This has also affected antimicrobials' use as antiseptics. These high levels of residues and AMR found in food animals consumed by humans and discharged into environments sustain the AMR pool in addition to the observed resistance by chemicals commonly used as antiseptics to control infection at the point of entrance. This portends a high potential risk to public health management and necessitates the establishment of an institutionalized system that will establish, monitor, control, and promote good antimicrobial stewardship using a one health approach to reduce the current spread of antimicrobial resistance.

Finally, high levels of multiple antibiotic resistance have been observed against many microbial organisms affecting humans and animals. However, most of the studies conducted to date do not use international standards in the delivery of the research results. Future research, in this regard, must carefully consider global standards as part of their methods to engage in carrying out research in Nigeria.

5. Limitations

It was difficult to harmonize our results based on the various AMRS approaches used in studies available for assessment. Moreover, in many of the studies, the Kirby-Baeur method was used, but adequate reference to standards from either Clinical & Laboratory Standards Institute (CLSI) or the European Committee on Antimicrobial Susceptibility Testing (EUCAST) was not provided.

Studies on antibiotics residues were scarce in Nigeria during the years under consideration, with limited studies available for analysis. The methods in most of the studies considered commercial kits, with a dearth of in-depth information on the procedures, which should have given ample opportunity to compares the biases in the methods used in the various residue studies. Relevant equipment that should support such studies on residue testing was wanting at the time of these studies in Nigeria.

6. Conclusions

Multidrug resistance has shown a heightened rise in Nigeria based on this study. The need to use international standards to evaluate most studies on AMR nationwide, in view of the variance of these standards, is necessary. Most of the antimicrobials observed in this study are on the WHO 2017 list of essential antimicrobials and are also listed in OIE 2017 Terrestrial animal health code has, thus, necessitated the evaluation of the situation of AMR in humans [76,80].

It is necessary to design a carefully planned, multi-sectoral, surveillance plan, which can be adopted for research and diagnostic purposes in various aspect of AMR. The need for standardization in all studies in the future and, possibly, the development of guidelines that should harmonize studies across platforms using the "One Health Approach" is imperative. This should target the promotion of good practices and antimicrobial stewardship, which should be enforced by the government, with the cooperation of all stakeholders

The relevant ministries and government departments should enforce: Registration and monitoring of animal production premises, especially, food producing animals; improvement of biosecurity compliance of food animal environments; prohibition of the use of antibiotics for growth promotion and prophylactic treatment; and putting in place a system to implement drug withdrawal periods in food animals.

More detailed descriptions of the results (figures) are available in the Supplementary materials, which are available online.

Supplementary Materials: The following are available online at http://www.mdpi.com/1660-4601/15/6/1284/s1, Figure S1: Flow chart of the methodological strategy (PRISMA 2009 Flow Diagram), Figure S2: Nigeria geopolitical zonal spread of the AMRS reports, Figure S3: Geopolitical zonal spread of the Antimicrobial Residue reports, Figure S4a: Level of resistance within generation of antimicrobials tested, Figure S4b: Proportional (%) pattern of resistance levels within generation of antimicrobials tested, Figure S5a: Frequency of Antimicrobial Resistance levels of classes of antibiotics, Figure S5b: Antimicrobial resistance patterns within classes along generation of antibiotics, Figure S6: Antimicrobial resistance patterns of β-lactam derivatives antibiotics, Figure S7: Antimicrobial resistance patterns of Quinolones, Figure S8: Antimicrobial resistance patterns of Aminoglycosides, Figure S9: Antimicrobial resistance patterns of Macrolide, Phenicol, and Tetracycline, Figure S10: Antimicrobial resistance patterns of Sulfonamides derivatives, Figure S11: Frequency of antimicrobial resistance levels of other classes of antibiotics, Figure S12: Antimicrobial resistance patterns of other classes of antibiotics, Figure S13: Pattern of antimicrobial resistance of *Escherichia coli*, Figure S14: Pattern of antimicrobial resistance of *Salmonella*, Figure S15: Pattern of antimicrobial resistance of *Staphylococcus*, Figure S16: Pattern of antimicrobial resistance of *Pseudomonas*, Figure S17: Pattern of antimicrobial resistance of *Klebsiella*, Figure S18: Pattern of antimicrobial resistance of other bacteria, Table excel S1: Raw data AMRS, S2: Comprehensive AMRS data, S3: Categorized AMRS data analytical.

Author Contributions: The study was conceived by all the authors; designed and supervised by S.F. and F.O.F.; search and review were done by M.G., S.O.O., E.J.A., M.K.A., H.A., and N.O.O.; analysis of the data and writing of the manuscript was done by I.A.O. and N.O.O. F.O.F. provided the overall supervision from design, data gathering, analysis, manuscript preparation, review, journal selection and proof reading of the final manuscript.

Funding: The corresponding author acknowledged the support of University of Pretoria for the Doctoral Research Support Scholarship for 2016 and 2017 funding years for the partial funding.

Conflicts of Interest: The authors declare no conflict of interest. The funding sponsors had no role in the design of the study; in the collection, analyses, or interpretation of data; in the writing of the manuscript, and in the decision to publish the results. The funding was part of regular funding block by the University of Pretoria to support research works of their PhD students.

Abbreviations

AMR	antimicrobial resistance
AMRS	antimicrobial resistance studies
AMX	Amoxicillin
AR	Antimicrobial residue
ARS	antimicrobial residue studies
B.S 2	Bacillus stearothermophilus
BS 1	Bacillus subtilis
CHL (CAP)	Chloramphenicol
ELISA	Enzyme-linked immune sorbent assay
FAO	Food and Agriculture Organization of the United Nations
FPT	Four plate agar diffusion test
HPLC	High Performance Liquid Chromatography
MDR	multidrug resistance
NC	North Central
NCDC	Nigerian Centre for Disease Control
NE	North East
NGC	no generational classification
NW	North West
PEN	Penicillin
SDA	Surface disinfectants and antiseptics
SE	South East
SS	South South
SW	South West
TET	Tetracycline
WHO	World Health Organization

References

1. Kingston, W. Antibiotics, invention and innovation. *Res. Policy* **2000**, *29*, 679–710. [CrossRef]
2. O'Neill, J. *Antimicrobials in Agriculture and the Environment: Reducing Unnecessary Use and Waste*; Wellcome Trust: London, UK, 2015; Available online: https://amr-review.org/sites/default/files/Antimicrobials%20in%20agriculture%20and%20the%20environment%20-%20Reducing%20unnecessary%20use%20and%20waste.pdf (accessed on 29 November 2017).
3. O'Neill, J. *Antimicrobial Resistance: Tackling a Crisis for the Health and Wealth of Nations*; Wellcome Trust: London, UK, 2014; Available online: http://www.amr-review.org (accessed on 29 November 2017).
4. World Health Organization. *Worldwide Country Situation Analysis: Response to Antimicrobial Resistance*; World Health Organization: Geneva, Switzerland, 2015; Available online: http://apps.who.int/iris/bitstream/10665/163468/1/9789241564946_eng.pdf (accessed on 29 November 2017).
5. O'Neill, J. *Tackling Drug-Resistant Infections Globally: Final Report and Recommendations: Final Report*; Wellcome Trust: London, UK, 2016; Available online: https://amr-review.org/sites/default/files/160518_Final%20paper_with%20cover.pdf (accessed on 29 November 2017).
6. United Nations. Press Release: High-Level Meeting on Antimicrobial Resistance. In Proceedings of the 71st General Assembly of the United Nations, New York, NY, USA, 21 September 2016; Available online: http://www.un.org/pga/71/2016/09/21/press-release-hl-meeting-on-antimicrobial-resistance/ (accessed on 29 November 2017).

7. World Health Organization. *Global Action Plan on Antimicrobial Resistance*; World Health Organization: Geneva, Switzerland, 2015; Available online: http://www.wpro.who.int/entity/drug_resistance/resources/global_action_plan_eng.pdf (accessed on 29 November 2017).

8. Food and Agricultural Organization of the United Nation. The FAO Action Plan on Antimicrobial Resistance (2016–2020). 2016. Available online: http://www.fao.org/3/a-i5996e.pdf (accessed on 22 September 2017).

9. Food and Agriculture Organization of the United Nations. Monitoring and Evaluation of the Global Action Plan on Antimicrobial Resistance. Produced Jointly by FAO, OIE and WHO. 2016. Available online: http://www.fao.org/3/a-i7711e.pdf (accessed on 29 November 2017).

10. Marshall, B.M.; Levy, S.B. Food Animals and Antimicrobials: Impacts on Human Health. *Clin. Microbiol. Rev.* **2011**, *4*, 718–733. [CrossRef] [PubMed]

11. Parmley, J.; Leung, Z.; Léger, D.; Finley, R.; Irwin, R.; Pintar, K.; Pollari, P.; Reid-Smith, R.; Waltner-Toews, D.; Karmali, M.; et al. A holistic approach toward enteric bacterial pathogens and antimicrobial resistance surveillance. In *Improving Food Safety through a One Health Approach: Workshop Summary*; National Academies Press: Washington, DC, USA, 2012. Available online: https://www.ncbi.nlm.nih.gov/books/NBK114511/ (accessed on 29 November 2017).

12. Leonard, C.T.; Ward, D.; Longson, C. Antimicrobial resistance: A light at the end of the tunnel? *Lancet* **2017**, *389*, 803. [CrossRef]

13. Nigeria Centre for Disease Control. Antimicrobial Use and Resistance in Nigeria. Situation Analysis and Recommendations. Produced by Federal Ministries of Agriculture, Environment and Health. Coordinated by Nigeria Centre for Disease Control. 2017. Available online: http://www.ncdc.gov.ng/themes/common/docs/protocols/56_1510840387.pdf (accessed on 29 November 2017).

14. Mzungu, I. Isolation, Antibiotic and Heavy Metal Susceptibility Patterns of Some Pathogens from Domestic Dumpsites and Waste Water. Master's Thesis, Ahmadu Bello University, Zaria, Nigeria, November 2007.

15. David, O.M.; Odeyemi, A.T. Antibiotic resistant pattern of environmental isolates of *Listeria monocytogenes* from Ado-Ekiti, Nigeria. *Afr. J. Biotechnol.* **2007**, *6*, 2135–2139. [CrossRef]

16. Adetunji, V.O.; Adegoke, G.O. Formation of biofilm by strains of *Listeria monocytogenes* isolated from soft cheese 'wara' and its processing environment. *Afr. J. Biotechnol.* **2008**, *7*, 2893–2897. [CrossRef]

17. Adelowo, O.O.; Fagade, O.E. The tetracycline resistance gene tet39 is present in both Gram-negative and Gram-positive bacteria from a polluted river, Southwestern Nigeria. *Lett. Appl. Microbiol.* **2009**, *48*, 167–172. [CrossRef] [PubMed]

18. Raufu, I.; Hendriksen, R.S.; Ameh, J.A.; Aarestrup, F.M. Occurrence and characterization of Salmonella Hiduddify from chickens and poultry meat in Nigeria. *Foodborne Pathog. Dis.* **2009**, *6*, 425–430. [CrossRef] [PubMed]

19. Muhammad, M.; Muhammad, L.U.; Ambali, A.G.; Mani, A.U.; Azard, S.; Barco, L. Prevalence of Salmonella associated with chick mortality at hatching and their susceptibility to antimicrobial agents. *Vet. Microbiol.* **2010**, *140*, 131–135. [CrossRef] [PubMed]

20. Ayeni, F.A.; Adeniyi, B.A.; Ogunbanwo, S.T.; Tabasco, R.; Paarup, T.; Peláez, C.; Requena, T. Inhibition of uropathogens by lactic acid bacteria isolated from dairy foods and cow's intestine in western Nigeria. *Arch. Microbiol.* **2009**, *191*, 639–648. [CrossRef] [PubMed]

21. Garba, I.; Tijjani, M.B.; Aliyu, M.S.; Yakubu, S.E.; Wada-Kura, A.; Olonitola, O.S. Prevalence of *Escherichia coli* in some public water sources in Gusau municipal, North-western Nigeria. *Bayero J. Pure Appl. Sci.* **2009**, *2*, 134–137. [CrossRef]

22. Fashae, K.; Ogunsola, F.; Aarestrup, F.M.; Hendriksen, R.S. Antimicrobial susceptibility and serovars of Salmonella from chickens and humans in Ibadan, Nigeria. *J. Infect. Dev. Ctries.* **2010**, *4*, 484–494. [CrossRef] [PubMed]

23. Odeyemi, A.T.; Dada, A.C.; Ogunbanjo, O.R.; Ojo, M.A. Bacteriological, physicochemical and mineral studies on Awedele spring water and soil samples in Ado Ekiti, Nigeria. *Afr. J. Environ. Sci. Technol.* **2010**, *4*, 319–327. [CrossRef]

24. Ojo, O.E.; Ajuwape, A.T.; Otesile, E.B.; Owoade, A.A.; Oyekunle, M.A.; Adetosoye, A.I. Potentially zoonotic shiga toxin-producing *Escherichia coli* serogroups in the faeces and meat of food-producing animals in Ibadan, Nigeria. *Int. J. Food Microbiol.* **2010**, *142*, 214–221. [CrossRef] [PubMed]

25. Adesiji, Y.O.; Alli, O.T.; Adekanle, M.A.; Jolayemi, J.B. Prevalence of *Arcobacter*, *Escherichia coli*, *Staphylococcus aureus* and *Salmonella* species in Retail Raw Chicken, Pork, Beef and Goat meat in Osogbo, Nigeria. *Sierra Leone J. Biomed. Res.* **2011**, *3*, 8–12. [CrossRef]

26. Fortini, D.; Fashae, K.; García-Fernández, A.; Villa, L.; Carattoli, A. Plasmid-mediated quinolone resistance and β-lactamases in *Escherichia coli* from healthy animals from Nigeria. *J. Antimicrob. Chemother.* **2011**, *66*, 1269–1272. [CrossRef] [PubMed]

27. Amosun, E.A.; Olatoye, I.O.; Adetosoye, A.I. Antimicrobial resistance in *Escherichia coli*, *Klebsiella pneumonia* and *Pseudomonas aeruginosa* isolated from the milk of dairy cows in three Nigerian cities. *Niger. Vet. J.* **2012**, *33*, 617–623.

28. Damian. U.I. Antibiotics Susceptibility Studies of Some Bacterial Isolates from Packaged Milk Marketed in Zaria, Nigeria. Master's Thesis, Ahmadu Bello University, Zaria, Nigeria, 2012.

29. Igbinosa, E.O.; Odjadjare, E.E.; Igbinosa, I.H.; Orhue, P.O.; Omoigberale, M.N.; Amhanre, N.I. Antibiotic synergy interaction against multidrug-resistant *Pseudomonas aeruginosa* isolated from an abattoir effluent environment. *Sci. World J.* **2012**, *2012*. [CrossRef] [PubMed]

30. Akobi, B.; Aboderin, O.; Sasaki, T.; Shittu, A. Characterization of *Staphylococcus aureus* isolates from faecal samples of the straw-coloured fruit bat (*Eidon helvum*) in Obafemi Awolowo University (OAU), Nigeria. *BMC Microbiol.* **2012**, *12*, 279. [CrossRef] [PubMed]

31. Ojo, O.E.; Awosile, B.; Agbaje, M.; Sonibare, A.O.; Oyekunle, M.A.; Kasali, O.B. Quinolone resistance in bacterial isolates from chicken carcasses in Abeokuta, Nigeria: A retrospective study from 2005–2011. *Niger. Vet. J.* **2012**, *33*, 483–491.

32. Fashae, K.; Hendriksen, R.S. Diversity and antimicrobial susceptibility of Salmonella enterica serovars isolated from pig farms in Ibadan, Nigeria. *Folia Microbiol. (Praha)* **2013**, *59*, 69–77. [CrossRef] [PubMed]

33. Mailafia, S.; Michael, O.; Kwaja, E. Evaluation of microbial contaminants and antibiogram of Nigerian paper currency notes (Naira) circulation in Gwagwalada, Abuja, Nigeria. *Niger. Vet. J.* **2013**, *34*, 726–735.

34. Oluduro, A.O. Antibiotic-resistant commensal *Escherichia coli* in faecal droplets from bats and poultry in Nigeria. *Vet. Ital.* **2012**, *48*, 297–308. [PubMed]

35. Kawo, A.H.; Musa, A.M. Enumeration, isolation and antibiotic susceptibility profile of bacteria associated with mobile cellphones in a university environment. *Niger. J. Basic Appl. Sci.* **2013**, *21*, 39–44. [CrossRef]

36. Ogunleye, A.O.; Ajuwape, A.T.P.; Adetosoye, A.I.; Carlson, S.A. Characterization of *Salmonella enterica* Ituri isolated from diseased poultry in Nigeria. *Afr. J. Biotechnol.* **2013**, *12*, 2125–2128. [CrossRef]

37. Oviasogie, F.E.; Agbonlahor, D.E. The burden, antibiogram and pathogenicity of bacteria found in municipal waste dumpsites and on waste site workers in Benin City. *J. Biomed. Sci.* **2013**, *12*, 115–130.

38. Raufu, I.; Bortolaia, V.; Svendsen, C.A.; Ameh, J.A.; Ambali, A.G.; Aarestrup, F.M.; Hendriksen, R.S. The first attempt of an active integrated laboratory-based Salmonella surveillance programme in the North-eastern region of Nigeria. *J. Appl. Microbiol.* **2013**, *115*, 1059–1067. [CrossRef] [PubMed]

39. Suleiman, A.; Zaria, L.T.; Grema, H.A.; Ahmadu, P. Antimicrobial resistant coagulase positive *Staphylococcus aureus* from chickens in Maiduguri, Nigeria. *Sokoto J. Vet. Sci.* **2013**, *11*, 51–55. [CrossRef]

40. Adelowo, O.O.; Fagade, O.E.; Agersø, Y. Antibiotic resistance and resistance genes in *Escherichia coli* from poultry farms, southwest Nigeria. *J. Infect. Dev. Ctries.* **2014**, *8*, 1103–1112. [CrossRef] [PubMed]

41. Adefarakan, T.A.; Oluduro, A.O.; David, O.M.; Ajayi, A.O.; Ariyo, A.B.; Fashina, C.D. Prevalence of antibiotic resistance and molecular characterization of *Escherichia coli* from faeces of apparently healthy rams and goats in Ile-Ife, Southwest, Nigeria. *Ife J. Sci.* **2014**, *16*, 447–460.

42. Adeyanju, G.T.; Ishola, O. *Salmonella* and *Escherichia coli* contamination of poultry meat from a processing plant and retail markets in Ibadan, Oyo State, Nigeria. *Springerplus* **2014**, *3*, 139. [CrossRef] [PubMed]

43. Raufu, I.A.; Fashae, K.; Ameh, J.A.; Ambali, A.; Ogunsola, F.T.; Coker, A.O.; Hendriksen, R.S. Persistence of fluoroquinolone-resistant Salmonella enterica serovar Kentucky from poultry and poultry sources in Nigeria. *J. Infect. Dev. Ctries.* **2014**, *8*, 384–388. [CrossRef] [PubMed]

44. Raufu, I.A.; Odetokun, I.A.; Oladunni, F.S.; Adam, M.; Kolapo, U.T.; Akorede, G.J.; Ghali, I.M.; Ameh, J.A.; Ambali, A. Serotypes, antimicrobial profiles, and public health significance of Salmonella from camels slaughtered in Maiduguri central abattoir, Nigeria. *Vet. World* **2015**, *8*, 1068–1072. [CrossRef] [PubMed]

45. Adenipekun, E.O.; Jackson, C.R.; Oluwadun, A.; Iwalokun, B.A.; Frye, J.G.; Barrett, J.B.; Hiott, L.M.; Woodley, T.A. Prevalence and antimicrobial resistance in *Escherichia coli* from food animals in Lagos, Nigeria. *Microb. Drug Resist.* **2015**, *21*, 358–365. [CrossRef] [PubMed]

46. Olowe, O.A.; Adewumi, O.; Odewale, G.; Ojurongbe, O.; Adefioye, O.J. Phenotypic and molecular characterisation of extended-spectrum Beta-lactamase producing *Escherichia coli* obtained from animal fecals Samples in Ado Ekiti, Nigeria. *J. Environ. Public Health* **2015**, *2015*. [CrossRef] [PubMed]

47. Olonitola, O.S.; Fahrenfeld, N.; Pruden, A. Antibiotic resistance profiles among mesophilic aerobic bacteria in Nigerian chicken litter and associated antibiotic resistance genes1. *Poult. Sci.* **2015**, *94*, 867–874. [CrossRef] [PubMed]

48. Ugwu, C.C.; Gomez-Sanz, E.; Agbo, I.C.; Torres, C.; Chah, K.F. Characterization of mannitol-fermenting methicillin-resistant staphylococci isolated from pigs in Nigeria. *Braz. J. Microbiol.* **2015**, *46*, 885–892. [CrossRef] [PubMed]

49. Ayeni, F.A.; Odumosu, B.T.; Oluseyi, A.E.; Ruppitsch, W. Identification and prevalence of tetracycline resistance in enterococci isolated from poultry in Ilishan, Ogun State, Nigeria. *J. Pharm. Bioallied Sci.* **2016**, *8*, 69–73. [CrossRef] [PubMed]

50. Eghomwanre, A.F.; Obayagbona, N.O.; Osarenotor, O.; Enagbonma, B.J. Evaluation of antibiotic resistance patterns and heavy metals tolerance of some bacteria isolated from contaminated soils and sediments from Warri, Delta State, Nigeria. *J. Appl. Sci. Environ. Manag.* **2016**, *20*, 287–291. [CrossRef]

51. Igbinosa, E.O. Detection and antimicrobial resistance of Vibrio isolates in aquaculture environments: Implications for public health. *Microb. Drug Resist.* **2016**, *22*, 238–245. [CrossRef] [PubMed]

52. Ishola, O.O.; Mosugu, J.I.; Adesokan, H.K. Prevalence and antibiotic susceptibility profiles of *Listeria monocytogenes* contamination of chicken flocks and meat in Oyo State, south-western Nigeria: Public health implications. *J. Prev. Med. Hyg.* **2016**, *57*, E157–E163. [PubMed]

53. Ngbede, E.O.; Raji, M.A.; Kwanashie, C.N.; Kwaga, J.K. Antimicrobial resistance and virulence profile of enterococci isolated from poultry and cattle sources in Nigeria. *Trop. Anim. Health Prod.* **2017**, *49*, 451–458. [CrossRef] [PubMed]

54. Odumosu, B.T.; Ajetumobi, O.; Adegbola, H.D.; Odutayo, I. Antibiotic susceptibility pattern and analysis of plasmid profiles of *Pseudomonas aeruginosa* from human, animal and plant sources. *SpringerPlus* **2016**, *5*, 1381. [CrossRef] [PubMed]

55. Umaru, G.A.; Kwaga, J.K.P.; Bello, M.; Raji, M.A.; Maitala, Y.S. Antibiotic resistance of *Staphylococcus aureus* isolated from fresh cow milk in settled Fulani herds in Kaduna State, Nigeria. *Bull. Anim. Health Prod. Afr.* **2016**, *64*, 173–182.

56. Dipeolu, M.A.; Alonge, D.O. Residues of tetracycline antibiotic in cattle meat marketed in Ogun and Lagos States of Nigeria. *ASSET* **2001**, *1*, 31–36.

57. Kabir, J.; Umoh, J.U.; Umoh, V.J. Characterisation and screening for antimicrobial substances of slaughtered cattle in Zaria, Nigeria. *Meat Sci.* **2002**, *61*, 435–439. [CrossRef]

58. Dipeolu, M.A. Residues of tetracycline antibiotic in marketed goats and pigs in Lagos and Ogun States Nigeria. *Niger. J. Anim. Sci.* **2002**, *5*. [CrossRef]

59. Dipeolu, M.A.; Dada, K.O. Residues of tetracycline in imported frozen chickens in South West Nigeria. *Trop. Vet.* **2005**, *23*, 1–4. [CrossRef]

60. Olatoye, I.O.; Ehinmowo, A.A. Oxytetracycline residues in edible tissues of cattle slaughtered in Akure, Nigeria. *Niger. Vet. J.* **2010**, *31*, 93–102. [CrossRef]

61. Ezenduka, E.V.; Oboegbulem, S.I.; Nwanta, J.A.; Onunkwo, J.I. Prevalence of antimicrobial residues in raw table eggs from farms and retail outlets in Enugu State, Nigeria. *Trop. Anim. Health Prod.* **2011**, *43*, 557–559. [CrossRef] [PubMed]

62. Adetunji, V.O.; Olaoye, O.O. Detection of β-Lactam antibiotics (Penicillin and Amoxicillin) residues in Goat milk. *Nat. Sci.* **2012**, *10*, 60–64.

63. Olatoye, O.; Kayode, S.T. Oxytetracycline residues in retail chicken eggs in Ibadan, Nigeria. *Food Addit. Contam. Part B Surveill.* **2012**, *5*, 255–259. [CrossRef] [PubMed]

64. Ezenduka, E.V.; Ugwumba, C. Antimicrobial residues screening in pigs and goats slaughtered in Nsukka Municipal abattoir, Southeast Nigeria. *Afr. J. Biotechnol.* **2012**, *11*, 12138–12140. [CrossRef]

65. Omeiza, K.G.; Otopa, E.A.; Okwoche, J.O. Assessment of drug residues in beef in Abuja, the Federal Capital Territory, Nigeria. *Vet. Ital.* **2012**, *48*, 283–289. [PubMed]

66. Olatoye, I.O.; Oyelakin, E.F.; Adeyemi, I.G.; Call, D.R. Chloramphenicol use and prevalence of its residues in broiler chickens and eggs in Ibadan, Nigeria. *Niger. Vet. J.* **2012**, *33*, 643–650.

67. Olusola, A.V.; Folashade, P.A.; Ayoade, O.I. Heavy metal (lead, Cadmium) and antibiotic (Tetracycline and Chloramphenicol) residues in fresh and frozen fish types (*Clarias gariepinus*, *Oreochromis niloticus*) in Ibadan, Oyo State, Nigeria. *Pak. J. Biol. Sci.* **2012**, *15*, 895–899. [CrossRef] [PubMed]

68. Olusola, A.V.; Diana, B.E.; Ayoade, O.I. Assessment of tetracycline, lead and cadmium residues in frozen chicken vended in Lagos and Ibadan, Nigeria. *Pak. J. Biol. Sci.* **2012**, *15*, 839–844. [CrossRef] [PubMed]

69. Adesokan, H.K.; Agada, C.A.; Adetunji, V.O.; Akanbi, I.M. Oxytetracycline and penicillin-G residues in cattle slaughtered in south-western Nigeria: Implications for livestock disease management and public health. *J. S. Afr. Vet. Assoc.* **2013**, *84*, E1–E5. [CrossRef] [PubMed]

70. Mbodi, F.E.; Nguku, P.; Okolocha, E.; Kabir, J. Determination of chloramphenicol residues in commercial chicken eggs in the Federal Capital Territory, Abuja, Nigeria. *Food Addit. Contam. Part A Chem. Anal. Control Expo. Risk Assess.* **2014**, *31*, 1834–1839. [CrossRef] [PubMed]

71. Olatoye, I.O.; Daniel, O.F.; Ishola, S.A. Screening of antibiotics and chemical analysis of penicillin residue in fresh milk and traditional dairy products in Oyo state, Nigeria. *Vet. World* **2015**, *9*, 948–954. [CrossRef] [PubMed]

72. Adesida, S.A.; Coker, A.O.; Smith, S.I. Resistotyping of *Campylobacter jejuni*. *Niger. Postgrad. Med. J.* **2003**, *10*, 211–215. [PubMed]

73. Moher, D.; Liberati, A.; Tetzlaff, J.; Altman, D.G.; The PRISMA Group. Preferred reporting items for systematic reviews and meta-analyses: The PRISMA statement. *PLoS Med.* **2009**, *6*, e1000097. [CrossRef] [PubMed]

74. Coates, A.R.; Halls, G.; Hu, Y. Novel classes of antibiotics or more of the same? *Br. J. Pharmacol.* **2011**, *163*, 184–194. [CrossRef] [PubMed]

75. World Organization for Animal Health. List of Antimicrobial Agents of Veterinary Importance. 2015. Available online: http://www.oie.int/fileadmin/Home/eng/Our_scientific_expertise/docs/pdf/Eng_OIE_List_antimicrobials_May2015.pdf (accessed on 29 November 2017).

76. World Health Organization. Critically Important Antimicrobials for Human Medicine. 5th Revision 2016. Ranking of Medically Important Antimicrobials for Risk Management of Antimicrobial Resistance Due to Non-Human Use. Updated June 2017. Available online: http://apps.who.int/iris/bitstream/10665/255027/1/9789241512220-eng.pdf (accessed on 29 November 2017).

77. Israel, E.U.; Emmanuel, E.G.; Sylvester, E.G.; Chukwuma, E. Self-medication with antibiotics amongst Civil Servants in Uyo, Southern Nigeria. *J. Adv. Med. Pharm. Sci.* **2015**, *2*, 89–97. [CrossRef]

78. Federal Ministries of Agriculture and Rural Development, Environment, and Health. Antimicrobial Use and Resistance in Nigeria: Situation Analysis and Recommendations. 2017. Available online: http://www.ncdc.gov.ng/themes/common/docs/protocols/56_1510840387.pdf (accessed on 14 May 2018).

79. Adesokan, H.; Akanbi, I.; Akanbi, I.; Obaweda, R. Pattern of antimicrobial usage in livestock animals in south-western Nigeria: The need for alternative plans. *Onderstepoort J. Vet. Res.* **2015**, *82*, 1–6. [CrossRef] [PubMed]

80. World Organization for Animal Health. OIE Terrestrial Animal Health Code. 2017. Available online: http://www.rr-africa.oie.int/docspdf/en/Codes/en_csat-vol1.pdf (accessed on 29 November 2017).

International Journal of
***Environmental Research
and Public Health***

Article

Prevalence of Integrons and Insertion Sequences in ESBL-Producing *E. coli* Isolated from Different Sources in Navarra, Spain

Lara Pérez-Etayo [1,*], Melibea Berzosa [1], David González [1,2] and Ana Isabel Vitas [1,2]

[1] Department of Microbiology and Parasitology, University of Navarra, 31008 Pamplona, Spain;
mberzosa.1@alumni.unav.es (M.B.); dgonzalez@unav.es (D.G.); avitas@unav.es (A.I.V.)
[2] IDISNA, Navarra Health Research Institute, 31008 Pamplona, Spain
* Correspondence: lperez.13@alumni.unav.es; Tel.: +34-948-425-600

Received: 14 September 2018; Accepted: 18 October 2018; Published: 20 October 2018

Abstract: Mobile genetic elements play an important role in the dissemination of antibiotic resistant bacteria among human and environmental sources. Therefore, the aim of this study was to determine the occurrence and patterns of integrons and insertion sequences of extended-spectrum β-lactamase (ESBL)-producing *Escherichia coli* isolated from different sources in Navarra, northern Spain. A total of 150 isolates coming from food products, farms and feeds, aquatic environments, and humans (healthy people and hospital inpatients), were analyzed. PCRs were applied for the study of class 1, 2, and 3 integrons (*intI1*, *intI2*, and *intI3*), as well as for the determination of insertion sequences (IS*26*, IS*Ecp1*, IS*CR1*, and IS*903*). Results show the wide presence and dissemination of *intI1* (92%), while *intI3* was not detected. It is remarkable, the prevalence of *intI2* among food isolates, as well as the co-existence of class 1 and class 2 (8% of isolates). The majority of isolates have two or three IS elements, with the most common being IS*26* (99.4%). The genetic pattern IS*26*–IS*Ecp1* (related with the pathogen clone ST131) was present in the 22% of isolates (including human isolates). In addition, the combination IS*Ecp1*–IS*26*–IS*903*–IS*CR1* was detected in 11 isolates being, to our knowledge, the first study that describes this genetic complex. Due to the wide variability observed, no relationship was determined among these mobile genetic elements and β-lactam resistance. More investigations regarding the genetic composition of these elements are needed to understand the role of multiple types of integrons and insertion sequences on the dissemination of antimicrobial resistance genes among different environments.

Keywords: ESBL-producing *E. coli*; β-lactamase genes; antimicrobial resistance; integrons; insertion sequences

---

1. Introduction

Antimicrobial resistance (AMR) has become a public health problem, reaching alarming levels in many parts of the world [1,2]. In recent years, resistances in the Enterobacteriaceae family have increased significantly because of the extensive use of antibiotics in human treatment, veterinary, and agriculture, leading to the selection and global spread of resistant clones [3,4]. In particular, the dissemination of extended-spectrum β-lactamases (ESBLs) have increased dramatically in the recent years, becoming a serious global threat [5,6].

Several genetic mechanisms have been involved in the acquisition and dispersion of antimicrobial resistances. The commonly called "mobilome" [7,8] is composed of a variety of mobile genetic elements (MGEs), including plasmids, transposons (Tn), insertion sequences (IS), integrons (*intI*), and introns. Conjugation, transformation, and transduction are the main mechanisms for the horizontal transfer of MGEs [9,10].

Integrons are DNA elements capable of capturing gene cassettes (including antimicrobial resistance genes) and disseminating them using an MGE [11]. Integrons are usually composed of two conserved segments (termed 5′-conserved region (5′-CS) and 3′-conserved region (3′-CS)) separated by a variable region which contains the gene cassettes. The 5′-CS end includes (i) the *int* gene coding for an integrase, that belongs to a distinct family of the tyrosine-recombinase; (ii) a primary recombination site (*attI*); and (iii) a promoter (Pc), which ensures the transcription of the cassette genes. On the other hand, the 3′-CS region is formed by (i) a truncated gene of resistance to quaternary ammonium compounds (*qacEΔ1*); (ii) a sulfonamide resistance gene (*sul1*); and (iii) an unknown function sequence (orf5) [12]. Class 1 (*intI1*) and class 2 (*intI2*) integrons are the most commonly involved in antibiotic resistances [13–17], while limited work has shown the presence of class 3 (*intI3*) in Enterobacteriaceae. The gene *intI3* was reported for the first time in a carbapenem-resistant *Serratia marcescens* strain [18] and has been also detected in *Klebsiella pneumoniae* isolates [19] and other Enterobacteriaceae [20]. In addition, *bla* ESBL genes have been associated with insertion sequences. These IS are the smallest transposable elements (<2.5 kb), and are classified into families according to different characteristics, with transposases (enzymes that catalyze the IS movement) being the main classification system used [21,22]. It has been well documented that IS*26*, IS*Ecp1*, IS*CR1*, and IS*903*, in association with class 1 integrons, are the most involved elements in the antimicrobial resistance to β-lactamics [23–27].

Therefore, the investigation of these elements might be critical, in order to predict the potential spread of ESBL-producing strains. In this context, the aim of this study was to evaluate the presence of different types of integrons (*intI1*, *intI2*, and *intI3*) and insertion sequences (IS*Ecp1*, IS*26*, IS*CR1*, and IS*903*) in a collection of 150 ESBL-producing *E. coli* isolated from different sources in Navarra, Spain.

2. Materials and Methods

2.1. Sample Collection

A total of 150 ESBL-producing *Escherichia coli* were selected from a wide collection of ESBL-producing Enterobacteriaceae, isolated in Navarra from different environments: food products (*n* = 48), farms and feeds (*n* = 20), rivers and wastewater treatment plants (WWTPs) (*n* = 33) and human origins, including healthy volunteers (*n* = 13) and hospital inpatients (*n* = 36). Clinical isolates from hospital inpatients were provided by Clínica Universidad de Navarra, and were collected from January 2009 to December 2012 [5]. Food and environmental samples were collected from different locations in Navarra in two sampling periods (2010–2013 [5,28,29]; 2015–2016 [30]) and, finally, isolates from healthy people were collected from September 2015 to September 2016 (data not published). All samples were already identified, and phenotypically and genotypically characterized, in order to know the antimicrobial susceptibility pattern, the types of β-lactamase genes, and the phylogenetic group [28–30]. Isolates were selected according to the following criteria: they must show multidrug-resistant phenotype (to at least three different classes of antimicrobials) and must carry at least one ESBL gene. The main characteristics of the selected isolates, regarding type of ESBL, is shown in Table 1. Resistance profiles and complete information of each isolate is presented in the Supplementary Material, Tables S1–S6).

Table 1. Genotypic characteristics of extended-spectrum β-lactamase (ESBL)-producing *E. coli* according to their origin [5,28–30].

Sample Origin	Percentages of Detected *bla* Genes					
	*bla*CTX-M-14	*bla*CTX-M-15	*bla*CTX-M-1	*bla*TEM-42	*bla*TEML-171	*bla*SHV-12
Hospital inpatients	41.7	61.1	8.3	11.1	NA	5.5
Healthy people	46.2	30.8	15.4	0	46.2	0
WWTP and rivers	33.3	30.3	18.2	6	NA	6
Food	32.7	4.1	18.3	12.3	31.8	35.6
Farms and feeds	31.6	5.26	47.4	26.3	5	21

NA: Not analyzed.

2.2. DNA Extraction and Detection of Integrons

DNA extraction was performed with DNeasy® Blood & Tissue kit (Qiagen, Barcelona, Spain), using a pre-treatment protocol for Gram-negative bacteria, and following the manufacturer's instruction. The quantity and quality of the DNA was analyzed using a Nanodrop ND-1000 spectrophotometer (NanoDrop Technologies, Wilmington, DE, USA).

Detection of class 1, class 2, and class 3 integrons in ESBL-producing *E. coli* was performed according to PCR, as described by Mazel et al. [31], and using only the primers shown in Table 2.

Table 2. Primers used for the detection of integrons.

Primer	Sequence (5′-3′)	Amplicon Size (pb)	T (°C) [3]	GenBank Accession No	Reference
intI1-Fw [1]	GGTCAAGGATCTGGATTTCG	483	62	U49101	[31]
intI1-Rv [2]	ACATGCGTGTAAATCATCGTC	483	62	U49101	[31]
intI2-Fw [1]	CACGGATATGCGACAAAAAGGT	789	62	L10818	[31]
intI2-Rv [2]	TAGCAAACGAGTGACGAAATG	789	62	L10818	[31]
intI3-Fw [1]	AGTGGGTGGCGAATGAGTG	600	60	D50438	[31]
intI3-Rv [2]	TGTTCTTGTATCGGCAGGTG	600	60	D50438	[31]

[1] Fw: forward; [2] Rv: reverse; [3] T (°C): annealing temperature.

DNA amplification was performed in a DNA thermal cycler GeneAmp® PCR system 2700 (Applied Biosystems Division, Foster City, CA, USA) in a final volume of 25 µL containing 2 µL of DNA extract mixed with 2.5 µL of 10× buffer (Bioline, London, UK), 5 µL of dNTPs (Bioline, London, UK), 1.5 µL of $MgCl_2$ 50 mM (Bioline, London, UK), 2 µL of each primer Sigma-Aldrich, Steinheim, Germany), and 1.5 U of Inmolase™ DNA polymerase (Bioline, London, UK). The conditions of the amplification were as follows: initial denaturation at 94 °C for 10 min, followed by 30 cycles of DNA denaturation at 94 °C for 45 s, primer annealing at 62 °C (*intI1* and *intI2*) or 60 °C (*intI3*) for 35 s, primer extension at 72 °C for 2 min, and a final elongation at 72 °C for 7 min. Positive and negative controls [17] were included in all PCR assays, and 1 kb ladder (Invitrogen) was used as a molecular size standard. After amplification, PCR products were separated by electrophoresis on 1% agarose gel in 1× TBE buffer, stained with ethidium bromide and visualized by UV transillumination. *E. coli* C828, *K. pneumoniae* C933 (provided both by Centro de Investigación Biomédica de la Rioja) and *E. coli* isolated from hospital inpatients, confirmed as carrying *intI2* by DNA sequencing, were used as positive controls for *intI1*, *intI3*, and *intI2*, respectively.

2.3. Detection of Insertion Sequences

DNA extracts were examined for the detection of different insertion sequences associated with ESBL genes, performing PCRs assays using the specific primers and conditions showed in Table 3 [27,32,33].

The PCRs were performed in a final volume of 25 µL containing 2 µL of DNA extract mixed with 2.5 µL of 10× buffer (Bioline, London, UK), 5 µL of dNTPs (Bioline, London, UK), 1.5 µL of

MgCl$_2$ 50 mM (Bioline, London, UK), 2 µL of each primer (Sigma-Aldrich, Madrid, Spain), and 1.5 U of Inmolase™ DNA polymerase (Bioline, London, UK), in a DNA thermal cycler GeneAmp® PCR system 2700 (Applied Biosystems Division, Foster City, CA, USA). Amplification conditions were modified in order to improve the specificity using an initial denaturation at 94 °C for 12 min, followed by 35 cycles of DNA denaturation at 94 °C for 1 min, and primer annealing temperature depending on the IS (Table 3), primer extension at 72 °C for 2 min, and a final elongation at 72 °C for 10 min. PCR products were separated by electrophoresis on 1% agarose gels and were visualized under UV light after staining with ethidium bromide.

Table 3. Primers and conditions used for the amplification of insertion sequences.

Primer [1]	Sequence (5′-3′)	Amplicon Size (pb)	T (°C) [3]	GenBank Accession No.	Reference
IS*Ecp1*-Fw [1]	ATCTAACATCAAATGCAGG	1381	60	AJ972954	[27]
IS*Ecp1*-Rv [2]	AGACTGCTTCTCACACAT	1381	60	AJ972954	[27]
IS*26*-Fw [1]	TCACTCCACGATTTACCGCT	557	61	AF205943	[27]
IS*26*-Rv [2]	CTTACCAGGCGCATTTCGCC	557	61	AF205943	[27]
IS*CR1*-Fw [1]	TCGCTGCGAGGATTGTCATC	1100	60	AF174129	[32]
IS*CR1*-Rv [2]	CTCGCTTGAGGCGTTGCAT	1100	60	AF174129	[32]
IS*903*-Fw [1]	CATATGAAATCATCTGCGC	473	56	EU056266	[33]
IS*903*-Rv [2]	CCGTAGCGGGTTGTGTTTTC	473	56	EU056266	[33]

[1] Fw: forward; [2] Rv: reverse; [3] T (°C): annealing temperature.

2.4. Sequence Analysis

Amplicons obtained in the different PCRs were sequenced to confirm the presence of integrons and insertion sequences. Bidirectional DNA sequence analysis was performed by the Macrogen EZ-Seq purification service (Macrogen Europe, Amsterdam, The Netherlands). Searches for DNA and protein homologies were carried out using the National Center for Biotechnology Information (http://www.ncbi.nlm.nih.gov/) using the BLAST program and the alignment of DNA and amino acids sequences were performed using Clustal Omega (http://www.ebi.ac.uk/Tools/msa/clustalo/).

2.5. Statistical Analysis

The results were subjected to statistical processing with the SPSS 15 software (SPSS Inc., Chicago, IL, USA), applying the chi-square test with a level of significance of $p < 0.05$.

3. Results and Discussion

3.1. Distribution of Integrons in ESBL-Producing E. coli

The occurrence and types of integrons, according to the isolate origin, is presented in Figure 1. As expected, class 1 showed the highest dissemination, being present in 92% of the isolates ($n = 138$) and in all environments, without significant differences among them ($p < 0.05$). Class 1 integrons have been reported as the most ubiquitous type among enteric bacteria [34–36]. In a similar way, Solberg et al. [37] reported the presence of class 1 integron in 70% of *E. coli* causing community-acquired infections. According to Roe et al. [38], the occurrence of class 1 integrons in healthy people suggests a possible acquisition of resistance genes circulating in different environments by a constant horizontal exchange of these genes. By contrast, class 2 integron was found in only 13 strains (8.5%), in accordance with the study of Ozgumus et al. [39], who found this class of integron in pathogenic, environmental, and commensal *E. coli* with a lower frequency than class 1. Finally, *intI3* was not detected, similarly to the report by Vinué et al. [40]. In fact, limited studies describe the presence of class 3 integron in *E. coli* [14,41] and, to date, there are no published data reporting the presence of this integron in ESBL-producing *E. coli* isolated from Spain.

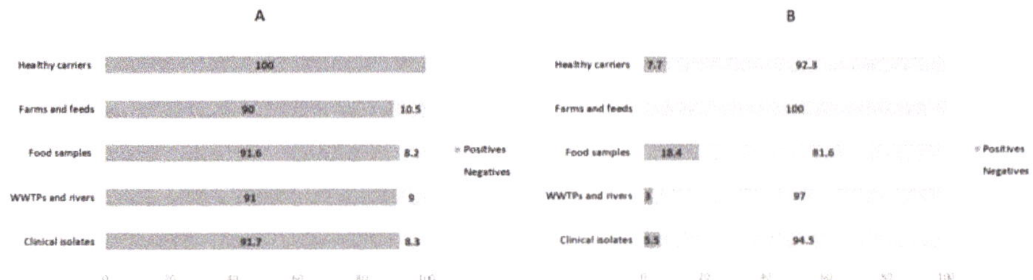

Figure 1. Prevalence (percentages) and distribution of (**A**)class 1 (*intI1*) and (**B**) class 2 (*intI2*) integrons in ESBL-producing *E. coli* according to their origin. ESBL: Extended spectrum beta-lactamases.

It should be noted that *intI2* was mainly detected in food isolates (18.4%), but not in farming environments (p = 0.044). This situation seems a little bit contradictory, but it could be due to the low number of isolates coming from farms and feed (n = 20), compared with food (n = 48). Probably, if we extended the study by increasing n, we would find positive results for this type of integron, as shown in the literature. In any case, our results are comparable to those obtained by Goldstein et al. [42], who demonstrated the presence of class 1 and class 2 integrons in food, livestock, and water contaminated with farm animal feces. In a similar way, *intI2* has been detected in poultry products [38].

In addition, it is remarkable that *intI1* and *intI2* coexist in 8% of the isolates (92.1% of those carrying *intI2*). Rizk et al. [20] reported the co-existence of more than one type of integron in 36.9% of isolates, and a prevalence of 38% was reported by Kargar et al. [41] in a study performed in 69 multidrug-resistant (MDR) *E. coli*. By contrast, Kor et al. [43] found only one isolate carrying both integrons among clinical isolates, and Odetoyin et al. [16] reported a prevalence of 2.4% in fecal *E. coli* isolated from mother–child pairs in Nigeria. The simultaneous existence of multiple integrons represents a great threat for the dissemination of antimicrobial resistance genes among Enterobacteriaceae [31].

3.2. Analysis of Insertion Sequences

The prevalent type was IS26 (99.4%), followed by IS*Ecp1* (68%) and IS903 (65.3%), while ISCR1 was detected only in 19 isolates (12.6%). The wide presence of IS26 in almost all multidrug-resistant isolates is a hint that IS26 is not associated with multidrug resistance, but only with ESBL-producing isolates.

The four insertion sequences were present in all environments (Figure 2), except ISCR1, that was not detected in farm and feeds. This latest result contrasts with the reported by Ali et al. [44], that showed the connection between ISCR1 and *intI1* in strains isolated from diverse dairy farms in China. However, the wide dissemination of IS among different niches has been reported by other authors. For instance, Cullik et al. [25] showed the association between *bla*CTX-M with the common elements IS*Ecp1*, IS26, and IS903, in ESBL-producing *E. coli* isolated in a German Hospital. The IS*Ecp1* type has been detected in clinical isolates from Korea, and in isolates from healthy or diseased food-producing animals, including swine and avian [45,46].

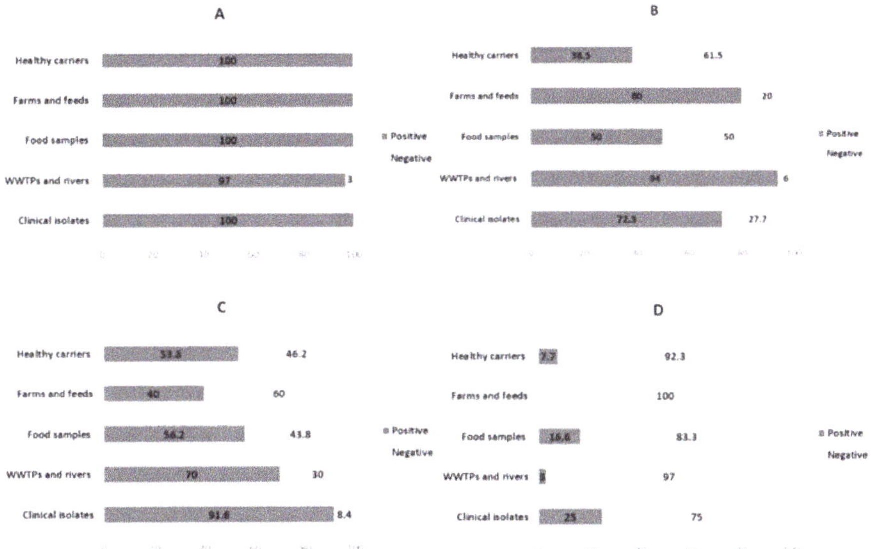

Figure 2. Prevalence (percentages) and distribution of insertion sequences in ESBL-producing *E. coli* according to their origin. (**A**) IS*26*; (**B**) IS*Ecp1*; (**C**) IS*903*; (**D**) IS*CR1*.

In addition, the frequent co-existence of several insertion sequences in the same strain has been detected, in agreement with other studies [25–27,47]. Genetic patterns are presented in Figure 3, showing that the majority of isolates carried two or three IS (42% and 40.7%, respectively), whereas 10% of them carried only one. The three prevalent genetic patterns were IS*26*–IS*Ecp1*–IS*903* (*n* = 55), IS*26*–IS*Ecp1* (*n* = 33), and IS*26*–IS*903* (*n* = 28) (37%, 22%, and 19%, respectively). The combination IS*26*–IS*Ecp1* has been related with the pathogen clone ST131 [25,48], and it was present in 4 isolates coming from healthy people (*n* = 2) and hospital inpatients (*n* = 2), that supposes a possible risk situation for the healthy population. Finally, it is remarkable that 7.3% of isolates contain the four IS (IS*Ecp1*–IS*26*–IS*903*–IS*CR1*), a situation that, to our best knowledge, is being described in the literature for the first time. These isolates come mainly from hospital inpatients (*n* = 9), but we also found the genetic patterns in isolates from a river (*n* = 1) and from a chicken hamburger (*n* = 1). In summary, these results show the complexity of mobile genetic elements, and suggest the facility to acquire different mechanisms to disseminate resistance genes through all environments.

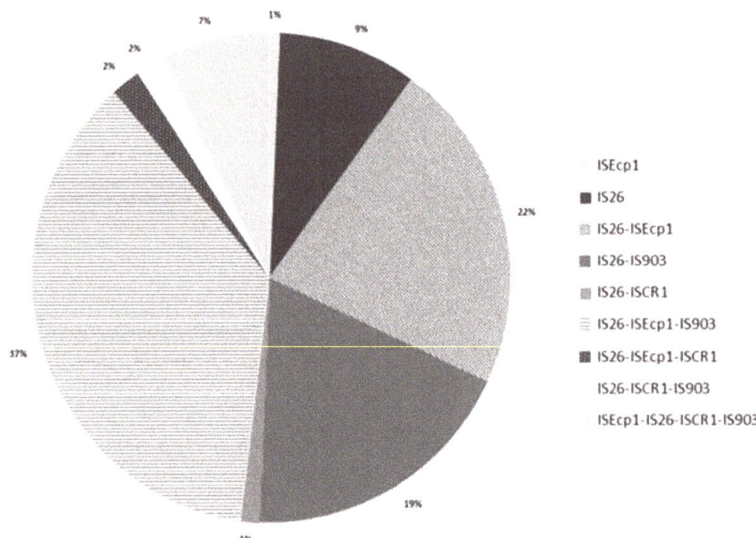

Figure 3. Genetic patterns and prevalences among the studied ESBL-producing *E. coli*.

3.3. The Important Role of Horizontal Genetic Elements in the Dissemination of ESBLs

Correlation between the presence of genetic elements and ESBL has been reported by several authors [25,49,50], and our results support this fact (Table 4). IS26 have been observed flanking the open reading frame (orf) regions of β-lactamase genes [51], and prevalences higher than 94% in all ESBL types were observed in this study. Similar results were detected in a study carried out in Kenya with 27 *E. coli* strains obtained from hospitalized patients, in which over 40% of isolates carrying $bla_{\text{TEM-52}}$, $bla_{\text{SHV-5}}$, or $bla_{\text{CTX-M-14}}$, were linked to the IS26 [50]. Otherwise, Billard-Pomares et al. [52] reported the characterization of a P1 bacteriophage from an ESBL-*E. coli* strain which had acquired two foreign DNA fragments, one of them being a fragment mobilized by two IS26 elements containing a $bla_{\text{SHV-2}}$ gene. Finally, Doi et al. [53] reported the relation between OXA (Beta lactamase product of blaOXA genes) and IS26 downstream of a class 1 integron in a *K. pneumoniae* strain. In summary, as Cullik et al. affirm [25], IS26 have an important role in the spread of resistance genes.

Table 4. Prevalences of insertion sequences and integrons among the different types of ESBL-*E. coli* producers.

bla Genes	IS26	IS903	IS*Ecp1*	ISCR1	*intI1*	*intI2*
$bla_{\text{CTX-M}}$	99.2	90.3	79	11.3	94	6.5
bla_{TEM}	100	89.9	88.5	16	94	7.3
$bla_{\text{OXA-1}}$	94.5	83.3	50	5.5	100	0
bla_{SHV}	100	56.5	26	0	95.7	21.8

Similarly, IS*Ecp1*-like insertion sequences have been observed upstream of orfs encoding members belonging to the CTX-M-1, CTX-M-2, and CTX-M-9 clusters. Kim et al. [45] found the association of IS*Ecp1* and CTX-M in clinical isolates, especially in strains containing CTX-M-14 (in agreement with the 37% observed in our study). A similar association was found in China by Sun et al. [54] in healthy and sick pets. In addition, Tamang et al. [55] reported that 97.6% $bla_{\text{CTX-M}}$ genes (isolated from cattle, farm workers, and the farm environment) possessed the insertion sequence IS*Ecp1* upstream of bla_{CTX}. On the other hand, our results show that 9 out of 102 isolates carrying IS*Ecp1* (isolated from WWTP, river, farm soil and feed) were disrupted by IS26. Similar findings have been reported in a German

University Hospital [25], where cases of IS*Ecp1* disrupted by an intact IS*26* were detected. In the same way, Wang et al. [48] detected a truncated copy of IS*Ecp1* gene with an IS*26* gene being located upstream in 3 out of 9 ESBL-producing *E. coli* isolated from fecal samples of food producing animals and healthy humans. Finally, despite the lower prevalence of IS*CR1* observed in the present study (12.6%), the aforementioned IS is another important element in the genetic platforms associated with the dissemination of CTX-M genes [22,56,57]. In general, IS*CR1* has been associated with CTX-M-2 and CTX-M-9 subtypes [57–59], but the majority of our strains carrying this IS were CTX-M-14 and CTX-M-15 producers. That could explain the low number of strains carrying IS*CR1*. Moreover, IS*CR1* mediates the formation of a complex with class 1 integrons [23,57]. From the total of isolates carrying IS*CR1*, 94.7% contain *intI1*, and even one of them contained both integrons (*intI1* and *intI2*). However, we have not found a specific association between the isolates containing *intI1* and the different ESBLs, due to its wide presence (92% of isolates). On the other hand, CTX-M-14 was present in the 46% of the isolates containing *intI2* (the same as SHV-12), whereas TEM and CTX-M-1 were detected in 38.5% and 15.4%, respectively, of *intI2* carriers.

Moreover, it must be pointed out that IS*Ecp1*-IS*903* is known as one of the major genetic platforms [22,27,54]. Our results showed that IS*903* and IS*Ecp1* were present in 55 isolates in co-existence with IS*26* (IS*26*-IS*Ecp1*-IS*903*). Similarly, a recent report detected this genetic platform [46] in CTX-M-14-producing *E. coli* isolated from animals [48]. Furthermore, all the analyzed strains show multidrug-resistant (MDR) phenotype, which means that they are resistant to at least three different classes of antimicrobials [28,29]. Similar results were reported by Woodford et al. [60] that found the plasmid pEK499 harboring 10 genes that confer resistance to eight antibiotic classes and also carrying IS (IS*26* and IS*Ecp1*).

Finally, Table 5 summarizes the relationship between the number of IS present in the same isolate, and the number of ESBL types produced by each microorganism. It can be seen that as the number of ESBL enclosed in the same genetic environment increases, the number of insertion sequences present also increases.

Table 5. Relationship between number of IS in each isolate and the number of expressed ESBLs.

Number of IS in Each Isolate	N Isolates	N Isolates (%) Producing			
		1 ESBL	2 ESBL	3 ESBL	4 ESBL
1	15	46.6	40	0	0
2	63	46	46	8	0
3	61	49	41	8.2	1.6
4	11	0	81.8	18.2	0

To sum up, the MDR ESBL-producing *E. coli* analyzed in the present study carried at least one genetic element (integron and IS). Since the strains were isolated from different sources (clinical isolates, healthy carriers, farms and feeds, food samples, WWTPs and rivers), these data revealed the potential risk for the dissemination of antimicrobial resistances among environmental and human bacteria.

4. Conclusions

In conclusion, this study highlights the high prevalence of different horizontal genetic elements among ESBL-producing *E. coli* isolates from food, environmental, and human samples. The analysis of integrons, showed that *intI1* was present in the majority of strains and in all sources, while the prevalence of *intI2* was lower but remarkable in the food isolates. Concerning insertion sequences, the multiple associations, like IS*26*-IS*Ecp1*, are relevant. Thus, the co-existence of diverse types of integrons and insertion sequences suggest possible risk for the dissemination of resistance genes among different environments and, therefore, additional investigations regarding the genetic composition of these integrons and insertion sequences are encouraged, to understand the role of these mobile elements in the spread of multidrug-resistant bacteria.

Supplementary Materials: Table S1: Antimicrobial profiles of ESBL-producing *E. coli* according to their origin; Table S2: Phenotypic and genotypic characteristics of strains isolated from hospital inpatients (n = 36); Table S3: Phenotypic and genotypic characteristics of aquatic strains included in the study (n = 33); Table S4: Phenotypic and genotypic characteristics of strains isolated from food (n = 48); Table S5: Phenotypic and genotypic characteristics of strains from farm origin included in the study (n = 20); Table S6: Phenotypic and genotypic characteristics of strains isolated from healthy people (n = 13).

Author Contributions: L.P.-E. performed the experiments, analyzed the data and wrote the paper. M.B. performed some PCRs. D.G. and A.I.V. conceived, designed the experiments, supervised data analysis and reviewed the manuscript.

Funding: This research was funded by a grant of the "Asociación de Amigos de la Universidad de Navarra", obtained in September 2016.

Acknowledgments: We are particularly grateful to Yolanda Saénz from the "Centro de Investigación Biomédica de la Rioja" for providing the positives controls for the PCRs.

Conflicts of Interest: The authors declare no conflict of interest. The funders had no role in the design of the study; in the collection, analyses, or interpretation of data; in the writing of the manuscript, and in the decision to publish the results.

References

1. WHO. *Antimicrobial Resistance. Global Report on Surveillance*; WHO: Geneva, Switzerland, 2014; pp. 383–394.
2. Roca, I.; Akova, M.; Baquero, F.; Carlet, J.; Cavaleri, M.; Coenen, S.; Cohen, J.; Findlay, D.; Gyssens, I.; Heure, O.E.; et al. The global threat of antimicrobial resistance: Science for intervention. *New Microbes New Infect.* **2015**, *6*, 22–29. [CrossRef] [PubMed]
3. Liu, X.J.; Lyu, Y.; Li, Y.; Xue, F.; Liu, J. Trends in antimicrobial resistance against Enterobacteriaceae strains isolated from blood: A 10-year epidemiological study in mainland China (2004–2014). *Chin. Med. J.* **2017**, *130*, 2050–2055. [CrossRef] [PubMed]
4. Shaikh, S.; Fatima, J.; Shakil, S.; Rizvi, S.M.D.; Kamal, M.A. Antibiotic resistance and extended spectrum beta-lactamases: Types, epidemiology and treatment. *Saudi J. Biol. Sci.* **2015**, *22*, 90–101. [CrossRef] [PubMed]
5. Ojer-Usoz, E.; González, D.; Vitas, A.I. Clonal diversity of ESBL-producing *Escherichia coli* isolated from environmental, human and food samples. *Int. J. Environ. Res. Public Health* **2017**, *14*, 676. [CrossRef] [PubMed]
6. Sikkema, R.; Koopmans, M. One Health training and research activities in Western Europe. *Infect. Ecol. Epidemiol.* **2016**, *6*, 33703. [CrossRef] [PubMed]
7. Medini, D.; Donati, C.; Tettelin, H.; Masignani, V.; Rappuoli, R. The microbial pan-genome. *Curr. Opin. Genet. Dev.* **2005**, *15*, 589–594. [CrossRef] [PubMed]
8. Tettelin, H.; Riley, D.; Cattuto, C.; Medini, D. Comparative genomics: The bacterial pan-genome. *Curr. Opin. Microbiol.* **2008**, *11*, 472–477. [CrossRef] [PubMed]
9. Norman, A.; Hansen, L.H.; Sørensen, S.J. Conjugative plasmids: Vessels of the communal gene pool. *Philos. Trans. R. Soc. B Biol. Sci.* **2009**, *364*, 2275–2289. [CrossRef] [PubMed]
10. Woodford, N.; Turton, J.F.; Livermore, D.M. Multiresistant Gram-negative bacteria: The role of high-risk clones in the dissemination of antibiotic resistance. *FEMS Microbiol. Rev.* **2011**, *35*, 736–755. [CrossRef] [PubMed]
11. Stokes, H.W.; Hall, R.M. A novel family of potentially mobile DNA elements encoding site-specific gene-integration functions: Integrons. *Mol. Microbiol.* **1989**, *3*, 1669–1683. [CrossRef] [PubMed]
12. Galani, I.; Souli, M.; Koratzanis, E.; Chryssouli, Z.; Giamarellou, H. Molecular characterization of an *Escherichia coli* clinical isolate that produces both metallo-β-lactamase VIM-2 and extended-spectrum β-lactamase GES-7: Identification of the In8 integron carrying the *bla*VIM-2 gene. *J. Antimicrob. Chemother.* **2006**, *58*, 432–433. [CrossRef] [PubMed]
13. Fluit, A.C.; Schmitz, F.J. Resistance integrons and super-integrons. *Clin. Microbiol. Infect.* **2004**, *10*, 272–288. [CrossRef] [PubMed]
14. Kaushik, M.; Kumar, S.; Kapoor, R.K.; Virdi, J.S.; Gulati, P. Integrons in Enterobacteriaceae: Diversity, distribution and epidemiology. *Int. J. Antimicrob. Agents* **2018**, *51*, 167–176. [CrossRef] [PubMed]

15. Machado, E.; Canto, R.; Baquero, F.; Gala, J.; Coque, T.M. Integron Content of Extended-Spectrum β-Lactamase-Producing *Escherichia coli* Strains over 12 Years in a Single Hospital in *Pseudomonas aeruginosa*. *Antimicrob. Agents Chemother.* **2005**, *49*, 1823–1829. [CrossRef] [PubMed]

16. Odetoyin, B.W.; Labar, A.S.; Lamikanra, A.; Aboderin, A.O.; Okeke, I.N. Classes 1 and 2 integrons in faecal *Escherichia coli* strains isolated from mother-child pairs in Nigeria. *PLoS ONE* **2017**, *12*. [CrossRef] [PubMed]

17. Saenz, Y.; Brinas, L.; Dominguez, E.; Ruiz, J.; Zarazaga, M.; Vila, J.; Torres, C. Mechanisms of Resistance in Multiple-Antibiotic-Resistant *Escherichia coli* Strains of Human, Animal, and Food Origins. *Antimicrob. Agents Chemother.* **2004**, *48*, 3996–4001. [CrossRef] [PubMed]

18. Arakawa, Y.; Murakami, M.; Suzuki, K.; Ito, H.; Wacharotayankun, R.; Kato, N.; Ohta, M. A Novel Integron-Like Element Carrying the Metallo-β-Lactamase Gene *bla* IMP. *Microbiology* **1995**, *39*, 1612–1615. [CrossRef]

19. Correia, M.; Boavida, F.; Grosso, F.; Salgado, M.J.; Lito, L.M.; Cristino, J.M.; Mendo, S.; Duarte, A. Molecular characterization of a new class 3 integron in *Klebsiella pneumoniae*. *Antimicrob. Agents Chemother.* **2003**, *47*, 2838–2843. [CrossRef] [PubMed]

20. Rizk, D.E.; El-Mahdy, A.M. Emergence of class 1 to 3 integrons among members of Enterobacteriaceae in Egypt. *Microb. Pathog.* **2017**, *112*, 50–56. [CrossRef] [PubMed]

21. Mahillon, J.; Chandler, M. Insertion sequences. *Microbiol. Mol. Biol. Rev.* **1998**, *62*. [CrossRef]

22. Zhao, W.H.; Hu, Z.Q. Epidemiology and genetics of CTX-M extended-spectrum β-lactamases in Gram-negative bacteria. *Crit. Rev. Microbiol.* **2013**, *39*, 79–101. [CrossRef] [PubMed]

23. Arduino, S.M.; Roy, P.H.; Jacoby, G.A.; Betina, E.; Pineiro, S.A.; Centron, D.; Arduino, S.M.; Roy, P.H.; Jacoby, G.A.; Orman, B.E.; et al. *bla*CTX-M-2 Is Located in an Unusual Class 1 Integron (In35) Which Includes Orf513 *bla*CTX-M-2 Is Located in an Unusual Class 1 Integron (In35) Which Includes Orf513. *Antimicrob. Agents Chemother.* **2002**, *46*, 2303–2306. [CrossRef] [PubMed]

24. Cheng, C.; Sun, J.; Zheng, F.; Lu, W.; Yang, Q.; Rui, Y. New structures simultaneously harboring class 1 integron and IS*CR1*-linked resistance genes in multidrug-resistant Gram-negative bacteria. *BMC Microbiol.* **2016**, *16*. [CrossRef] [PubMed]

25. Cullik, A.; Pfeifer, Y.; Prager, R.; Von Baum, H.; Witte, W. A novel IS*26* structure surrounds *bla*CTX-M genes in different plasmids from German clinical *Escherichia coli* isolates. *J. Med. Microbiol.* **2010**, *59*, 580–587. [CrossRef] [PubMed]

26. Diestra, K.; Juan, C.; Curiao, T.; Moya, B.; Miro, E.; Oteo, J.; Coque, T.M.; Perez-Vazquez, M.; Campos, J.; Canton, R.; et al. Characterization of plasmids encoding *bla*ESBL and surrounding genes in Spanish clinical isolates of *Escherichia coli* and *Klebsiella pneumoniae*. *J. Antimicrob. Chemother.* **2008**, *63*, 60–66. [CrossRef] [PubMed]

27. Eckert, C.; Gautier, V.; Arlet, G. DNA sequence analysis of the genetic environment of various *bla*CTX-M genes. *J. Antimicrob. Chemother.* **2006**, *57*, 14–23. [CrossRef] [PubMed]

28. Ojer-Usoz, E.; González, D.; Vitas, A.I.; Leiva, J.; García-Jalón, I.; Febles-Casquero, A.; de la Escolano, M.S. Prevalence of extended-spectrum β-lactamase-producing Enterobacteriaceae in meat products sold in Navarra, Spain. *Meat Sci.* **2013**, *93*, 316–321. [CrossRef] [PubMed]

29. Ojer-Usoz, E.; González, D.; García-Jalón, I.; Vitas, A.I. High dissemination of extended-spectrum β-lactamase-producing Enterobacteriaceae ineffluents from wastewater treatment plants. *Water Res.* **2014**, *56*, 37–47. [CrossRef] [PubMed]

30. Vitas, A.I.; Naik, D.; Pérez-Etayo, L.; González, D. Increased exposure to extended-spectrum β-lactamase-producing multidrug-resistant Enterobacteriaceae through the consumption of chicken and sushi products. *Int. J. Food Microbiol.* **2018**, *269*, 80–86. [CrossRef] [PubMed]

31. Mazel, D.; Dychinco, B.; Webb, V.; Davies, J. Antibiotic resistance in the ECOR collection: Integrons and identification of a novel aad gene. *Antimicrob. Agents Chemother.* **2000**, *44*, 1568–1574. [CrossRef] [PubMed]

32. Novais, Â.; Cantón, R.; Valverde, A.; Machado, E.; Galán, J.C.; Peixe, L.; Carattoli, A.; Baquero, F.; Coque, T.M. Dissemination and persistence of *bla*CTX-M-9 are linked to class 1 integrons containing CR1 associated with defective transposon derivatives from Tn402 located in early antibiotic resistance plasmids of IncHI2, IncP1-α, and IncFI groups. *Antimicrob. Agents Chemother.* **2006**, *50*, 2741–2750. [CrossRef] [PubMed]

33. Poirel, L.; Decousser, J.; Nordmann, P. Insertion Sequence IS*Ecp1*B Is Involved in Expression and Mobilization of a *bla*CTX-M β-Lactamase Gene. *Antimicrob. Agents Chemother.* **2003**, *47*, 2938–2945. [CrossRef] [PubMed]

34. Deng, Y.; Bao, X.; Ji, L.; Chen, L.; Liu, J.; Miao, J.; Chen, D.; Bian, H.; Li, Y.; Yu, G. Resistance integrons: Class 1, 2 and 3 integrons. *Ann. Clin. Microbiol. Antimicrob.* **2015**, *14*. [CrossRef] [PubMed]

35. Hall, R.M.; Collis, C.M. Mobile gene cassettes and integrons: Capture and spread of genes by site-specific recombination. *Mol. Microbiol.* **1995**, *15*, 593–600. [CrossRef] [PubMed]

36. Machado, E.; Coque, T.M.; Cantón, R.; Sousa, J.C.; Peixe, L. Antibiotic resistance integrons and extended-spectrum β-lactamases among Enterobacteriaceae isolates recovered from chickens and swine in Portugal. *J. Antimicrob. Chemother.* **2008**, *62*, 296–302. [CrossRef] [PubMed]

37. Solberg, O.; Ajiboye, R.; Riley, L. Origin of class 1 and 2 integrons and gene cassettes in a population-based sample of uropathogenic *Escherichia coli*. *J. Clin. Microbiol.* **2006**, *44*, 1347–1351. [CrossRef] [PubMed]

38. Roe, M.T.; Vega, E.; Pillai, S.D. Antimicrobial Resistance Markers of Class 1 and Class 2 Integron-bearing *Escherichia coli* from Irrigation Water and Sediments. *Emerg. Infect. Dis.* **2003**, *9*, 5–9. [CrossRef] [PubMed]

39. Ozgumus, O.B.; Sandalli, C.; Sevim, A.; Celik-Sevim, E.; Sivri, N. Class 1 and class 2 integrons and plasmid-mediated antibiotic resistance in coliforms isolated from ten rivers in northern Turkey. *J. Microbiol.* **2009**, *47*, 19–27. [CrossRef] [PubMed]

40. Vinué, L.; Sáenz, Y.; Somalo, S.; Escudero, E.; Moreno, M.Á.; Ruiz-Larrea, F.; Torres, C. Prevalence and diversity of integrons and associated resistance genes in faecal *Escherichia coli* isolates of healthy humans in Spain. *J. Antimicrob. Chemother.* **2008**, *62*, 934–937. [CrossRef] [PubMed]

41. Kargar, M.; Mohammadalipour, Z.; Doosti, A.; Lorzadeh, S.; Japoni-Nejad, A. High prevalence of class 1 to 3 integrons among multidrug-resistant diarrheagenic *Escherichia coli* in southwest of Iran. *Osong Public Health Res. Perspect.* **2014**, *5*, 193–198. [CrossRef] [PubMed]

42. Goldstein, C.; Lee, M.D.; Sanchez, S.; Phillips, B.; Register, B.; Grady, M.; Liebert, C.; Summers, A.O.; White, D.G.; Maurer, J.J.; et al. Incidence of Class 1 and 2 Integrases in Clinical and Commensal Bacteria from Livestock, Companion Animals, and Exotics. *Antimicrob. Agents Chemother.* **2001**, *45*, 723–726. [CrossRef] [PubMed]

43. Kor, S.B.; Choo, Q.C.; Chew, C.H. New integron gene arrays from multiresistant clinical isolates of members of the Enterobacteriaceae and *Pseudomonas aeruginosa* from hospitals in Malaysia. *J. Med. Microbiol.* **2013**, *62*, 412–420. [CrossRef] [PubMed]

44. Ali, T.; Ur Rahman, S.; Zhang, L.; Shahid, M.; Zhang, S.; Liu, G.; Gao, J.; Han, B. ESBL-producing *Escherichia coli* from cows suffering mastitis in China contain clinical class 1 integrons with CTX-M linked to ISCR1. *Front. Microbiol.* **2016**, *7*. [CrossRef] [PubMed]

45. Kim, J.; Bae, I.K.; Jeong, S.H.; Chang, C.L.; Lee, C.H.; Lee, K. Characterization of IncF plasmids carrying the *bla*CTX-M-14 gene in clinical isolates of *Escherichia coli* from Korea. *J. Antimicrob. Chemother.* **2011**, *66*, 1263–1268. [CrossRef] [PubMed]

46. Liao, X.P.; Xia, J.; Yang, L.; Li, L.; Sun, J.; Liu, Y.H.; Jiang, H.X. Characterization of CTX-M-14-producing *Escherichia coli* from food-producing animals. *Front. Microbiol.* **2015**, *6*. [CrossRef] [PubMed]

47. Bae, I.K.; Lee, Y.H.; Jeong, H.J.; Hong, S.G.; Lee, S.H.; Jeong, S.H. A novel *bla*CTX-M-14 gene-harboring complex class 1 integron with an In4-like backbone structure from a clinical isolate of *Escherichia coli*. *Diagn. Microbiol. Infect. Dis.* **2008**, *62*, 340–342. [CrossRef] [PubMed]

48. Wang, J.; Stephan, R.; Zurfluh, K.; Hächler, H.; Fanning, S. Characterization of the genetic environment of *bla*ESBL genes, integrons and toxin-antitoxin systems identified on large transferrable plasmids in multi-drug resistant *Escherichia coli*. *Front. Microbiol.* **2015**, *6*. [CrossRef]

49. Van Leverstein Hall, M.A.; Dierikx, C.M.; Cohen Stuart, J.; Voets, G.M.; van den Munckhof, M.P.; van Essen-Zandbergen, A.; Platteel, T.; Fluit, A.C.; van de Sande-Bruinsma, N.; Scharinga, J.; et al. Dutch patients, retail chicken meat and poultry share the same ESBL genes, plasmids and strains. *Clin. Microbiol. Infect.* **2011**, *17*, 873–880. [CrossRef] [PubMed]

50. Kiiru, J.; Butaye, P.; Goddeeris, B.M.; Kariuki, S. Analysis for prevalence and physical linkages amongst integrons, ISEcp 1, ISCR 1, Tn 21 and Tn 7 encountered in *Escherichia coli* strains from hospitalized and non-hospitalized patients in Kenya during a 19-year period (1992–2011). *BMC Microbiol.* **2013**, *13*. [CrossRef] [PubMed]

51. Porse, A.; Schønning, K.; Munck, C.; Sommer, M.O.A. Survival and Evolution of a Large Multidrug Resistance Plasmid in New Clinical Bacterial Hosts. *Mol. Biol. Evol.* **2016**, *33*, 2860–2873. [CrossRef] [PubMed]

52. Billard-Pomares, T.; Fouteau, S.; Jacquet, M.E.; Roche, D.; Barbe, V.; Castellanos, M.; Bouet, J.Y.; Cruveiller, S.; Médigue, C.; Blanco, J.; et al. Characterization of a P1-like bacteriophage carrying an SHV-2 extended-spectrum β-lactamase from an *Escherichia coli* strain. *Antimicrob. Agents Chemother.* **2014**, *58*, 6550–6557. [CrossRef] [PubMed]

53. Doi, Y.; Hazen, T.H.; Boitano, M.; Tsai, Y.C.; Clark, T.A.; Korlach, J.; Rasko, D.A. Whole-genome assembly of *Klebsiella pneumoniae* coproducing NDM-1 and OXA-232 carbapenemases using single-molecule, real-time sequencing. *Antimicrob. Agents Chemother.* **2014**, *58*, 5947–5953. [CrossRef] [PubMed]

54. Sun, Y.; Zeng, Z.; Chen, S.; Ma, J.; He, L.; Liu, Y.; Deng, Y.; Lei, T.; Zhao, J.; Liu, J.H. High prevalence of *bla*CTX-M extended-spectrum β-lactamase genes in *Escherichia coli* isolates from pets and emergence of CTX-M-64 in China. *Clin. Microbiol. Infect.* **2010**, *16*, 1475–1481. [CrossRef] [PubMed]

55. Tamang, M.D.; Nam, H.M.; Gurung, M.; Jang, G.C.; Kim, S.R.; Jung, S.C.; Park, Y.H.; Lim, S.K. Molecular characterization of CTX-M β-lactamase and associated addiction systems in *Escherichia coli* circulating among cattle, farm workers, and the farm environment. *Appl. Environ. Microbiol.* **2013**, *79*, 3898–3905. [CrossRef] [PubMed]

56. Rodriguez-Martinez, J.M.; Poirel, L.; Canton, R.; Nordmann, P. Common region CR1 for expression of antibiotic resistance genes. *Antimicrob. Agents Chemother.* **2006**, *50*, 2544–2546. [CrossRef] [PubMed]

57. Toleman, M.A.; Bennett, P.M.; Walsh, T.R. Common regions e.g. orf513 and antibiotic resistance: IS91-like elements evolving complex class 1 integrons. *J. Antimicrob. Chemother.* **2006**, *58*. [CrossRef] [PubMed]

58. Toleman, M.A.; Bennett, P.M.; Walsh, T.R. ISCR Elements: Novel Gene-Capturing Systems of the 21st Century? *Microbiol. Mol. Biol. Rev.* **2006**, *70*, 296–316. [CrossRef] [PubMed]

59. Millan, B.; Castro, D.; Araque, M.; Ghiglione, B.; Gutkind, G. ISCR1 associated with *bla*CTX-M-1 y *bla*CTX-M-2 genes in IncN and IncFIIA plasmids isolated from *Klebsiella pneumoniae* of nosocomial origin in Mérida, Venezuela. *Biomédica* **2012**, *33*, 268–275. [CrossRef]

60. Woodford, N.; Carattoli, A.; Karisik, E.; Underwood, A.; Ellington, M.J.; Livermore, D.M. Complete nucleotide sequences of plasmids pEK204, pEK499, and pEK516, encoding CTX-M enzymes in three major *Escherichia coli* lineages from the United Kingdom, all belonging to the international O25:H4-ST131 clone. *Antimicrob. Agents Chemother.* **2009**, *53*, 4472–4482. [CrossRef] [PubMed]

International Journal of
*Environmental Research
and Public Health*

Article

Antibiotic Susceptibility, Genetic Diversity, and the Presence of Toxin Producing Genes in *Campylobacter* Isolates from Poultry

Jeeyeon Lee [1,2], Jiyeon Jeong [1], Heeyoung Lee [2], Jimyeong Ha [1,2], Sejeong Kim [1,2], Yukyung Choi [1,2], Hyemin Oh [1,2], Kunho Seo [3], Yohan Yoon [1,2,*] and Soomin Lee [2,*]

[1] Department of Food and Nutrition, Sookmyung Women's University, Seoul 04310, Korea; fmjylee02@naver.com (J.L.); jjy5654@naver.com (J.J.); hayan29@naver.com (J.H.); 3337119@hanmail.net (S.K.); lab_yukyung@naver.com (Y.C.); odry0731@naver.com (H.O.)
[2] Risk Analysis Research Center, Sookmyung Women's University, Seoul 04310, Korea; hylee06@sookmyung.ac.kr
[3] Department of Veterinary Medicine, Konkuk University, Seoul 05029, Korea; bracstu3@konkuk.ac.kr
* Correspondence: yyoon@sookmyung.ac.kr (Y.Y.); slee0719@naver.com (S.L.); Tel.: +82-2-2077-7585 (Y.Y. & S.L.)

Received: 27 September 2017; Accepted: 14 November 2017; Published: 17 November 2017

Abstract: This study examined antibiotic susceptibility, genetic diversity, and characteristics of virulence genes in *Campylobacter* isolates from poultry. Chicken (n = 152) and duck (n = 154) samples were collected from 18 wet markets in Korea. *Campylobacter* spp. isolated from the carcasses were identified by PCR. The isolated colonies were analyzed for antibiotic susceptibility to chloramphenicol, amikacin, erythromycin, tetracycline, ciprofloxacin, nalidixic acid, and enrofloxacin. The isolates were also used to analyze genetic diversity using the DiversiLab™ system and were tested for the presence of cytolethal distending toxin (*cdt*) genes. *Campylobacter* spp. were isolated from 45 poultry samples out of 306 poultry samples (14.7%) and the average levels of *Campylobacter* contamination were 22.0 CFU/g and 366.1 CFU/g in chicken and duck samples, respectively. Moreover, more than 90% of the isolates showed resistance to nalidixic acid and ciprofloxacin. Genetic correlation analysis showed greater than 95% similarity between 84.4% of the isolates, and three *cdt* genes (*cdtA*, *cdtB*, and *cdtC*) were present in 71.1% of *Campylobacter* isolates. These results indicate that *Campylobacter* contamination should be decreased to prevent and treat *Campylobacter* foodborne illness.

Keywords: *Campylobacter*; poultry; antibiotic susceptibility; Rep-PCR; *cdt* toxin

1. Introduction

Campylobacter spp. are Gram-negative, microaerophilic bacteria, and the most common cause of bacterial foodborne illness in the world [1–4]. Among 17 *Campylobacter* species, *Campylobacter jejuni* and *Campylobacter coli* are the major causative agents of foodborne illness in human [5–7]. Animal species such as chicken, cattle and wild birds are reservoirs for *Campylobacter* [8,9]. *Campylobacter* infection causes watery diarrhea, fever, bloody stools, abdominal pain, and some complications such as Guillain-Barré syndrome (GBS) and Reiter's syndrome in severe case [10]. Facciolà et al. [10] suggested that it is difficult to find the contamination sources because *Campylobacter* outbreaks were sporadic and caused by cross-contamination.

Recently, campylobacteriosis have increased dramatically in South Korea. Until 2002, there were no *Campylobacter* outbreaks, but 831 people were infected by *Campylobacter* in 2016 [11]. In Switzerland, campylobacteriosis have also been increased, and healthcare cost for the patients was $7.5 million per year, expected to increase steadily [12]. *Campylobacter* have several virulence factors such as flagellin, capsular polysaccharides, and cytotoxins [13]. Regarding cytotoxin production, *Campylobacter* can produce cytolethal distending toxin (CDT), which is encoded by *cdtA*, *cdtB*, and *cdtC* genes [14–16]. This toxin can induce the host cell distension, then lead to cell death [17]. In severe cases, antibiotic (erythromycin, ciprofloxacin, tetracycline, etc.) treatment is necessary to treat *Campylobacter* infection, but *Campylobacter* spp. have recently begun to show resistance to several antibiotics [18–20]. In a previous study, 159 *Campylobacter* isolates from poultry samples in China were examined for antibiotic resistance and 94% (149 isolates) of *Campylobacter* isolates were resistant to tetracycline, doxycycline, and erythromycin [18]. Thus, *Campylobacter* isolates need to be investigated for antibiotic susceptibility.

To analyze the genetic correlation among bacterial isolates, restriction-based, amplification-based, and sequencing-based methods have been used [21]. Restriction-based methods include plasmid analysis, restriction fragment length polymorphism (RFLP) analysis, and pulsed-field gel electrophoresis (PFGE). Amplification-based methods are amplified fragment length polymorphisms (AFLP), random amplified polymorphic DNA PCR (RAPD-PCR), and repetitive element PCR (Rep-PCR). Sequencing-based methods include multilocus sequence typing (MLST) and single-nucleotide polymorphism (SNP) analysis. Rep-PCR can assign molecular fingerprints according to the repetitive sequences in bacterial genomes [22,23]. Compared to other PCR typing methods, Rep-PCR has advantages: processing is rapid and it has the ability to analyze small amounts of DNA [21,24]. Abay et al. [25] also suggested that Rep-PCR was more powerful for typeability of *Campylobacter* than PFGE.

The objective of this study was to investigate the prevalence of *Campylobacter* in poultry carcasses in wet markets, determine antibiotic susceptibility patterns, the presence of *cdt* genes, and analyzed the genetic diversity between the *Campylobacter* isolates.

2. Materials and Methods

2.1. Sample Collection

Chicken (*n* = 152) and duck (*n* = 154) carcasses were purchased from 18 wet markets throughout Korea during the summer (June–August, in 2014) and winter seasons (December in 2014 to February in 2015) (Figure 1). Three to ten samples for both chicken and duck carcasses were collected per market and per visit, and each market was visited twice for summer and winter. The samples were placed in a cooler on ice and transported to a laboratory. They were analyzed within 24 h.

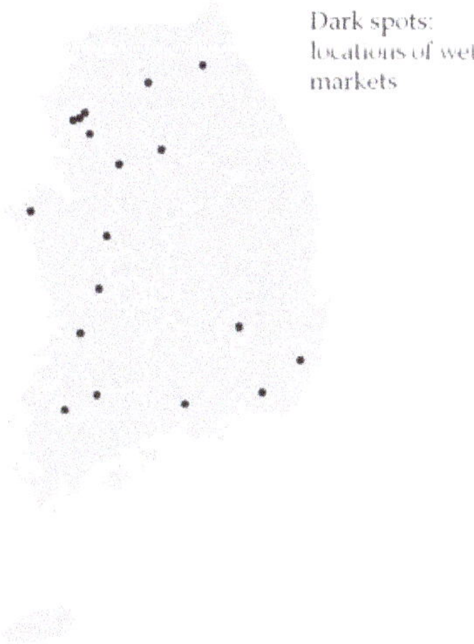

Figure 1. The locations of wet markets for poultry samples collected in Korea.

2.2. Campylobacter Isolation, Enumeration, and Identification

Each poultry sample was placed into a sample bag containing 400 mL 0.1% buffered peptone water (BPW, Becton, Dickinson and Company, Sparks, MD, USA) and gently shaken for 60 s. For *Campylobacter* isolation, the rinsate (27 mL) was mixed with 27 mL 2 × blood-free Bolton broth (Oxoid Ltd., Basingstoke, Hampshire, UK) and the mixture was enriched at 42 °C for 48 h. Loopful portions (10 μL) of the enrichments were streaked on modified charcoal-cefoperazone-deoxycholate agar (mCCDA; Oxoid Ltd., Basingstoke, UK) and incubated at 42 °C for 48 h in a microaerobic environment (5% O_2, 10% CO_2, and 85% N_2) created by CampyGen^{TM} gas packs (Oxoid Ltd., Basingstoke, UK). The two presumptive *Campylobacter* colonies (gray, mucoid, and flat) on a plate were selected and each colony of them was streaked on two Colombia agar plates (bioMérieux, Marcy-l'Étoile, France) for aerobic and microaerobic conditions at 42 °C for 48 h under both aerobic and microaerobic conditions. The colonies grown under microaerobic conditions were further analyzed to identify *Campylobacter* by PCR using the primers listed in Table 1. To extract *Campylobacter* DNA, the presumptive colonies at plate were suspended in 0.2 mL of sterilized distilled water, and heated at 99 °C for 10 min. The suspensions were centrifuged at 14,000 rpm for 3 min, and supernatants were then used for PCR amplification. The program was as follows: pre-denaturation at 95 °C for 15 min, 25 cycles of denaturation at 95 °C for 0.5 min, annealing at 58 °C for 1.5 min, and extension at 72 °C for 1 min. A final extension step at 72 °C for 7 min was performed [26]. The PCR products were visualized by electrophoresis and UV-transillumination. The isolates were used in further experiments for analysis of antibiotic resistance, genetic diversity and *cdt* genes. To enumerate *Campylobacter* cells, 1 mL of the rinsate was serially diluted using 0.1% BPW, and 0.1 mL of aliquots were plated on mCCDA (Oxoid Ltd., Basingstoke, UK). The plates were then incubated at 42 °C for 48 h under microaerobic conditions. Five presumptive colonies on each plate were then analyzed by PCR using the conditions described above. The contamination levels of *Campylobacter* were determined by multiplying the number of positive colonies per five presumptive colonies to the total number of colonies. Additionally, each carcass was weighted to calculate the colony forming units per g (CFU/g).

Table 1. Primer sequences used to identify the *Campylobacter* genus and species.

Species	Target Gene	Primer	Sequence (5′→3′)	Size (bp)	Reference
Genus *Campylobacter*	16S rRNA	C412F C1228R	GGATGACACTTTTCGGAGC CATTGTAGCACGTGTGTC	816	[27]
Campylobacter jejuni	*cj0414*	C-1 C-3	CAAATAAAGTTAGAGGTAGAATGT CCATAAGCACTAGCTAGCTGAT	161	[28]
Campylobacter coli	*ask*	CC18F CC519R	GGTATGATTTCTACAAAGCGAG ATAAAAGACTATCGTCGCGTG	502	[27]

2.3. Antibiotic Susceptibility Testing

The isolated colonies were further analyzed for antibiotic susceptibility to chloramphenicol, amikacin, erythromycin, tetracycline, ciprofloxacin, nalidixic acid, and enrofloxacin (Sigma-Aldrich, St Louis, MO, USA), according to the guidelines of the Clinical & Laboratory Standards Institute [29]. To determine antibiotic resistance, the breakpoints suggested by CLSI [29], CDC [30], Hong et al. [31], and Kang et al. [32] were used as follows: chloramphenicol at 32 μg/mL, amikacin at 64 μg/mL, erythromycin at 32 μg/mL, tetracycline at 16 μg/mL, ciprofloxacin at 4 μg/mL, nalidixic acid at 64 μg/mL, and enrofloxacin at 4 μg/mL. The *Campylobacter* isolates on Colombia agar (bioMérieux, Marcy-l'Étoile, France) were suspended in Mueller-Hinton broth (MHB; Becton, Dickinson and Company, Sparks, MD, USA) to obtain a McFarland 0.5 standard, and further diluted 10-fold. Using needles, *Campylobacter* isolates were spotted on Mueller-Hinton agar (MHA; Becton, Dickinson and Company, Sparks, MD, USA) with 5% lysed horse blood plates (Oxoid Ltd., Basingstoke, UK), formulated at 0.5–128 μg/mL with seven antibiotics. The plates were incubated under microaerobic conditions at 37 °C for 48 h. MIC was determined by colony formation on the plates and the reference strain used was *Campylobacter jejuni* ATCC33560.

2.4. Analysis of Genetic Diversity

To analyze the genetic diversity, 45 *Campylobacter* isolates from poultry were streaked on Colombia agar (bioMérieux, Marcy-l'Étoile, France), followed by microaerobic incubation at 42 °C for 48 h. DNA was extracted from *Campylobacter* isolates using a commercial kit (UltraClean™ Microbial DNA Isolation Kit, MoBio Laboratories, Solana Beach, CA, USA). The extracted DNA was amplified using DiversiLab *Campylobacter* Kit (bioMérieux, Marcy-l'Étoile, France). The amplified products were separated by electrophoresis on microfluidics chips (Agilent Technologies, Palo Alto, CA, USA) and analyzed with the Agilent 2100 Bioanalyzer (Agilent Technologies, Palo Alto, CA, USA). The peak and band data were analyzed by DiversiLab™ software version 2.1.66 (bioMérieux, Marcy-l'Étoile, France) using Pearson's correlation coefficient and unweighted pair group method with the arithmetic mean, followed by dendrogram generation. The cutoff value was 95% for determining genetic similarity [33,34].

2.5. Analysis of Cytolethal Distending Toxin Genes

To observe the presence of *cdt* genes (*cdtA*, *cdtB*, and *cdtC*) from isolates, the extracted DNA was amplified using the primers listed in Table 2 [14]. The PCR products were visualized by gel electrophoresis and UV-transillumination.

Table 2. PCR primers and amplification conditions used to analysis of *cdt* genes for *Campylobacter* isolates.

Genus	Gene	Sequence (5′→3′)	Amplification [1] Condition	Size (bp)
Campylobacter jejuni	*cdtA*	F: AGGACTTGAACCTACTTTTC R: AGGTGGAGTAGTTAAAAACC	94 °C, 30 s −55 °C, 30 s −72 °C, 30 s	631
	cdtB	F: ATCTTTTAACCTTGCTTTTGC R: GCAAGCATTAAAATCGCAGC	94 °C, 30 s −56 °C, 30 s −72 °C, 30 s	714
	cdtC	F: TTTAGCCTTTGCAACTCCTA R: AAGGGGTAGCAGCTGTTAA	94 °C, 30 s −55 °C, 30 s −72 °C, 30 s	524
Campylobacter coli	*cdtA*	F: ATTGCCAAGGCTAAAATCTC R: GATAAAGTCTCCAAAACTGC	94 °C, 30 s −55 °C, 30 s −72 °C, 30 s	329
	cdtB	F: TTTAATGTATTATTTGCCGC R: TCATTGCCTATGCGTATG	94 °C, 30 s −56 °C, 30 s −72 °C, 30 s	413
	cdtC	F: TAGGGATATGCACGCAAAAG R: GCTTAATACAGTTACGATAG	94 °C, 30 s −55 °C, 30 s −72 °C, 30 s	313

[1] Amplification: denaturation-annealing-extension.

2.6. Statistical Analysis

The data for the prevalence and contamination levels of *Campylobacter* between chicken and duck were statistically analyzed by SAS version 9.3 (SAS Institute Inc., Cary, NC, USA), and Chi-square test and *t*-test were used for prevalence and contamination levels, respectively, to determine significance at $\alpha = 0.05$.

3. Results and Discussion

3.1. Prevalence and Contamination Levels of Campylobacter

Of 306 poultry samples, *Campylobacter* spp. were identified from 45 samples (14.7%, 15 chicken samples and 30 duck samples) after enrichment (qualitative), but the number of positive samples was higher in quantitative results than in qualitative samples (Table 3). Since other bacteria may also be enriched with *Campylobacter*, resulting in disturbing the identification, the prevalence rate was lower in qualitative results than in quantitative results. The mean contamination levels of the isolated *Campylobacter* spp. in chicken and duck samples were 22.0 ± 36.3 CFU/g and 366.1 ± 733.6 CFU/g, respectively (Table 3).

Table 3. Prevalence and contamination levels of *Campylobacter* in chicken and duck carcasses at wet markets in Korea during summer and winter.

Seasons	Sample	Prevalence (No. of Positive Samples/No. of Samples (%))	Contamination Level	
			No. of Positive Samples/No. of Samples (%)	Mean ± SD (CFU/g)
Summer	Chicken	7/80 (8.8)	3/80 (3.8)	32.1 ± 21.0
	Duck	15/80 (18.8)	7/80 (8.8)	15.7 ± 14.2
	Subtotal	22/160 (13.8)	10/160 (6.3)	20.6 ± 17.2
Winter	Chicken	8/72 (11.1)	19/72 (26.4)	20.4 ± 38.8
	Duck	15/74 (20.3)	38/74 (51.4)	427.4 ± 780.2
	Subtotal	23/146 (15.8)	57/146 (39.0)	301.1 ± 673.1
Total	Chicken	15/152 (9.9) [A]	22/152 (14.5)	22.0 ± 36.6 [b]
	Duck	30/154 (19.5) [A]	45/154 (29.2)	366.1 ± 733.6 [a]
	Total	45/306 (14.7)	67/306 (21.9)	259.8 ± 628.9

Different upper letters (A, a, and b) in the same column indicate a difference ($p < 0.05$).

These results suggest that a quantitative method may be appropriate to investigate *Campylobacter* prevalence rather than a qualitative method, and duck samples have a higher contamination frequency and have higher levels of contamination significantly ($p = 0.0210$) than those in chicken samples in the Korean markets. *Campylobacter* was isolated regardless of the season; however, the contamination levels of *Campylobacter* were higher in the winter than in the summer. Of the 45 *Campylobacter* spp. isolates, 29 isolates were *C. jejuni* and 16 isolates were *C. coli*. In France, 372 of 425 chicken samples (87.5%) were *Campylobacter* positive, and their mean contamination level was 2.4 log CFU/g [35]. Also, Garin et al. [36] showed that *Campylobacter* spp. were detected from 491 of 750 chicken carcasses (65.5%) in five countries (Senegal, Cameroon, Madagascar, New Caledonia and Vietnam), and the mean value of contamination level was 3.2 log CFU/g. Additionally, Zhu et al. [37] analyzed 1587 chicken carcasses collected from seven provinces in China, and 716 carcasses (45.1%) were contaminated to Campylobacter, and the contamination level was 2.1 log CFU/g (median value). These studies indicate that *Campylobacter* contamination levels were similar among countries, however, the prevalence of *Campylobacter* can be considered low in wet markets in Korea. *Campylobacter* are microaerophilic bacteria. Thus, the bacterial cell counts can be gradually decreased under aerobic condition during distribution. Hence, long exposure time to aerobic condition during distribution to wet markets may induce low prevalence of *Campylobacter* in poultry in Korea.

3.2. Antimicrobial Resistance Patterns

Because antimicrobial resistance patterns were not different between *C. jejuni* and *C. coli*, the data were combined in Table 4. The *Campylobacter* isolates were resistant to nalidixic acid (93.3%), ciprofloxacin (91.1%), and tetracycline (71.1%) (Table 4). The isolates showed especially strong resistance to antibiotics such as nalidixic acid ciprofloxacin, tetracycline. However, *Campylobacter* isolates were sensitive to chloramphenicol (others), enrofloxacin (fluoroquinolones), erythromycin (macrolides), and amikacin (aminoglycosides) (Table 4). In Italy, *Campylobacter* isolates also showed high resistance rates to ciprofloxacin, tetracycline, and nalidixic acid [38]. Similarly, in the USA, the rate of antimicrobial resistance to tetracycline was very high, at 99.1% in *Campylobacter* isolates from broiler carcasses, followed by resistance to nalidixic acid and ciprofloxacin [39].

Table 4. Percentage of susceptibility and resistance of seven antibiotics for *Campylobacter* isolates from poultry.

Class	Antibiotics	Susceptibility		Resistance	
		No. of Isolates	Ratio (%)	No. of Isolates	Ratio (%)
A [1]	Amikacin	25	55.6	20	44.4
M	Erythromycin	43	95.6	2	4.4
T	Tetracycline	13	28.9	32	71.1
F	Ciprofloxacin	4	8.9	41	91.1
F	Enrofloxacin	38	84.4	7	15.6
Q	Nalidixic acid	3	6.7	42	93.3
Others	Chloramphenicol	45	100.0	0	0.0

[1] A: Aminoglycosides; M: Macrolides; T: Tetracyclines; F: Fluoroquinolones; Q: Quinolones.

Raeisi et al. [40] showed that *Campylobacter* isolates from poultry were resistant to ciprofloxacin, tetracycline and nalidixic acid. Also, 100% of *C. jejuni* isolates (n = 31) from chicken in China had resistance to ciprofloxacin and nalidixic acid [41]. In Poland, *Campylobacter* isolates were susceptible to erythromycin and resistant to tetracycline and ciprofloxacin [42]. Taken together, we can conclude that both poultry and human isolates of *Campylobacter* spp. are generally resistant to quinolone and fluoroquinolone antibiotics, such as nalidixic acid and ciprofloxacin. This may be caused by the use of these antibiotics in veterinary and human medicine. Therefore, this result suggests that antibiotics used for humans should not be used in poultry.

3.3. Genetic Diversity between Isolates

Campylobacter isolates were group according to the Rep-PCR dendrogram patterns (Figure 2). In genetic diversity, more than 95% similarity was shown in 38 isolates (84.4%) and these isolates were grouped into 10 groups (Figure 2). When comparing the 10 groups, obvious geographic correlations were not observed (Figure 2). For instance, key numbers 21–23 in group 6 were isolated from same location (Ulsan). Although 26–27 in group 7, and 39–41 in group 9 were isolated from same location (Cheongju), they were placed in different genetic group. However, Hiett et al. [43] subtyped for 50 *Campylobacter* isolates, and the most isolates from same location were genetically very similar. Like this result, very close genetic similarity can be expected for the isolates from same locations, but it was not observed in Korea as discussed above. This result indicates that chicken and duck in different wet markets in Korea may be distributed from only few slaughterhouses.

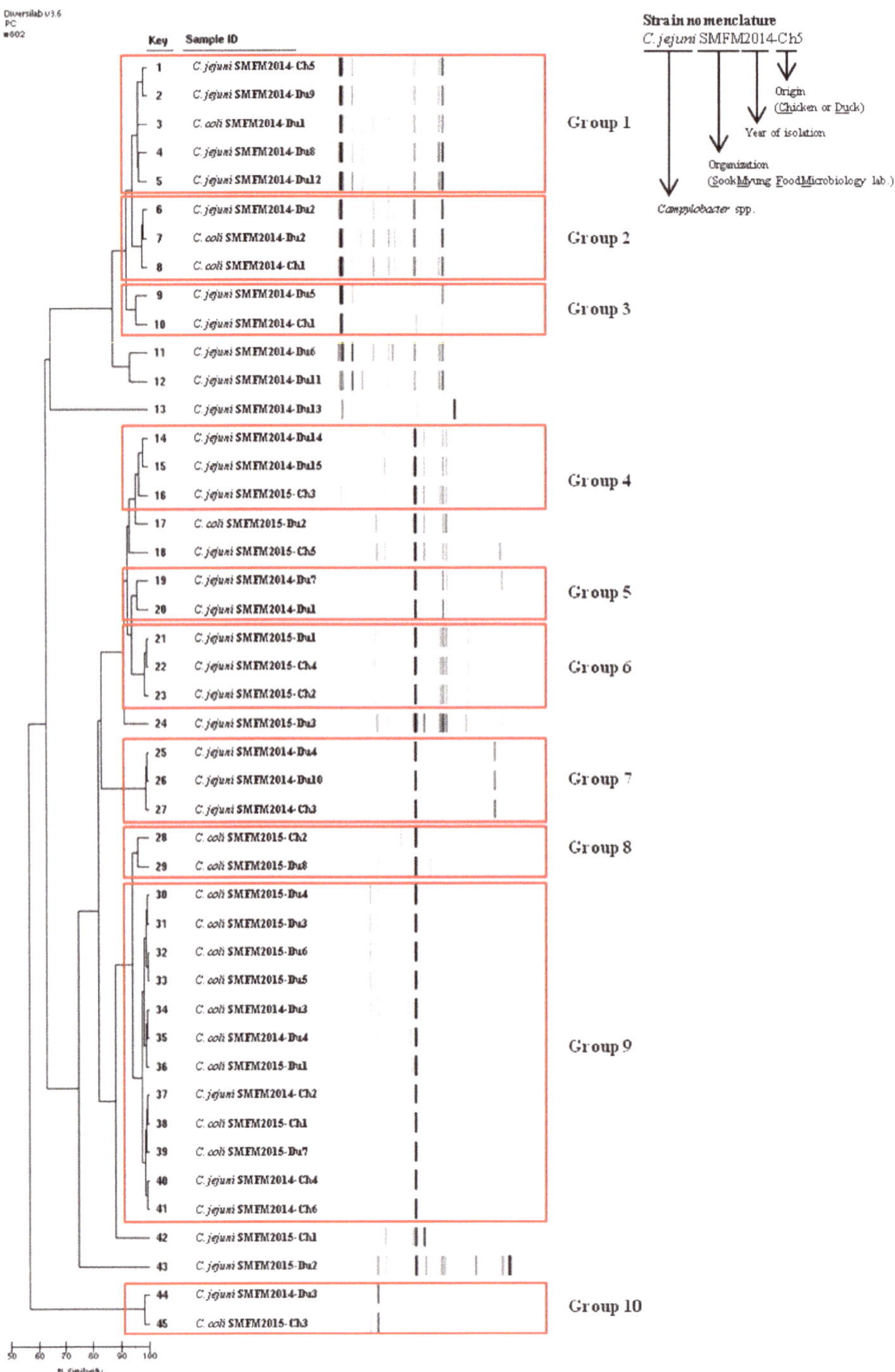

Figure 2. Dendrogram and gel-like image of the DiversiLab systems for *Campylobacter* isolates from poultry samples.

3.4. Distribution of cdtA, cdtB, and cdtC

Campylobacter can produce CDT, composed of A, B, and C subunits, which are encoded by *cdtA*, *cdtB*, and *cdtC* genes [44]. The 71.1% of the *Campylobacter* isolates had these three genes (Table 5). Nine of 15 chicken *Campylobacter* isolates and 23 of 30 duck *Campylobacter* isolates had the three *cdt* genes. Four isolates were found to be without any *cdt* genes and nine isolates had two *cdt* genes (*cdtA+/cdtB+*, *cdtA+/cdtC+*, or *cdtB+/cdtC+*). There was no relationship between the distribution of *cdt* genes and the regions the isolates had been obtained from. Oh et al. [45] showed that 37 *C. jejuni* isolates out of 38 chicken samples had all *cdt* genes. Findik et al. [5] found that 75.6% of *C. jejuni* isolates (127 isolates out of 168) from various sources, including human, poultry, cattle, sheep, and dog, had all *cdt* genes and five isolates were without *cdt* genes. In Brazil, all *cdt* genes were detected in 66.7% of *Campylobacter* isolates [46]. These results indicate that most *Campylobacter* isolates from our study have the potential to produce CDT.

Table 5. Cytolethal Distending Toxin (CDT) gene profiles of *Campylobacter* isolated from chicken and duck carcasses at wet markets.

Toxin Profile	Number of Isolates				Total (%)
	Chicken		Duck		
	Summer	Winter	Summer	Winter	
Negative	1	-	2	1	4 (4.3)
cdtA+	-	-	-	-	-
cdtB+	-	-	-	-	-
cdtC+	-	-	-	-	-
cdtA+/cdtB+	-		1	-	1 (2.2)
cdtA+/cdtC+	-	-	1	-	1 (2.2)
cdtB+/cdtC+	1	4	1	1	7 (15.6)
cdtA+/cdtB+/cdtC+	5	4	10	13	32 (71.1)
Total	7	8	15	15	45 (100.0)

4. Conclusions

In this study, the prevalence of the pathogen, antibiotic resistance, genetic diversity, and the presence of *cdt* genes in *Campylobacter* isolates were identified from poultry in Korean wet markets. Although the prevalence of *Campylobacter* in poultry was relatively low compared to that in other countries, antibiotic resistance patterns of the isolates were similar to those in other countries. In addition, geographic genetic diversity was not observed and a high proportion of *cdt* genes were present in *Campylobacter* isolates. Therefore, *Campylobacter* contamination should be decreased in order to prevent and treat the *Campylobacter* foodborne illness.

Acknowledgments: This research was supported by a grant (14162MFDS077) from the Ministry of Food and Drug Safety in 2015.

Author Contributions: Jeeyeon Lee participated in the design and coordination of the study, performed the experiments, analyzed the data, and wrote the manuscript. Jiyeon Jeong participated in the design of the study, performed the experiments and analyzed the data. Heeyoung Lee participated in the design of the study and helped draft the manuscript. Jimyeong Ha, Sejeong Kim, Yukyung Choi, and Hyemin Oh performed the experiments and helped the draft the manuscript. Kunho Seo participated in the design and coordination of the study. Yohan Yoon participated in the design of the study, oversaw the data collection in the study and contributed to the manuscript revision process. Soomin Lee also participated in the design of the study and contributed to the manuscript revision process. All authors read and approved the final manuscript.

Conflicts of Interest: The authors declare no conflict of interest.

References

1. Han, K.; Jang, S.S.; Choo, E.; Heu, S.; Ryu, S. Prevalence, genetic diversity, and antibiotic resistance patterns of *Campylobacter jejuni* form retail raw chickens in Korea. *Int. J. Food Microbiol.* **2007**, *114*, 50–59. [CrossRef] [PubMed]
2. Miflin, J.K.; Templeton, J.M.; Blackall, P.J. Antibiotic resistance in *Campylobacter jejuni* and *Campylobacter coli* isolated from poultry in the South-East Queensland region. *J. Antimicrob. Chemother.* **2007**, *59*, 775–778. [CrossRef] [PubMed]
3. Qin, S.-S.; Wu, C.-M.; Wang, Y.; Jeon, B.; Shen, Z.-Q.; Wang, Y.; Zhang, Q.; Shen, J.-Z. Antimicrobial resistance in *Campylobacter coli* isolated from pigs in two provinces of China. *Int. J. Food Microbiol.* **2011**, *146*, 94–98. [CrossRef] [PubMed]
4. Suzuki, H.; Yamamoto, S. *Campylobacter* contamination in retail poultry meats and by-products in the world: A literature survey. *J. Vet. Med. Sci.* **2009**, *71*, 255–261. [CrossRef] [PubMed]
5. Findik, A.; Ica, T.; Onuk, E.E.; Percin, D.; Kevenk, T.O.; Ciftci, A. Molecular typing and *cdt* genes prevalence of *Campylobacter jejuni* isolates from various sources. *Trop. Anim. Health Prod.* **2011**, *43*, 711–719. [CrossRef] [PubMed]
6. Moore, J.E.; Corcoran, D.; Dooley, J.S.F.; Fanning, S.; Lucey, B.; Matsuda, M.; McDowell, D.A.; Mégraud, F.; Millar, B.C.; O'Mahony, R.; et al. Campylobacter. *Vet. Res.* **2005**, *36*, 351–382. [CrossRef] [PubMed]
7. On, S.L. Taxonomy of *Campylobacter*, *Arcobacter*, *Helicobacter* and related bacteria: Current status, future prospects and immediate concerns. *J. Appl. Microbiol.* **2001**, *90*, 1S–15S. [CrossRef]
8. Refregier-Petton, J.; Rose, N.; Denis, M.; Salvat, G. Risk factors for *Campylobacter* spp. contamination in French broiler-chicken flocks at the end of the rearing period. *Prev. Vet. Med.* **2001**, *50*, 89–100. [CrossRef]
9. Skarp, C.P.A.; Hänninen, M.L.; Rautelin, H.I.K. Campylobacteriosis: The role of poultry meat. *Clin. Microbiol. Inf.* **2016**, *22*, 103–109. [CrossRef] [PubMed]
10. Facciolà, A.; Riso, R.; Avventuroso, E.; Visalli, G.; Delia, S.A.; Laganà, P. *Campylobacter*: From microbiology to prevention. *J. Prev. Med. Hyg.* **2017**, *58*, E79–E92. [PubMed]
11. Ministry of Food and Drug Safety. Statistics of Foodborne Illness. Available online: http://www.foodsafetykorea.go.kr/portal/healthyfoodlife/foodPoisoningStat.do?menu_no=519&menu_grp=MENU_GRP02 (accessed on 17 October 2017).
12. Schmutz, C.; Mäusezahl, D.; Bless, P.J.; Hatz, C.; Schwenkglenks, M.; Urbinello, D. Estimating healthcare costs of acute gastroenteritis and human campylobacteriosis in Switzerland. *Epidemiol. Infect.* **2017**, *145*, 627–641. [CrossRef] [PubMed]
13. Zilbauer, M.; Dorrell, N.; Wren, B.W.; Bajaj-Elliott, M. *Campylobacter jejuni*-mediated disease pathogenesis: An update. *Trans. R. Soc. Trop. Med. Hyg.* **2008**, *102*, 123–129. [CrossRef] [PubMed]
14. Asakura, M.; Samosornsuk, W.; Hinenoya, A.; Misawa, N.; Nishimura, K.; Matsuhisa, A.; Yamasaki, S. Development of a cytolethal distending toxin (*cdt*) gene-based species-specific multiplex PCR assay for the detection and identification of *Campylobacter jejuni*, *Campylobacter coli* and *Campylobacter fetus*. *FEMS Immunol. Med. Microbiol.* **2008**, *52*, 260–266. [CrossRef] [PubMed]
15. Bolton, D.J. *Campylobacter* virulence and survival factors. *Food Microbiol.* **2015**, *48*, 99–108. [CrossRef] [PubMed]
16. Yamasaki, S.; Asakura, M.; Tsukamoto, T.; Faruque, S.M.; Deb, R.; Ramamurthy, T. Cytolethal distending toxin (CDT): Genetic diversity, structure and role in diarrheal disease. *Toxin Rev.* **2006**, *25*, 61–88. [CrossRef]
17. Weis, A.M.; Miller, W.A.; Byrne, B.A.; Chouicha, N.; Boyce, W.M.; Townsend, A.K. Prevalence and pathogenic potential of *Campylobacter* isolates from free-living, human-commensal American crows. *Appl. Environ. Microbiol.* **2014**, *80*, 1639–1644. [CrossRef] [PubMed]
18. Ge, B.; White, D.G.; McDermott, P.F.; Girard, W.; Zhao, S.; Hubert, S.; Meng, J. Antimicrobial-resistant *Campylobacter* species from retail raw meats. *Appl. Environ. Microbiol.* **2003**, *69*, 3005–3007. [CrossRef] [PubMed]
19. Gibreel, A.; Tracz, D.M.; Nonaka, L.; Ngo, T.M.; Connell, S.R.; Taylor, D.E. Incidence of antibiotic resistance in *Campylobacter jejuni* isolated in Alberta, Canada, from 1999 to 2002, with special reference to *tet*(O)-mediated tetracycline resistance. *Antimicrob. Agents Chemother.* **2004**, *48*, 3442–3450. [CrossRef] [PubMed]
20. Zhong, X.; Wu, Q.; Zhang, J.; Shen, S. Prevalence, genetic diversity and antimicrobial susceptibility of *Campylobacter jejuni* isolated from retail food in China. *Food Control* **2016**, *62*, 10–15. [CrossRef]

21. Foley, S.L.; Lynne, A.M.; Nayak, R. Molecular typing methodologies for microbial source tracking and epidemiological investigations of Gram-negative bacterial foodborne pathogens. *Infect. Genet. Evol.* **2009**, *9*, 430–440. [CrossRef] [PubMed]

22. Behringer, M.; Miller, W.G.; Oyarzabal, O.A. Typing of *Campylobacter jejuni* and *Campylobacter coli* isolated from live broilers and retail broiler meat by flaA-RFLP, MLST, PFGE and REP-PCR. *J. Microbiol. Meth.* **2011**, *84*, 194–201. [CrossRef] [PubMed]

23. Hulton, C.S.J.; Higgins, C.F.; Sharp, P.M. ERIC sequences, a novel family of repetitive elements in the genome of *Escherichia coli, Salmonella typhimurium* and other enterobacterial. *Mol. Microbiol.* **1991**, *5*, 825–834. [CrossRef] [PubMed]

24. Appuhamy, S.; Parton, R.; Coote, J.G.; Gibbs, H.A. Genomic fingerprinting of *Haemophilus somnus* by a combination of PCR methods. *J. Clin. Microbiol.* **1997**, *35*, 288–291. [PubMed]

25. Abay, S.; Kayman, T.; Otlu, B.; Hizlisoy, H.; Aydin, F.; Ertas, N. Genetic diversity and antibiotic resistance profiles of *Campylobacter jejuni* isolates from poultry and humans in Turkey. *Int. J. Food Microbiol.* **2014**, *178*, 29–38. [CrossRef] [PubMed]

26. Yamazaki-Matsune, W.; Taguchi, M.; Seto, K.; Kawahara, R.; Kawatsu, K.; Kumeda, Y.; Kitazato, M.; Nukina, M.; Misawa, N.; Tsukamoto, T. Development of a multiplex PCR assay for identification of *Campylobacter coli, Campylobacter fetus, Campylobacter hyointestinalis* subsp. *hyointestinalis, Campylobacter jejuni, Campylobacter lari* and *Campylobacter upsaliensis. J. Med. Microbiol.* **2007**, *56*, 1467–1473. [CrossRef] [PubMed]

27. Linton, D.; Owen, R.J.; Stanley, J. Rapid identification by PCR of the genus *Campylobacter* and five *Campylobacter* species enteropathogenic for man and animals. *Res. Microbiol.* **1996**, *147*, 707–718. [CrossRef]

28. Wang, R.F.; Salvic, M.F.; Cao, W.W. A rapid PCR method for direct detection of low numbers of *Campylobacter jejuni. J. Rapid Methods Autom. Microbiol.* **1992**, *1*, 101–108. [CrossRef]

29. Clinical and Laboratory Standards Institute. *Performance Standards for Antimicrobial Susceptibility Testing; 20th Informational Supplement*; M100-S20; CLSI: Pennsylvania, PA, USA, 2010.

30. Centers for Disease Control and Prevention. *National Antimicrobial Resistance Monitoring System for Enteric Bacteria (NARMS): 2005 Human Isolates Final Report*; Department of Health and Human Services, CDC: Atlanta, GA, USA, 2008.

31. Hong, J.; Kim, J.M.; Jung, W.K.; Kim, S.H.; Bae, W.; Koo, H.C.; Gil, J.; Kim, M.; Ser, J.; Park, Y.H. Prevalence and antibiotic resistance of *Campylobacter* spp. isolated from chicken meat, pork, and beef in Korea, from 2001 to 2006. *J. Food Prot.* **2007**, *70*, 860–866. [CrossRef] [PubMed]

32. Kang, Y.S.; Cho, Y.S.; Yoon, S.K.; Yu, M.A.; Kim, C.M.; Lee, J.O.; Pyun, Y.R. Prevalence and antimicrobial resistance of *Campylobacter jejuni* and *Campylobacter coli* isolated from raw chicken meat and human stools in Korea. *J. Food Prot.* **2006**, *69*, 2915–2923. [CrossRef] [PubMed]

33. Herbold, N.M.; Clotilde, L.M.; Anderson, K.M.; Kase, J.; Hartman, G.L.; Himathongkham, S.; Lin, A.; Lauzon, C.R. Clustering of clinical and environmental *Escherichia coli* O104 isolates using the DiversiLab™ repetitive sequence-based PCR system. *Curr. Microbiol.* **2015**, *70*, 436–440. [CrossRef] [PubMed]

34. Ross, T.L.; Merz, W.G.; Farkosh, M.; Carroll, K.C. Comparison of an automated repetitive sequence-based PCR microbial typing system to pulsed-field gel electrophoresis for analysis of outbreaks of Methicillin-resistant *Staphylococcus aureus. J. Clin. Microbiol.* **2005**, *43*, 5642–5647. [CrossRef] [PubMed]

35. Hue, O.; Le Bouquin, S.; Laisney, M.J.; Allain, V.; Lalande, F.; Petetin, I.; Rouxel, S.; Quesne, S.; Gloaguen, P.Y.; Picherot, M.; et al. Prevalence of and risk factors for *Campylobacter* spp. contamination of broiler chicken carcasses at the slaughterhouse. *Food Microbiol.* **2010**, *27*, 992–999. [CrossRef] [PubMed]

36. Garin, B.; Gouali, M.; Wouafo, M.; Perchec, A.M.; Thu, P.M.; Ravaonindrina, N.; Urbès, F.; Gay, M.; Diawara, A.; Leclercp, A.; et al. Prevalence, quantification and antimicrobial resistance of *Campylobacter* spp. on chicken neck-skins at points of slaughter in 5 major cities located on 4 continents. *Int. J. Food Microbiol.* **2012**, *157*, 102–107. [CrossRef] [PubMed]

37. Zhu, J.; Yao, B.; Song, X.; Wang, Y.; Cui, S.; Xu, H.; Yang, B.; Huang, J.; Liu, G.; Yang, X.; et al. Prevalence and quantification of *Campylobacter* contamination on raw chicken carcasses for retail sale in China. *Food Control* **2017**, *75*, 196–202. [CrossRef]

38. Di Giannatale, E.; Di Serafino, G.; Zilli, K.; Alessiani, A.; Sacchini, L.; Garofolo, G.; Aprea, G.; Marotta, F. Characterization of antimicrobial resistance patterns and detection of virulence genes in *Campylobacter* isolates in Italy. *Sensors* **2014**, *14*, 3308–3322. [CrossRef] [PubMed]

39. Son, I.; Englen, M.D.; Berrang, M.E.; Fedorka-Cray, P.J.; Harrison, M.A. Antimicrobial resistance of *Arcobacter* and *Campylobacter* from broiler carcasses. *Int. J. Antimicrob. Agents* **2007**, *29*, 451–455. [CrossRef] [PubMed]

40. Raeisi, M.; Khoshbakht, R.; Ghaemi, E.A.; Bayani, M.; Hashemi, M.; Seyedghasemi, N.S.; Shirzad-Aski, H. Antimicrobial resistance and virulence-associated genes of *Campylobacter* spp. isolated from raw milk, fish, poultry, and red meat. *Microb. Drug Resist.* **2017**, *23*, 925–933. [CrossRef] [PubMed]

41. Ma, H.; Su, Y.; Ma, L.; Ma, L.; Li, P.; Du, X.; Gölz, G.; Wang, S.; Lu, X. Prevalence and characterization of *Campylobacter jejuni* isolated from retail chicken in Tianjin, China. *J. Food Prot.* **2017**, *80*, 1032–1040. [CrossRef] [PubMed]

42. Szczepanska, B.; Andrzejewska, M.; Spica, D.; Klawe, J.J. Prevalence and antimicrobial resistance of *Campylobacter jejuni* and *Campylobacter coli* isolated from children and environmental sources in urban and suburban areas. *BMC Microbiol.* **2017**, *17*, 80. [CrossRef] [PubMed]

43. Hiett, K.L.; Seal, B.S.; Siragusa, G.R. *Campylobacter* spp. subtype analysis using gel-based repetitive extragenic palindromic-PCR discriminates in parallel fashion to *fla A* short variable region DNA sequence analysis. *J. Appl. Microbiol.* **2006**, *101*, 1249–1258. [CrossRef] [PubMed]

44. Mortensen, N.P.; Schiellerup, P.; Boisen, N.; Klein, B.M.; Locht, H.; Abuoun, M.; Newell, D.; Krogfelt, K.A. The role of *Campylobacter jejuni* cytolethal distending toxin in gastroenteritis: Toxin detection, antibody production, and clinical outcome. *APMIS* **2011**, *119*, 626–634. [CrossRef] [PubMed]

45. Oh, J.Y.; Kwon, Y.K.; Wei, B.; Jang, H.K.; Lim, S.K.; Kim, C.H.; Jung, S.C.; Kang, M.S. Epidemiological relationships of *Campylobacter jejuni* strains isolated from humans and chickens in South Korea. *J. Microbiol.* **2017**, *55*, 13–20. [CrossRef] [PubMed]

46. Carvalho, A.F.D.; Silva, D.M.D.; Azevedo, S.S.; Piatti, R.M.; Genovez, M.E.; Scarcelli, E. Detection of CDT toxin genes in *Campylobacter* spp. strains isolated from broiler carcasses and vegetables in São Paulo, Brazil. *Braz. J. Microbiol.* **2013**, *44*, 693–699. [CrossRef] [PubMed]

International Journal of
*Environmental Research
and Public Health*

Article

Diversity and Antimicrobial Resistance Genotypes in Non-Typhoidal *Salmonella* Isolates from Poultry Farms in Uganda

Terence Odoch [1,*], Camilla Sekse [2], Trine M. L'Abee-Lund [3], Helge Christoffer Høgberg Hansen [3], Clovice Kankya [1] and Yngvild Wasteson [3]

[1]   Department of Bio-security, Ecosystems and Veterinary Public Health, College of Veterinary Medicine, Animal Resources and Biosecurity (COVAB), Makerere University, P.O. Box 7062, Kampala, Uganda; clokankya@yahoo.com

[2]   Norwegian Veterinary Institute, 0106 Oslo, Norway; camilla.sekse@vetinst.no

[3]   Department of Food Safety and Infection Biology, Faculty of Veterinary Medicine, Norwegian University of Life Sciences (NMBU), 0454 Oslo, Norway; trine.labee-lund@nmbu.no (T.M.L.-L.); helge.hansen@nmbu.no (H.C.H.H.); yngvild.wasteson@nmbu.no (Y.W.)

*   Correspondence: odochterence@gmail.com or odoch@covab.mak.ac.ug; Tel.: +256-772-360168

Received: 14 December 2017; Accepted: 9 February 2018; Published: 13 February 2018

Abstract: Non-typhoidal *Salmonella* (NTS) are foodborne pathogens of global public health significance. The aim of this study was to subtype a collection of 85 NTS originating from poultry farms in Uganda, and to evaluate a subgroup of phenotypically resistant isolates for common antimicrobial resistance genes and associated integrons. All isolates were subtyped by pulsed-field gel electrophoresis (PFGE). Phenotypically resistant isolates ($n = 54$) were screened by PCR for the most relevant AMR genes corresponding to their phenotypic resistance pattern, and all 54 isolates were screened by PCR for the presence of integron class 1 and 2 encoding genes. These genes are known to commonly encode resistance to ampicillin, tetracycline, ciprofloxacin, trimethoprim, sulfonamide and chloramphenicol. PFGE revealed 15 pulsotypes representing 11 serotypes from 75 isolates, as 10 were non-typable. Thirty one (57.4%) of the 54 resistant isolates carried at least one of the seven genes (bla_{TEM-1}, $cmlA$, $tetA$, $qnrS$, $sul1$, $dhfrI$, $dhfrVII$) identified by PCR and six (11%) carried class 1 integrons. This study has shown that a diversity of NTS-clones are present in Ugandan poultry farm settings, while at the same time similar NTS-clones occur in different farms and areas. The presence of resistance genes to important antimicrobials used in human and veterinary medicine has been demonstrated, hence the need to strengthen strategies to combat antimicrobial resistance at all levels.

Keywords: antimicrobial resistance; genotypes; non-typhoidal *Salmonella*; poultry; genes; integrons; subtyping

---

1. Introduction

Salmonella enterica subsp. *enterica* include serotypes that are global foodborne pathogens significantly affecting public health and economy [1–3]. In humans, salmonellosis is classified into typhoid and non-typhoidal salmonellosis. Most cases of non-typhoidal *Salmonella* (NTS) disease are associated with consumption of contaminated foods of animal origin, particularly poultry, meat and in some instances vegetables [4–6]. Globally, NTS is estimated to cause 93.8 million cases of gastroenteritis annually, of which 80 million cases are foodborne and causing 155,000 deaths [7]. Although African countries have low estimated cases of NTS gastroenteritis compared to other parts of the world, they have a much higher level of invasive non-enteric NTS infections [7,8]. NTS bacteraemia is an emerging opportunistic infection in individuals infected with HIV and is reported to be highly correlated with malaria, especially in children and elderly persons [9–13].

In poultry, transmission of NTS can occur by direct contacts with infected birds, consumption of contaminated feeds and water, and contact with environmental reservoirs [13]. Transmission can also occur through cross contamination anywhere along the production chain, and for specific serotypes, vertical transmission is also possible [14,15]. However, NTS infections in poultry is mainly asymptomatic [14], and may therefore not get the necessary attention with regard to prevention and control. The diversity of NTS circulating in poultry and livestock production environment in most developing countries is poorly understood, as very limited studies have been undertaken. Molecular typing is important for characterization of bacteria to establish genetic relatedness between isolates in order to elucidate the dynamics of the bacterial populations. Although whole genome sequencing is getting more established, pulsed-field gel electrophoresis (PFGE) technique is still considered an adequate molecular method suitable for subtyping of serotypes of *Salmonella.*

The increasing development of antimicrobial resistance (AMR) in NTS is complicating treatment of bacteraemia cases and results in poorer treatment outcomes. Even more worrying is the emergence of multidrug resistance (MDR) in NTS against commonly used antibiotics in human and animal treatment, which has become a serious public health challenge [15–18]. Resistance is increasing not only against first line antibiotics, but also against clinically important antimicrobial agents like fluoroquinolones and third generation cephalosporins [19]. Inappropriate use of antimicrobials in agriculture is known to be a key factor contributing to the development of AMR, and the influence of livestock environment in the development of MDR in NTS has been demonstrated [20]. Increased intensification of production in agriculture, use of antibiotics as feed additives, and prophylactic treatment are some of the practices that influence development of AMR [21,22]. MDR NTS can be transferred from the poultry reservoirs to humans through the food chain, but AMR can also be transferred from one bacterium to another through resistance genes associated with integrons and mobile genetic elements such as plasmids and transposons. Most studies on AMR in poultry are done in developed countries while in most developing countries, including Uganda, there are no surveillance and monitoring programs for important foodborne pathogens and AMR in primary production units. To date in Africa, only a few limited studies have documented AMR and corresponding genes in NTS isolated from humans, animal products, and poultry farms [23–29]. Therefore, data is scarce and the extent of NTS and AMR remains poorly known. As a result, development of appropriate mitigation measures and control efforts is compromised. The aim of this study was to characterize a collection of NTS isolates from poultry by using PFGE for molecular subtyping and to investigate the presence of integrons and acquired antimicrobial resistance genes from the phenotypically resistant isolates. The NTS were isolated from faecal samples collected from poultry farms in three districts (Wakiso, Lira, and Masaka) in Uganda between 2015 and 2016 [30].

2. Materials and Methods

2.1. The NTS isolate collection

The majority (75/85) of the NTS isolates used in this study were from a previous study by Odoch et al. [30]. The remaining 10 isolates originated from additional sampling. However, all 85 isolates were from fecal samples collected from poultry houses in three districts with high numbers of commercial poultry farms (Wakiso, Lira, and Masaka) in Uganda between 2015 and 2016, according to a sampling design and procedure described in Odoch et al [30]. A map of the study area is provided as Supplementary Materials Figure S1. NTS were isolated, identified, serotyped and tested for antimicrobial sensitivity according to standard methods as earlier described [30]: Culture and isolation of NTS were done according to ISO 6579:2002/Amd 1:2007, Annex D: Detection of *Salmonella* spp. in animal faeces and in environmental samples from the primary production [31]. Biochemical confirmatory tests were done by using the API-20E (BioMerieux, Marcy l'Etoile, France) identification system. All isolates were serotyped according to the Kauffman–White–Le–Minor technique at the Norwegian Veterinary Institute. Phenotypic susceptibility testing of 13 antimicrobials (gentamicin, sulonamide, trimethoprim-sulfamethoxazole, ciprofloxacin, cefotaxime, meropenem, chloramphenicol, ceftazidime, ampicillin, amoxicillin/clavulanic acid, trimethoprim, tetracycline, and enrofloxacin) was performed by the disc diffusion test. The metadata, serotype and phenotypic resistance of the isolates are presented in the Supplementary Materials (Table S1).

2.2. Pulsed-Field Gel Electrophoresis (PFGE) and Bionumerics Analysis

The PulseNet standardized protocol for PFGE for molecular subtyping of *Salmonella* (https://www.cdc.gov/pulsenet/pathogens/pfge.html) was used on all the 85 isolates. Overnight cultures were used to prepare DNA templates according to the PulseNet protocol. DNA was digested with the restriction enzyme *XbaI* and *Salmonella* Braenderup H9812 was used as a molecular size standard in all PFGE investigations. Electrophoresis was performed with the CHEF-DR III system (Bio-Rad Laboratories, Hercules, CA, USA) with the following set parameters: initial switch time 2.2 s, final switch time 63.8 s, voltage-6 V, time-19 h and temperature 14 °C. The gels were stained with ethidium bromide and the bands visualized under UV transillumination and captured by GelDoc EQ system with Quantity One®software (Version 4.2.1; Bio-Rad Laboratories, Hercules, CA, USA). PFGE banding patterns were compared using a combination of visual inspection and the BioNumerics software vers. 6.6.11 (Applied Maths, Ghent, Belgium). A dendrogram was generated using band-based dice similarity coefficient and the unweighted pair group method using a geometric average (UPGMA) with 1.2% position tolerance and 1.2% optimization. A cutoff of 97% similarity was used to define a PFGE pulsotype (PT).

2.3. Bacterial DNA Extraction

Total DNA for PCR were extracted using the boiled lysate method [32]. This was done by taking 200 μL of an overnight culture, mixing with 800 μL of sterile distilled water and boiling for 10 minutes. The resultant solution was centrifuged at 13,000 rpm for five minutes and the supernatant was used as a DNA template. This was kept at −20 °C for subsequent use.

2.4. Detection of Integrons and Antibiotic Resistance Genes

The isolates that were classified as resistant according to the results of the disc diffusion test (n = 54) were screened by PCR for the most relevant AMR genes corresponding to their phenotypic resistance pattern. In addition, all resistant isolates were screened by PCR for the presence of integron class 1 and 2 encoding genes. The isolates tested were *S.* Newport (n = 18), *S.* Bolton (n = 8), *S.* Hadar (n = 6), *S.* Mbandaka (n = 4), *S.* Heidelberg (n = 8), *S.* Typhimurium (n = 2), and *S.* Zanzibar (n = 8) serotypes. The existence of class 1 integron was investigated by PCR for the detection of genes encoding the variable part between the 5′ conserved segment and the 3′ conserved segment of the variable region [33]. Presence of class 2 integron was investigated by detection of *hep74* and *hep51* genes using primers and following PCR conditions previously reported [33]. Presence of 22 AMR genes (Table 1) known to confer resistance to six commonly used classes of antimicrobials (β-lactams, tetracyclines, phenicols, fluoroquinolones, trimethoprim, and sulfonamides) were investigated by PCR. The primer sets used for detection of integrons and AMR genes are shown in Table 1. Ampicillin resistant isolates (n = 4) were screened for four β-lactamase resistance encoding genes, and ciprofloxacin resistant isolates (n = 40) were screened for four fluoroquinolone plasmid mediated quinolone resistance (PMQR) determinant genes. Chloramphenicol resistant isolates (n=4) were screened for four phenicol resistance genes, tetracycline resistant isolates (n=12) were screened for three genes. Sulfonamide resistant isolates (n = 21) were screened for two genes and six trimethoprim resistant isolates were screened for five trimethoprim resistance genes. These genes were selected because they are the most frequently detected genes associated with the corresponding phenotypes of the NTS isolates [34]. All the integron PCR products were purified and sequenced (GATC Biotech, Cologne, Germany) and the sequence results were analysed using BLAST and compared to GenBank database (http://blast.ncbi.nlm.nih.gov/blast.cgi). Similarly, one PCR product from each of the AMR PCRs was sequenced to confirm the PCR results. Negative controls were included in all PCR analyses.

Table 1. PCR primers used for amplification of genes encoding integrons and antimicrobial resistance in non-typhoidal *Salmonella* isolates.

Target Category	Target Gene	Primer Sequence	Amplicon Size (bp)	Annealing Temp (°C)	Reference
Integron	Class 1 integron 5′-CS 3′-CS	GGCATCCAAGCAGCAAG AAGCAGACTTGACCTGA	Variable size	55	[33]
	Class 2 integron hep74 hep51	CGGGATCCCGGACGGCATGCACGATTTGTA GATGCCATCGCAAGTACGAG	491	55	[33]
Resistance to ampicillin by detection of four β-lactamase genes	bla_{PSE-1}	CGCTTCCCGTTAACAAGTAC CTGGTTCATTTCAGATAGCG	419	57	[35]
	bla_{CMY-2}	TGGCCAGAACTGACAGGCAAA TTTCTCCTGAACGTGGCTGGC	462	64	[36]
	bla_{TEM-1}	AGGAAGAGTATGATTCAACA CTCGTCGTTTGGTATGGC	535	55	[37]
	bla_{OxA}	ACCAGATTCAACTTTCAA TCTTGGCTTTTATGCTTG	590	55	[38]
Resistance to ciprofloxacin by detection of four fluoroquinolone plasmid mediated quinolone resistance genes	$qnrA$	AGAGGATTTCTCACGCCAGG TGCCAGGCACAGATCTTGAC	580	54	[39]
	$qnrB$	GATCGTGAAAGCCAGAAAGG ATGAGCAACGATGCCTGGTA	476	53	[40]
	$qnrC$	GGGTTGTACATTTATTGAATCG CACCTACCCATTTATTTTCA	307	53	[40]
	$qnrS$	GCAAGTTCATTGAACAGGGT TCTAAACCGTCGAGTTCGGCG	428	54	[39]
Resistance to chloramphenicol by detection of four phenicol resistance genes	$floR$	AACCCGCCCTCTGGATCAAGTCAA CAAATCACGGGCCACGCTGTATC	548	60	[41]
	$cat1$	CTTGTCGCCTTGCGTATAAT ATCCCAATGGCATCGTAAAG	508	55	[42]
	$cat2$	AACGGCATGATGAACCTGAA ATCCCAATGGCATCGTAAAG	547	55	[42]
	$cmlA$	CGCCACGGTGTTGTTGTTAT GCGACCTGCGTAAATGTCAC	394	55	[42]

Table 1. *Cont.*

Target Category	Target Gene	Primer Sequence	Amplicon Size (bp)	Annealing Temp (°C)	Reference
Resistance to sulfonamide by detection of two dihydropteroate reductase genes	*sul1*	GCG CGG CGT GGG CTA CCT GATTTCCGCGACACCGAGACAA	350	65	[43]
	sul2	CGG CAT CGT CAA CAT AACC GTG TGC GGA TGA AGT CAG	720	52	[43]
Resistance to tetracycline by detection of three efflux pump genes	*tetA*	GCTACATCCTGCTTGCCTTC CATAGATCGCCGTGAAGAGG	210	55	[35]
	tetB	TTGGTTAGGGGCAAGTTTTG GTAATGGGCCAATAACACCG	659	55	[35]
	tetG	CAG CTTTCG GATTCT TACGG GAT TGGTGA GGCTCG TTAGC	844	55	[35]
Resistance to trimethoprim by detection of five dihydrofolate reductase genes	*dhfrI*	AAGAATGGAGTTATCGGGAATG GGGTAAAAACTGGCCTAAAATTG	391	50	[37]
	dhfrV	CTGCAAAAGCGAAAAACGG AGCAATAGTTAATGTTTGAGCTAAAG	432	50	[37]
	dhfrVII	GGTAATGGCCCTGATATCCC TGTAGATTTGACCGCCACC	265	50	[37]
	dhfrIX	TCTAAACATGATTGTCGCTGTC TTGTTTTCAGTAATGGTCGGG	452	50	[37]
	dhfrXIII	CAGGTGAGCAGAAGATTTTT CCTCAAAGGTTTGATGTACC	294	50	[37]

The β-lactamase encoding genes (*blaPSE-1, blaCMY-2, bla*TEM-1*, blaOxA*) encode production of β-lactamase enzyme that breaks the β-lactam antibiotic ring open and deactivates the molecule's antibacterial properties. The plasmid mediated quinolone resistance genes (*qnrA, qnrB, qnrC, qnrS*) encode pentapeptide repeat proteins that bind to and protects DNA gyrase and topoisomerases IV from the inhibition of quinolones. The phenicol resistance genes, (*cat1, cat2*) encode chloramphenicol acetyltransferase enzyme that inactivates chloramphenicol, chloramphenicol resistance gene, *cmlA* and florfenicol resistance gene *floR*, encode efflux pump proteins. Sulfonamide resistance genes *sul1* and *sul2* encode insensitive sulfonamide-resistant dihydropteroate synthase which cannot be inhibited by sulfonamide. Tetracycline resistance genes (*tetA, tetB, tetG*) encode membrane associated efflux pump proteins that export tetracycline from the cell and reduces drug concentration and thereby protecting ribosomes. Trimethoprim resistance genes (*dhfrI, dhfrV, dhfrVII, dhfrIX, dhfrXIII*) encode a drug-insensitive dihydrofolate reductase which cannot be inhibited by trimethoprim.

3. Results

3.1. Pulsed-Field Gel Electrophoresis Typing

A total of 75 *Salmonella* isolates were typable, and 15 PTs were identified (Figure 1) and the PFGE banding pattern of all isolates were included in a dendrogram as the Supplementary Materials (Figure S2). The 10 nontypable (NT) isolates belonged to different serotypes; *Salmonella* Bolton (*n* = 1), *S.* Newport (*n* = 3), *S.* Typhimurium (*n* = 1), *S.* Hadar (*n* = 4), and *S.* Heidelberg (*n* = 1). For the majority of the typable isolates, there was a complete association between serotype and PT. The 21 typable *S.* Newport isolates all belonged to PT (H), but were isolated from several farms in all districts (Figure 1). Ciprofloxacin resistant isolates were the majority and most diverse in terms of serotypes, pulsotypes and geographic distribution. Four *S.* Mbandaka isolates were characterized by the same PT (N) and phenotypic resistance pattern, but were isolated from three different farms in two districts. A similar distribution pattern was also observed for 10 *S.* Aberdeen isolates of PT (F); these were isolated from nine different farms from all districts. However, the isolates were fully sensitive in the disc diffusion test. The exceptions from the serotype-PT associations were *S.* Hadar and *S.* Heidelberg. A total of seven *S.* Hadar isolates were typable. Four of them with identical PT originated from the same district, but from two farms, and had same phenotypic resistance towards three antimicrobials. The other three *S.* Hadar isolates had three different PTs, however, two of these isolates were similar with only one band difference (Figure S2). The typable *S.* Heidelberg isolates consisted of two different PTs; one PT (A) with two isolates from the same district and one PT (B) with seven isolates from the other two districts. The isolates in PT (A) were fully susceptible in the disc diffusion test, while all in PT (B) expressed ciprofloxacin resistance and two also expressed sulfonamide resistance.

3.2. Detection of Integrons and Antibiotic Resistance Genes

Genes encoding class 1 integrons were only detected in six *S.* Hadar isolates, four belonging to PT (G) and two nontypable. The integrons were similar in size, with approximately 1700 bp. All the *S.* Hadar isolates that carried integrons originated from four farms in one district, Wakiso. Genes encoding class 2 integrons were not detected in any of the isolates. Sequencing of the six integron PCR products revealed the presence of *aadA1* and *dfrA15* genes that confer resistance to streptomycin/spectinomycin and trimethoprim, respectively.

AMR genes were detected in 31 (57.4%) of the 54 phenotypically resistant. Only seven genes (*bla*TEM-1*, cmlA, qnrS, tetA, sul1, dhfrI, dhfrVII*) of the 22 AMR genes were detected among the selected phenotypically resistant isolates. These genes are known to confer resistance to six categories of antimicrobials (β-lactams, chloramphenicol, fluoroquinolones, tetracyclines, sulfonamides, and trimethoprim).

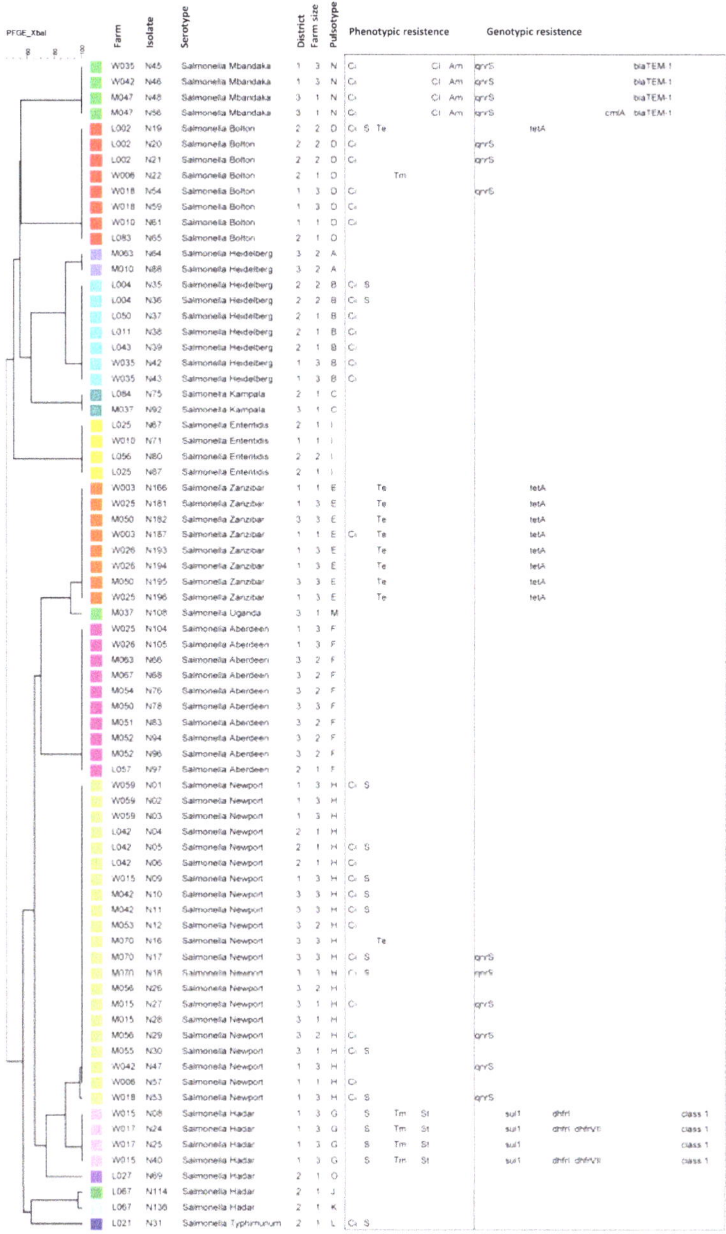

Figure 1. Dendrogram based on Pulsed-Field gel electrophoresis (PFGE) patterns of 75 non-typhoidal *Salmonella* from poultry from Uganda. A cutoff level of 97% similarity defines a PFGE profile. For each isolate the isolate number, PFGE profile, serotype, farm, size of farm, district, phenotypic resistance (Ci; ciprofloxacin, S; sulphonamide, Te; tetracycline, Tm; trimethoprim, St; sulphamethoxazole_trimethoprim, Cl; chloramphenicol, Am; ampicillin) and identified genotypic resistance genes (*qnrS, sul1, tetA, dhfrI, dhfrVII, cmlA, bla*TEM-1, *integrons, dfrA15, aadA1*) have been included.

All four ampicillin and chloramphenicol resistant *S.* Mbandaka strains harbored the bla_{TEM} gene that confers resistance to β-lactams, but only one of them was harboring the chloramphenicol resistance gene *cmlA*. The PMQR gene *qnrS* was detected in 16 (18.8%) out of the total 85 isolates. Forty of these displayed ciprofloxacin resistance, of which 16 (40%) carried *qnrS*. All 13 tetracycline resistant isolates were positive for the *tetA* gene. The sulfonamide resistant gene *sul1*, was the only one identified in six of the 21 sulfonamide resistant isolates (*sul2* was not detected). Out of the six trimethoprim resistant *S.* Hadar strains, four were resistant to sulfonamide/trimethoprim and they all harbored the *dhfr*1 gene (Table S1). Three of the six harbored both *dhfr1* and *dhfrVII* (Figure 1).

4. Discussion

The diversity of NTS circulating in poultry in most developing countries is poorly understood, as few studies have been undertaken [44–46]. In this study, 15 PTs from 11 different serotypes of NTS isolates were identified, with most of the identified serotypes having only one PT implying they are clonally related. The PFGE dendrogram combined with the geographical origin of the isolates indicate that many related clones are circulating in geographically diverse areas. For example *S.* Newport, the most prevalent serotype of all, belonged to the same PT and was isolated from all the districts. This situation is not surprising considering the uncontrolled movement of poultry and poultry products in Uganda. In addition, most commercial farms share sources of chicks, feeds, feed ingredients, and live bird markets and these are all potential common sources of NTS contamination. A similar situation has been reported in Senegal [46]. Because NTS is known to persist in the environments for months [47,48], they can easily be spread over large geographical areas. Some of the NTS serotypes represented in this study have caused foodborne illnesses and outbreaks globally [49]. There were isolates with similar PTs that varied with regard to their content of resistance genes, the AMR genes tested for are acquired genes, and not through mutations in chromosomally encoded genes, therefore the genes might be spread among isolates due to their location on plasmids, transposons and integrons. Integration of these elements does not necessarily result in changes in PT.

Through this study, the occurrence of AMR genes among a diversity of NTS isolates from poultry farms in the study districts have been unveiled. The isolates were screened for the genes conferring resistance to the antibiotics to which the isolate revealed a resistance phenotype. The genes detected confer resistance to some of the most important antimicrobials used for treatment of bacterial infections in humans and animals [50]. However, among the 22 AMR genes that are commonly occurring within the *Enterobacteriaceae* family, only seven genes were identified. Discordance was seen where observed phenotypic AMR was not reflected by the detection of corresponding AMR genes. For example, neither *sul1* nor *sul2* genes were detected in the nine phenotypically sulfonamide resistant *S.* Newport isolates. This discordance could be due to presence of other and more unusual resistance mechanisms encoded by genes not included in this study.

Previous investigations on the occurrence of integrons in NTS isolates from animal sources have yielded varying results [51–53]. Class 1 integrons are known for their roles in the dissemination of AMR, especially in the carrying of multiple AMR genes. In this study, integrons were identified in six *S.* Hadar isolates and all of them were identified with *aadA1* and *dfrA15* genes that confer resistance to streptomycin/spectinomycin and trimethoprim, respectively. It is in agreement with studies and reports that most of these genes are found in gene cassettes located within class 1 and 2 integrons [41,51]. In addition, PCR identified four of these *S.* Hadar isolates with *dhfrI* genes with three of the four carrying both *dhfrI* genes and *dhfrVII* genes. More than 30 gene variants encoding dihydrofolate reductase have been identified [38] and *dfrA* are the most commonly genes identified from NTS.

Class I integrons are always associated with *sul1* genes. In this study, *sul1* gene was the only sulfonamide resistance gene identified in six of the 21 phenotypically sulfonamide resistant isolates. Previous studies have reported that in NTS, *sul1* is more common than *sul2* and *sul3* and these genes encode the dihydropteroate synthase [54]. As reported earlier, increase in resistance to sulfonamides/trimethoprim in Uganda has serious public health implications as it is the main drug used to control opportunistic infections in HIV/AIDS patients [30].

The PMQR gene *qnrS* was the only PMQR gene detected from the NTS isolates that were phenotypically resistant to ciprofloxacin. This finding is in agreement with some similar studies undertaken previously [55–57]. It may, however, be noted that the detection of the *qnrS* genes was restricted to the serotypes *S*. Newport, *S*. Bolton and *S*. Mbandaka, while they were not detected in *S*. Zanzibar, *S*. Typhimurium, *S*. Heidelberg. PMQR genes are rapidly spreading globally, although their presence only mediate low levels of fluoroquinolone resistance, they can interact with genomic determinants to increase the minimum inhibitory concentrations of fluoroquinolones of the PMQR harboring bacteria [58]. Ciprofloxacin is an important fluoroquinolone used in Uganda and other countries for treatment of salmonellosis and other bacteraemic infections. It is often used as a last resort antimicrobial in the treatment of blood stream infections in children and is classified by World Health Organization (WHO) as critically important [50]. In the current study areas, a potential risk exits that ciprofloxacin resistance genes could get transferred to humans through contact with poultry, and consequently complicate the use of ciprofloxacin. The high occurrence of *qnrS* in NTS from poultry needs to be explored further to determine whether it could be associated with use of enrofloxacin in poultry. Enrofloxacin, also a fluoroquinolone, is sometimes used prophylactically and metaphylactically in combination with other drugs in some commercial poultry farms in Uganda [30]. As all fluoroquinolones have the same mechanism of inhibition of the topoisomerase genes, resistance to any one of them will confer resistance to all others. High presence of the plasmid-mediated quinolone resistance gene *qnrS* therefore shows the potential of horizontal transfer of resistance genes [59].

In this study, all the tetracycline resistant isolates carried *tetA* genes, they were all negative for *tetB* and *tetG* genes. This result is similar to what has been reported in previous studies undertaken in Thailand, Australia, Germany, Morocco, and Egypt [18,60–63]. However, the results is also in contrast to another study in Egypt [64]. Many genes responsible for tetracycline resistance have been identified and described [65]. The occurrence of *tetA* gene is known to be widespread in NTS and is associated with non-conjugative transposons. These genes are associated with efflux pump mechanisms implying that these are the predominant mechanisms for tetracycline resistance in NTS in these areas. High presence of *tetA* genes is not surprising as tetracycline is an extensively used drug in human and veterinary medicine, mainly because it is cheap and readily available [66].

All four *S*. Mbandaka isolates that were resistant to chloramphenicol were negative for phenicol resistance encoding genes *floR*, *cat1*, *cat2*, and only one was positive for *cmlA* genes. This finding is consistent with an earlier study [67]. The chloramphenicol exporter gene *cmlA* has been previously found in plasmid-located class 1 integrons in *S*. Typhimurium. Use of chloramphenicol for animal treatment is banned in many countries, including Uganda, due to health hazards associated with the persistence of residues in foods [68]. These same isolates of *S*. Mbandaka were identified with *qnrS* gene and bla_{TEM-1} gene but were negative for all the other screened β-lactamase encoding genes(bla_{PSE-1}, bla_{CMY-2}, bla_{OxA}). The gene $bla_{TEM-1,}$ is reported to be the most widely distributed of the β-lactamase genes worldwide [52] and is mainly known to be spread by plasmids. Not much information is available on the occurrence of beta-lactamase encoding genes in isolates from poultry in Uganda, but similar results have been reported in studies elsewhere [69–71]. Carriage of the bla_{TEM-1} gene is a threat to the potency of β-lactam antibiotics and in the case of Uganda, ampicillin is still widely used in human and veterinary medicine.

The interpretation of results from this study needs to be taken with a bit of caution, especially when looking at the bigger picture of the whole country. This study evaluated a limited number of resistance genes and only on phenotypically resistant isolates from a previous study [30], the sample size was quite small and samples were collected from only three districts that were purposively selected. However, as far as we are concerned, it is the first of its kind in Uganda and the data generated should make a significant contribution towards the national and international efforts to control antimicrobial resistance.

5. Conclusions

This study was a follow up of a previous study that determined prevalence, antimicrobial susceptibility and risk factors associated with NTS in Uganda [30]. The occurrence of AMR genes and integrons in *Salmonella enterica* isolates from Ugandan poultry has been unveiled, and through subtyping, the diversity of NTS isolates from three districts in Uganda has been explored.

The study has put into perspective the need to monitor use of antimicrobials and occurrence of AMR genes in farm ecosystems in developing countries, in order to institute measures to contain spread of AMR. Poultry keeping is predicted to continue growing in developing countries and in Uganda it will remain an important economic activity. However, as demonstrated, poultry farm environments remain a significant source of spread of AMR genes. Farmers have to be educated on the adoption of strict biosecurity measures, prudent use of antimicrobials and better management practices. More investigations need to be undertaken to further enhance understanding of the driving forces in farm ecosystems for the development of AMR in important foodborne pathogens like *Salmonella*. This study underscores the need for using the One Health approach to generate data on AMR in *Salmonella* organisms originating from humans, animals, and environmental samples.

Supplementary Materials: The following are available online at http://www.mdpi.com/1660-4601/15/2/324/s1, Figure S1: A map of the study areas, Figure S2: A PFGE dendrogram of all typable isolates including the PFGE banding pattern, Table S1: List of all Salmonella isolates with metadata.

Acknowledgments: Funding for this project was provided under the NORHED project No.UGA-13/0031, based at Makerere University and Norwegian University of Life Sciences (NMBU). Our sincere thanks and gratitude go to professor Eystein Skjerve for overall coordination and logistical support. We are grateful to Henning Sørum for providing the positive controls for the resistance genes. In addition, we acknowledge the technical support and contributions of Aud Kari Fauske, Gaute Skogtun, Kristina Borch-Pedersen, and Kristin O'Sullivan, all from NMBU.

Author Contributions: Terence Odoch, Camilla Sekse and Yngvild Wasteson conceptualized and designed the study; Terence Odoch and Clovice Kankya collected field data and samples, and were responsible for bacterial isolation procedure; Trine M. L'Abée-Lund, Yngvild Wasteson and Helge Christoffer Høgberg Hansen mobilized molecular laboratory reagents and supervised molecular laboratory procedures; Terence Odoch and Helge Christoffer Høgberg Hansen performed PCR and molecular subtyping techniques; Terence Odoch, Camilla Sekse and Yngvild Wasteson analyzed laboratory results; Terence Odoch, Yngvild Wasteson, Trine M. L'Abée-Lund and Terence Odoch wrote the paper with contributions from Clovice Kankya and Helge Christoffer Høgberg Hansen. All authors read and approved the final paper submitted.

Conflicts of Interest: The authors declare no conflict of interest.

References

1. Camps, N.; Dominguez, A.; Company, M.; Perez, M.; Pardos, J.; Llobet, T.; Usera, M.A.; Salleras, L.; Working Group for the Investigation of the Outbreak of Salmonellosis. A foodborne outbreak of Salmonella infection due to overproduction of egg-containing foods for a festival. *Epidemiol. Infect.* **2005**, *133*, 817–822. [CrossRef] [PubMed]

2. Shao, D.; Shi, Z.; Wei, J.; Ma, Z. A brief review of foodborne zoonoses in China. *Epidemiol. Infect.* **2011**, *139*, 1497–1504. [CrossRef] [PubMed]

3. Scallan, E.; Hoekstra, R.M.; Angulo, F.J.; Tauxe, R.V.; Widdowson, M.A.; Roy, S.L.; Jones, J.L.; Griffin, P.M. Foodborne Illness Acquired in the United States—Major Pathogens. *Emerg. Infect. Dis.* **2011**, *17*, 7–15. [CrossRef] [PubMed]

4. Dallal, M.M.S.; Taremi, M.; Gachkar, L.; Modarressi, S.; Sanaei, M.; Bakhtiari, R.; Yazdi, M.K.S.; Zali, M.R. Characterization of antibiotic resistant patterns of *Salmonella serotypes* isolated from beef and chicken samples in Tehran. *Jundishapur J. Microbiol.* **2009**, *2*, 124–131.

5. Painter, J.A.; Hoekstra, R.M.; Ayers, T.; Tauxe, R.V.; Braden, C.R.; Angulo, F.J.; Griffin, P.M. Attribution of foodborne illnesses, hospitalizations, and deaths to food commodities by using outbreak data, United States, 1998–2008. *Emerg. Infect. Dis.* **2013**, *19*, 407–415. [CrossRef] [PubMed]

6. Antunes, P.; Mourao, J.; Campos, J.; Peixe, L. Salmonellosis: The role of poultry meat. *Clin. Microbiol. Infect.* **2016**, *22*, 110–121. [CrossRef] [PubMed]

7. Majowicz, S.E.; Musto, J.; Scallan, E.; Angulo, F.J.; Kirk, M.; O'Brien, S.J.; Jones, T.F.; Fazil, A.; Hoekstra, R.M. International Collaboration on Enteric Disease 'Burden of Illness, S.; The global burden of nontyphoidal Salmonella gastroenteritis. *Clin. Infect. Dis.* **2010**, *50*, 882–889. [CrossRef] [PubMed]

8. Ao, T.T.; Feasey, N.A.; Gordon, M.A.; Keddy, K.H.; Angulo, F.J.; Crump, J.A. Global burden of invasive nontyphoidal Salmonella disease, 2010. *Emerg. Infect. Dis.* **2015**, *21*, 941–949. [CrossRef] [PubMed]

9. Graham, S.M.; Molyneux, E.M.; Walsh, A.L.; Cheesbrough, J.S.; Molyneux, M.E.; Hart, C.A. Nontyphoidal Salmonella infections of children in tropical Africa. *Pediatr. Infect. Dis. J.* **2000**, *19*, 1189–1196. [CrossRef] [PubMed]

10. Gordon, M.A.; Graham, S.M.; Walsh, A.L.; Wilson, L.; Phiri, A.; Molyneux, E.; Zijlstra, E.E.; Heyderman, R.S.; Hart, C.A.; Molyneux, M.E. Epidemics of invasive *Salmonella enterica* serovar enteritidis and S-enterica serovar typhimurium infection associated with multidrug resistance among adults and children in Malawi. *Clin. Infect. Dis.* **2008**, *46*, 963–969. [CrossRef] [PubMed]

11. Haeusler, G.M.; Curtis, N. Non-typhoidal Salmonella in Children: Microbiology, Epidemiology and Treatment. In *Hot Topics in Infection and Immunity in Children IX*; Curtis, N., Finn, A., Pollard, A.J., Eds.; Springer: New York, NY, USA, 2013.

12. Kariuki, S.; Gordon, M.A.; Feasey, N.; Parry, C.M. Antimicrobial resistance and management of invasive Salmonella disease. *Vaccine* **2015**, *33*, S21–S29. [CrossRef] [PubMed]

13. Kariuki, S.; Revathi, G.; Kariuki, N.; Kiiru, J.; Mwituria, J.; Hart, C.A. Characterisation of community acquired non-typhoidal Salmonella from bacteraemia and diarrhoeal infections in children admitted to hospital in Nairobi, Kenya. *BMC Microbiol.* **2006**, *6*, 101. [CrossRef] [PubMed]

14. Gaffga, N.H.; Behravesh, C.B.; Ettestad, P.J.; Smelser, C.B.; Rhorer, A.R.; Cronquist, A.B.; Comstock, N.A.; Bidol, S.A.; Patel, N.J.; Gerner-Smidt, P.; et al. Outbreak of Salmonellosis Linked to Live Poultry from a Mail-Order Hatchery. *N. Engl. J. Med.* **2012**, *366*, 2065–2073. [CrossRef] [PubMed]

15. Lynne, A.M.; Rhodes-Clark, B.S.; Bliven, K.; Zhao, S.H.; Foley, S.L. Antimicrobial resistance genes associated with *Salmonella enterica* serovar Newport isolates from food animals. *Antimicrob. Agents Chemother.* **2008**, *52*, 353–356. [CrossRef] [PubMed]

16. Yang, X.J.; Huang, J.H.; Wu, Q.P.; Zhang, J.M.; Liu, S.R.; Guo, W.P.; Cai, S.Z.; Yu, S.B. Prevalence, antimicrobial resistance and genetic diversity of Salmonella isolated from retail ready-to-eat foods in China. *Food Control* **2016**, *60*, 50–56. [CrossRef]

17. Tamang, M.D.; Nam, H.M.; Kim, T.S.; Jang, G.C.; Jung, S.C.; Lim, S.K. Emergence of Extended-Spectrum beta-Lactamase (CTX-M-15 and CTX-M-14)-Producing Nontyphoid Salmonella with Reduced Susceptibility to Ciprofloxacin among Food Animals and Humans in Korea. *J. Clin. Microbiol.* **2011**, *49*, 2671–2675. [CrossRef] [PubMed]

18. Chuanchuen, R.; Padungtod, P. Antimicrobial Resistance Genes in *Salmonella enterica* Isolates from Poultry and Swine in Thailand. *J. Vet. Med. Sci.* **2009**, *71*, 1349–1355. [CrossRef] [PubMed]

19. Lunguya, O.; Lejon, V.; Phoba, M.F.; Bertrand, S.; Vanhoof, R.; Glupczynski, Y.; Verhaegen, J.; Muyembe-Tamfum, J.J.; Jacobs, J. Antimicrobial Resistance in Invasive Non-typhoid Salmonella from the Democratic Republic of the Congo: Emergence of Decreased Fluoroquinolone Susceptibility and Extended-spectrum Beta Lactamases. *PLoS Negl. Trop. Dis.* **2013**, *7*, e2103. [CrossRef] [PubMed]

20. An, R.; Alshalchi, S.; Breimhurst, P.; Munoz-Aguayo, J.; Flores-Figueroa, C.; Vidovic, S. Strong influence of livestock environments on the emergence and dissemination of distinct multidrug-resistant phenotypes among the population of non-typhoidal Salmonella. *PLoS ONE* **2017**, *12*, e0179005. [CrossRef] [PubMed]

21. Hur, J.; Jawale, C.; Lee, J.H. Antimicrobial resistance of Salmonella isolated from food animals: A review. *Food Res. Int.* **2012**, *45*, 819–830. [CrossRef]

22. Van Nhiem, D.; Paulsen, P.; Suriyasathaporn, W.; Smulders, F.J.M.; Kyule, M.N.; Baumann, M.P.O.; Zessin, K.H.; Ngan, P.H. Preliminary analysis of tetracycline residues in marketed pork in Hanoi, Vietnam. In *Impact of Emerging Zoonotic Diseases on Animal Health*; Blouin, E.F., Maillard, J.C., Eds.; New York Academy of Sciences: New York, NY, USA, 2006.

23. Abdel-Maksoud, M.; Abdel-Khalek, R.; El-Gendy, A.; Gamal, R.F.; Abdelhady, H.M.; House, B.L. Genetic characterisation of multidrug-resistant *Salmonella enterica* serotypes isolated from poultry in Cairo, Egypt. *Afr. J. Lab. Med.* **2015**, *4*. [CrossRef]

24. Tabo, D.A.; Diguimbaye, C.D.; Granier, S.A.; Moury, F.; Brisabois, A.; Elgroud, R.; Millemann, Y. Prevalence and antimicrobial resistance of non-typhoidal Salmonella serotypes isolated from laying hens and broiler chicken farms in N'Djamena, Chad. *Vet. Microbiol.* **2013**, *166*, 293–298. [CrossRef] [PubMed]

25. Ahmed, A.M.; Shimamoto, T.; Shimamoto, T. Characterization of integrons and resistance genes in multidrug-resistant *Salmonella enterica* isolated from meat and dairy products in Egypt. *Int. J. Food Microbiol.* **2014**, *189*, 39–44. [CrossRef] [PubMed]

26. Fashae, K.; Ogunsola, F.; Aarestrup, F.M.; Hendriksen, R.S. Antimicrobial susceptibility and serovars of Salmonella from chickens and humans in Ibadan, Nigeria. *J. Infect. Dev. Ctries.* **2010**, *4*, 484–494. [PubMed]

27. Dione, M.M.; Ieven, M.; Garin, B.; Marcotty, T.; Geerts, S. Prevalence and Antimicrobial Resistance of Salmonella Isolated from Broiler Farms, Chicken Carcasses, and Street-Vended Restaurants in Casamance, Senegal. *J. Food Prot.* **2009**, *72*, 2423–2427. [CrossRef] [PubMed]

28. Elgroud, R.; Zerdoumi, F.; Benazzouz, M.; Bouzitouna-Bentchouala, C.; Granier, S.A.; Fremy, S.; Brisabois, A.; Dufour, B.; Millemann, Y. Characteristics of Salmonella Contamination of Broilers and Slaughterhouses in the Region of Constantine (Algeria). *Zoonoses Public Health* **2009**, *56*, 84–93. [CrossRef] [PubMed]

29. Ben Salem, R.; Abbassi, M.S.; Garcia, V.; Garcia-Fierro, R.; Fernandez, J.; Kilani, H.; Jaouani, I.; Khayeche, M.; Messadi, L.; Rodicio, M.R. Antimicrobial drug resistance and genetic properties of *Salmonella enterica* serotype Enteritidis circulating in chicken farms in Tunisia. *J. Infect. Public Health* **2017**, *10*, 855–860. [CrossRef] [PubMed]

30. Odoch, T.; Wasteson, Y.; L'Abée-Lund, T.; Muwonge, A.; Kankya, C.; Nyakarahuka, L.; Tegule, S.; Skjerve, E. Prevalence, antimicrobial susceptibility and risk factors associated with non-typhoidal Salmonella on Ugandan layer hen farms. *BMC Vet. Res.* **2017**, *13*, 365. [CrossRef] [PubMed]

31. International Organization for Standardization (ISO). *ISO 6579:2002/Amd 1:2007 Annex D: Detection of Salmonella spp. in Animal Faeces and in Environmental Samples from the Primary Production*; ISO: Geneva, Switzerland, 2007.

32. Ahmed, A.M.; Hussein, A.I.A.; Shimamoto, T. Proteus mirabilis clinical isolate harbouring a new variant of Salmonella genomic island 1 containing the multiple antibiotic resistance region. *J. Antimicrob. Chemother.* **2007**, *59*, 184–190. [CrossRef] [PubMed]

33. Ahmed, A.M.; Motoi, Y.; Sato, M.; Maruyama, M.; Watanabe, H.; Fukumoto, Y.; Shimamoto, T. Zoo animals as reservoirs of gram-negative bacteria harboring integrons and antimicrobial resistance genes. *Appl. Environ. Microbiol.* **2007**, *73*, 6686–6690. [CrossRef] [PubMed]

34. Michael, G.B.; Butaye, P.; Cloeckaert, A.; Schwarz, S. Genes and mutations conferring antimicrobial resistance in Salmonella: An update. *Microbes Infect.* **2006**, *8*, 1898–1914. [CrossRef] [PubMed]

35. Bacci, C.; Boni, E.; Alpigiani, I.; Lanzoni, E.; Bonardi, S.; Brindani, F. Phenotypic and genotypic features of antibiotic resistance in *Salmonella enterica* isolated from chicken meat and chicken and quail carcasses. *Int. J. Food Microbiol.* **2012**, *160*, 16–23. [CrossRef] [PubMed]

36. Perez-Perez, F.J.; Hanson, N.D. Detection of plasmid-mediated AmpC beta-lactamase genes in clinical isolates by using multiplex PCR. *J. Clin. Microbiol.* **2002**, *40*, 2153–2162. [CrossRef] [PubMed]

37. Maynard, C.; Fairbrother, J.M.; Bekal, S.; Sanschagrin, F.; Levesque, R.C.; Brousseau, R.; Masson, L.; Lariviere, S.; Harel, J. Antimicrobial resistance genes in enterotoxigenic Escherichia coli O149 : K91 isolates obtained over a 23-year period from pigs. *Antimicrob. Agents Chemother.* **2003**, *47*, 3214–3221. [CrossRef] [PubMed]

38. Wang, Y.; Yang, B.W.; Wu, Y.; Zhang, Z.F.; Meng, X.F.; Xi, M.L.; Wang, X.; Xia, X.D.; Shi, X.M.; Wang, D.P.; et al. Molecular characterization of *Salmonella enterica* serovar Enteritidis on retail raw poultry in six provinces and two National cities in China. *Food Microbiol.* **2015**, *46*, 74–80. [CrossRef] [PubMed]

39. Cattoir, V.; Poirel, L.; Rotimi, V.; Soussy, C.J.; Nordmann, P. Multiplex PCR for detection of plasmid-mediated quinolone resistance qnr genes in ESBL-producing enterobacterial isolates. *J. Antimicrob. Chemother.* **2007**, *60*, 394–397. [CrossRef] [PubMed]

40. Bin Kim, H.; Park, C.H.; Kim, C.J.; Kim, E.C.; Jacoby, G.A.; Hooper, D.C. Prevalence of Plasmid-Mediated Quinolone Resistance Determinants over a 9-Year Period. *Antimicrob. Agents Chemother.* **2009**, *53*, 639–645.

41. Randall, L.P.; Cooles, S.W.; Osborn, M.K.; Piddock, L.J.V.; Woodward, M.J. Antibiotic resistance genes, integrons and multiple antibiotic resistance in thirty-five serotypes of *Salmonella enterica* isolated from humans and animals in the UK. *J. Antimicrob. Chemother.* **2004**, *53*, 208–216. [CrossRef] [PubMed]

42. Chen, S.; Zhao, S.H.; White, D.G.; Schroeder, C.M.; Lu, R.; Yang, H.C.; McDermott, P.F.; Ayers, S.; Meng, J.H. Characterization of multiple-antimicrobial-resistant *Salmonella serovars* isolated from retail meats. *Appl. Environ. Microbiol.* **2004**, *70*, 1–7. [CrossRef] [PubMed]

43. Poppe, C.; Martin, L.; Muckle, A.; Archambault, M.; McEwen, S.; Weir, E. Characterization of antimicrobial resistance of Salmonella Newport isolated from animals, the environment, and animal food products in Canada. *Can. J. Vet. Res.* **2006**, *70*, 105–114. [PubMed]

44. Afema, J.A.; Byarugaba, D.K.; Shah, D.H.; Atukwase, E.; Nambi, M.; Sischo, W.M. Potential Sources and Transmission of Salmonella and Antimicrobial Resistance in Kampala, Uganda. *PLoS ONE* **2016**, *11*, e0152130. [CrossRef] [PubMed]

45. Andoh, L.A.; Dalsgaard, A.; Obiri-Danso, K.; Newman, M.J.; Barco, L.; Olsen, J.E. Prevalence and antimicrobial resistance of Salmonella serovars isolated from poultry in Ghana. *Epidemiol. Infect.* **2016**, *144*, 3288–3299. [CrossRef] [PubMed]

46. Cardinale, E.; Gros-Claude, J.D.P.; Rivoal, K.; Rose, V.; Tall, F.; Mead, G.C.; Salvat, G. Epidemiological analysis of *Salmonella enterica* ssp enterica serovars Hadar, Brancaster and Enteritidis from humans and broiler chickens in Senegal using pulsed-field gel electrophoresis and antibiotic susceptibility. *J. Appl. Microbiol.* **2005**, *99*, 968–977. [CrossRef] [PubMed]

47. Thomason, B.M.; Biddle, J.W.; Cherry, W.B. Detection of salmonellae in environment. *Appl. Microbiol.* **1975**, *30*, 764–767. [PubMed]

48. Winfield, M.D.; Groisman, E.A. Role of nonhost environments in the lifestyles of Salmonella and Escherichia coli. *Appl. Environ. Microbiol.* **2003**, *69*, 3687–3694. [CrossRef] [PubMed]

49. Jackson, B.R.; Griffin, P.M.; Cole, D.; Walsh, K.A.; Chai, S.J. Outbreak-associated *Salmonella enterica* Serotypes and Food Commodities, United States, 1998–2008. *Emerg. Infect. Dis.* **2013**, *19*, 1239–1244. [CrossRef] [PubMed]

50. Collignon, P.; Powers, J.H.; Chiller, T.M.; Aidara-Kane, A.; Aarestrup, F.M. World Health Organization Ranking of Antimicrobials According to Their Importance in Human Medicine: A Critical Step for Developing Risk Management Strategies for the Use of Antimicrobials in Food Production Animals. *Clin. Infect. Dis.* **2009**, *49*, 132–141. [CrossRef] [PubMed]

51. Ahmed, A.M.; Nakano, H.; Shimamoto, T. Molecular characterization of integrons in non-typhoid Salmonella serovars isolated in Japan: Description of an unusual class 2 integron. *J. Antimicrob. Chemother.* **2005**, *55*, 371–374. [CrossRef] [PubMed]

52. Peirano, G.; Agerso, Y.; Aarestrup, F.M.; dos Reis, E.M.F.; Rodrigues, D.D. Occurrence of integrons and antimicrobial resistance genes among *Salmonella enterica* from Brazil. *J. Antimicrob. Chemother.* **2006**, *58*, 305–309. [CrossRef] [PubMed]

53. Guerri, M.L.; Aladuena, A.; Echeita, A.; Rotger, R. Detection of integrons and antibiotic-resistance genes in *Salmonella enterica* serovar Typhimurium, isolates with resistance to ampicillin and variable susceptibility to amoxicillin-clavulanate. *Int. J. Antimicrob. Agents* **2004**, *24*, 327–333. [CrossRef] [PubMed]

54. Antunes, P.; Machado, J.; Sousa, J.C.; Peixe, L. Dissemination of sulfonamide resistance genes (sul1, sul2, and sul3) in Portuguese *Salmonella enterica* strains and relation with integrons. *Antimicrob. Agents Chemother.* **2005**, *49*, 836–839. [CrossRef] [PubMed]

55. Ata, Z.; Yibar, A.; Arslan, E.; Mustak, K.; Gunaydin, E. Plasmid-mediated quinolone resistance in Salmonella serotypes isolated from chicken carcasses in Turkey. *Acta Vet BRNO* **2014**, *83*, 281–286. [CrossRef]

56. Strahilevitz, J.; Jacoby, G.A.; Hooper, D.C.; Robicsek, A. Plasmid-Mediated Quinolone Resistance: A Multifaceted Threat. *Clin. Microbiol. Rev.* **2009**, *22*, 664–689. [CrossRef] [PubMed]

57. Oh, J.Y.; Kwon, Y.K.; Tamang, M.D.; Jang, H.K.; Jeong, O.M.; Lee, H.S.; Kang, M.S. Plasmid-Mediated Quinolone Resistance in *Escherichia coli* Isolates from Wild Birds and Chickens in South Korea. *Microb. Drug Resist.* **2016**, *22*, 69–79. [CrossRef] [PubMed]

58. Geetha, V.K.; Yugendran, T.; Srinivasan, R.; Harish, B.N. Plasmid-mediated quinolone resistance in typhoidal Salmonellae: A preliminary report from South India. *Indian J. Med. Microbiol.* **2014**, *32*, 31–34. [PubMed]

59. Hopkins, K.L.; Davies, R.H.; Threlfall, E.J. Mechanisms of quinolone resistance in *Escherichia coli* and Salmonella: Recent developments. *Int. J. Antimicrob. Agents* **2005**, *25*, 358–373. [CrossRef] [PubMed]

60. Pande, V.V.; Gole, V.C.; McWhorter, A.R.; Abraham, S.; Chousalkar, K.K. Antimicrobial resistance of non-typhoidal *Salmonella isolates* from egg layer flocks and egg shells. *Int. J. Food Microbiol.* **2015**, *203*, 23–26. [CrossRef] [PubMed]

61. Miko, A.; Pries, K.; Schroeter, A.; Helmuth, R. Molecular mechanisms of resistance in multidrug-resistant serovars of Salmonella enterica isolated from foods in Germany. *J. Antimicrob. Chemother.* **2005**, *56*, 1025–1033. [CrossRef] [PubMed]

62. Murgia, M.; Bouchrif, B.; Timinouni, M.; Al-Qahtani, A.; Al-Ahdal, M.N.; Cappuccinelli, P.; Rubino, S.; Paglietti, B. Antibiotic resistance determinants and genetic analysis of *Salmonella enterica* isolated from food in Morocco. *Int. J. Food Microbiol.* **2015**, *215*, 31–39. [CrossRef] [PubMed]

63. Ahmed, H.A.; El-Hofy, F.I.; Shafik, S.M.; Abdelrahman, M.A.; Elsaid, G.A. Characterization of Virulence-Associated Genes, Antimicrobial Resistance Genes, and Class 1 Integrons in *Salmonella enterica* serovar Typhimurium Isolates from Chicken Meat and Humans in Egypt. *Foodborne Pathog. Dis.* **2016**, *13*, 281–288. [CrossRef] [PubMed]

64. El-Sharkawy, H.; Tahoun, A.; El-Gohary, A.; El-Abasy, M.; El-Khayat, F.; Gillespie, T.; Kitade, Y.; Hafez, H.M.; Neubauer, H.; El-Adawy, H. Epidemiological, molecular characterization and antibiotic resistance of *Salmonella enterica* serovars isolated from chicken farms in Egypt. *Gut Pathog.* **2017**, *9*, 8. [CrossRef] [PubMed]

65. Roberts, M.C. Update on acquired tetracycline resistance genes. *Fems Microbiol. Lett.* **2005**, *245*, 195–203. [CrossRef] [PubMed]

66. World Organisation for Animal Health (OIE). *Annual Report on the Use of Antimicrobial Agents in Animals-Better Understanding of the Global Situation*; OIE: Paris, France, 2015.

67. Abatcha, M.G.; Zakaria, Z.; Gurmeet, K.D.; Thong, K.T. Antibiograms, Resistance Genes, Class I Integrons and PFGE profiles of Zoonotic Salmonella in Malaysia. *Trop. Biomed.* **2015**, *32*, 573–586.

68. Berendsen, B.; Stolker, L.; de Jong, J.; Nielen, M.; Tserendorj, E.; Sodnomdarjaa, R.; Cannavan, A.; Elliott, C. Evidence of natural occurrence of the banned antibiotic chloramphenicol in herbs and grass. *Anal. Bioanal. Chem.* **2010**, *397*, 1955–1963. [CrossRef] [PubMed]

69. Qiao, J.; Zhang, Q.; Alali, W.Q.; Wang, J.W.; Meng, L.Y.; Xiao, Y.P.; Yang, H.; Chen, S.; Cui, S.H.; Yang, B.W. Characterization of extended-spectrum beta-lactamases (ESBLs)-producing Salmonella in retail raw chicken carcasses. *Int. J. Food Microbiol.* **2017**, *248*, 72–81. [CrossRef] [PubMed]

70. Eguale, T.; Birungi, J.; Asrat, D.; Njahira, M.N.; Njuguna, J.; Gebreyes, W.A.; Gunn, J.S.; Djikeng, A.; Engidawork, E. Genetic markers associated with resistance to beta-lactam and quinolone antimicrobials in non-typhoidal *Salmonella isolates* from humans and animals in central Ethiopia. *Antimicrob. Resist. Infect. Control* **2017**, *6*, 13. [CrossRef] [PubMed]

71. Giuriatti, J.; Stefani, L.M.; Brisola, M.C.; Crecencio, R.B.; Bitner, D.S.; Faria, G.A. Salmonella Heidelberg: Genetic profile of its antimicrobial resistance related to extended spectrum beta-lactamases (ESBLs). *Microb. Pathog.* **2017**, *109*, 195–199. [CrossRef] [PubMed]

International Journal of
*Environmental Research
and Public Health*

Article

Antibiotic Resistance and Antibiotic Resistance Genes in *Escherichia coli* Isolates from Hospital Wastewater in Vietnam

La Thi Quynh Lien [1,2,*] , Pham Thi Lan [3], Nguyen Thi Kim Chuc [3], Nguyen Quynh Hoa [4], Pham Hong Nhung [3,5], Nguyen Thi Minh Thoa [3], Vishal Diwan [1,6], Ashok J. Tamhankar [1,7] and Cecilia Stålsby Lundborg [1]

[1]   Health Systems and Policy (HSP): Improving the Use of Medicines, Department of Public Health Sciences, Karolinska Institutet, Tomtebodavägen 18A, 17177 Stockholm, Sweden; vishaldiwan@hotmail.com (V.D.); ejetee@gmail.com (A.J.T.); Cecilia.Stalsby.Lundborg@ki.se (C.S.L.)

[2]   Department of Pharmaceutical Management and Pharmaco-Economics, Hanoi University of Pharmacy, 13-15 Le Thanh Tong, Hoan Kiem District, Hanoi 110403, Vietnam

[3]   Department of Dermatology and Venereology, Department of Family Medicine, Department of Microbiology, Hanoi Medical University, 01 Ton That Tung, Dong Da District, Hanoi 116516, Vietnam; landhy2003@yahoo.com (P.T.L.); ntkchuc@yahoo.com (N.T.K.C.); hongnhung@hmu.edu.vn (P.H.N.); ntmthoa2017@gmail.com (N.T.M.T.)

[4]   National Centralized Drug Procurement Centre, Vietnam Ministry of Health, 138A Giang Vo Street, Ba Dinh district, Hanoi 118401, Vietnam; quynhhoa29@gmail.com

[5]   Department of Microbiology, Bach Mai Hospital, 78 Giai Phong, Dong Da District, Hanoi 116365, Vietnam

[6]   Department of Public Health & Environment, R.D. Gardi Medical College, Agar Road, Ujjain 456006, India

[7]   Indian Initiative for Management of Antibiotic Resistance, Department of Environmental Medicine, R.D. Gardi Medical College, Agar Road, Ujjain 456006, India

*   Correspondence: lienltq@hup.edu.vn; Tel.: +46-(0)-704-363-771

Received: 24 April 2017; Accepted: 23 June 2017; Published: 29 June 2017

Abstract: The environmental spread of antibiotic-resistant bacteria has been recognised as a growing public health threat for which hospitals play a significant role. The aims of this study were to investigate the prevalence of antibiotic resistance and antibiotic resistance genes (ARGs) in *Escherichia coli* isolates from hospital wastewater in Vietnam. Wastewater samples before and after treatment were collected using continuous sampling every month over a year. Standard disk diffusion and E-test were used for antibiotic susceptibility testing. Extended-spectrum beta-lactamase (ESBL) production was tested using combined disk diffusion. ARGs were detected by polymerase chain reactions. Resistance to at least one antibiotic was detected in 83% of isolates; multidrug resistance was found in 32%. The highest resistance prevalence was found for co-trimoxazole (70%) and the lowest for imipenem (1%). Forty-three percent of isolates were ESBL-producing, with the bla_{TEM} gene being more common than bla_{CTX-M}. Co-harbouring of the bla_{CTX-M}, bla_{TEM} and $qepA$ genes was found in 46% of isolates resistant to ciprofloxacin. The large presence of antibiotic-resistant *E. coli* isolates combined with ARGs in hospital wastewater, even post-treatment, poses a threat to public health. It highlights the need to develop effective processes for hospital wastewater treatment plants to eliminate antibiotic resistant bacteria and ARGs.

Keywords: antibiotic resistance; antibiotic resistance genes; bla_{CTX-M}; bla_{TEM}; $qepA$; hospital wastewater

---

1. Introduction

The environmental spread of antibiotic resistant bacteria has been recognized as a growing public health threat [1,2]. Hospitals are "hotspots" for antibiotic use and not only play an important role in antibiotic dissemination but also in the release of antibiotic resistant bacteria into the environment. Hospital wastewater treatment plants containing antibiotic residues can favour the development of antibiotic resistance due to the selective pressure placed on bacteria [3,4]. Moreover, antibiotic resistance genes (ARGs) carried by bacterial contaminants can be transferred to other bacterial populations including pathogenic bacteria found in hospital wastewater [1]. Hospital effluents can reach water bodies used in agriculture or for domestic purposes. From there, antibiotic resistant bacteria and/or ARGs can be transferred to humans.

In recent years, the presence of antibiotic resistant *Escherichia coli* (*E. coli*), particularly extended-spectrum beta-lactamase (ESBL)-producing isolates, in surface water has attracted attention [5]. A direct relationship between clinical *E. coli* isolates and the quantity of ESBL-producing *E. coli* strains found in hospital wastewater has been demonstrated [6]. Consequently, the existence of ESBL-producing *E. coli* carriers in hospitals may lead to their environmental spread [6].

Antibiotic resistance causes prolonged illness, excess mortality, and higher costs for patients and health systems [7–9]. Despite increased warnings and numerous efforts to contain it, antibiotic resistance has been increasing [10–13]. At the recent United Nations general assembly, it was highlighted that antibiotic resistance is among the greatest global health risks, requiring urgent attention [14].

The risks are potentially more serious in low- and middle-income countries where many hospitals either do not have wastewater treatment plants or they are ineffective. To make matters worse, in many places, but particularly rural areas, surface water is used for agriculture and domestic purposes or even consumed untreated. Most research on antibiotic resistant bacteria in hospital wastewater originates from high-income countries [15].

Therefore, this study sought to investigate the prevalence of resistant *E. coli* isolates to commonly used antibiotics, ESBL-producing isolates along with genes coding for cephalosporin resistance *bla*$_{CTX-M}$ and *bla*$_{TEM}$, and a gene coding for ciprofloxacin resistance *qepA*, in hospital wastewater in a rural and an urban hospital in Vietnam.

2. Materials and Methods

This is a repeated cross-sectional study with monthly data collection in one rural and one urban hospital in Vietnam, a lower middle-income country. The rural hospital has 220 beds and is situated 60 km northwest of central Hanoi. The 520-bedded urban hospital is located in central Hanoi. Both hospitals' wastewater is routed to wastewater treatment plants (WWTPs) where it is treated using filtering, microbiological, and biochemical mechanisms. After treatment, hospital effluents are discharged into sewer systems, which lead to nearby rivers.

2.1. Collection of Water Samples

The collection of water samples is described in detail elsewhere [4]. Briefly, samples of wastewater before treatment (WBT) as well as wastewater after treatment (WAT) were collected using 24 h continuous sampling on a weekday during the last week of every month in 2013. The water samples were stored in closed containers surrounded by ice and transferred to the microbiological laboratory in Bach Mai Hospital in central Hanoi within 6 h of testing. The urban hospital and its wastewater treatment plant was under reconstruction from June to August 2013, therefore sampling was ceased during this period.

2.2. Antibacterial Susceptibility Testing and Detection of ARGs

Coliforms were detected with the most probable number procedure [16]. A presumptive test involved three subsets of tubes containing different amounts of lactose or lauryl tryptose broth. Each subset contained five tubes with inverted Durham tubes to collect gas produced by fermentation. The three subsets were inoculated with water samples of 10, 1.0, and 0.1 mL, respectively. The tubes were then incubated for 24 h at 35–37 °C. A positive test for gas formation was presumptive evidence of coliforms. A confirmatory test for coliforms was made by inoculating another broth from one of the positive tubes. The test was completed by final isolation of the coliforms on selective and differential media, Gram staining the isolates, and reconfirming gas production. Coliform isolates were then sub-cultured on *Brilliance*™ UTI agar to collect presumptive *E. coli* isolates. Following biochemical confirmation using standard tests, identified *E. coli* isolates were tested for antibiotic susceptibility using the standard Kirby Bauer disc diffusion method for: (i) amoxicillin/clavulanic acid; (ii) ceftazidime; (iii) ceftriaxone; (iv) ciprofloxacin; (v) co-trimoxazole (trimethoprim/sulfamethoxazole); (vi) fosfomycin; (vii) gentamicin; and (viii) imipenem. The selected antibiotics were commonly used in the hospitals and routinely tested in clinical laboratories. Antibiotic susceptibility test results were interpreted as resistant, intermediate, and susceptible using Clinical and Laboratory Standard Institute guidelines (CLSI M100-2013) [17]. Minimum inhibitory concentrations (MICs) were determined for ciprofloxacin and ceftazidime using E-test. Disc diffusion zone diameters were also compared with epidemiological cut-off values (ECOFF) [18].

For *E. coli* isolates resistant to third generation cephalosporins, extended-spectrum beta-lactamase (ESBL) production was tested using combined disc diffusion. Genotypic confirmation was done through polymerase chain reactions. Genes coding for beta-lactam resistance, $bla_{CTX\text{-}M}$ and bla_{TEM}, were tested in ESBL-producing isolates and $qepA$ gene was tested for in ciprofloxacin-resistant isolates [19,20].

2.3. Data Analysis

Prevalence of resistance to at least one of the studied antibiotics, to each studied antibiotic, and multidrug resistance (MDR) in *E. coli* isolates were analysed. The definition of MDR reported by Magiorakos et al. (2012) from ECDC Joint Expert Meetings was applied; according to that, bacterial isolates were considered multidrug-resistant if they were non-susceptible to at least one agent in three or more antibiotic categories [21]. Fisher's exact test was applied to test the difference between the prevalence of antibiotic resistant *E. coli* isolates before and after wastewater treatment in each hospital using Stata 12 (StataCorp LP, College Station, TX, USA).

3. Results

In total, 265 *E. coli* isolates were collected from both the hospitals during the study period; 158 from the rural hospital (WBT = 84; WAT = 74) and 107 isolates from the urban hospital (WBT = 60; WAT = 47).

3.1. Resistance to Studied Antibiotics

In the rural hospital, 85% of *E. coli* isolates were resistant to at least one of the tested antibiotics (WBT = 94%; WAT = 74%). Resistance was most common towards co-trimoxazole, with 70% of isolates being resistant to it (WBT = 86%; WAT = 53%). Resistance to ceftriaxone was found in 49% of isolates (WBT = 55%; WAT = 42%), and resistance to ceftazidime, gentamicin, and amoxicillin/clavulanic acid was around 40%, respectively. Thirty percent of the isolates were resistant to ciprofloxacin (WBT = 25%; WAT = 35%), and 2% were resistant to fosfomycin (WBT = 1%; WAT = 3%). Resistance to imipenem was only detected in one isolate (1%). MDR was found in 35% of isolates (WBT = 44%; WAT = 26%). Prevalence of resistance to the studied antibiotics in *E. coli* isolates from WAT was less common than WBT, with the exception of ciprofloxacin and fosfomycin, for which enrichment of resistant *E. coli* isolates after wastewater treatment was found. The differences are statistically significant for amoxicillin/clavulanic acid, co-trimoxazole, gentamicin, resistance to at least one studied antibiotic, and MDR (Table 1).

In the urban hospital, 79% of *E. coli* isolates were resistant to at least one of the studied antibiotics (WBT = 88%; WAT = 68%). Co-trimoxazole resistance was again most common, with resistance found in 71% of isolates (WAT = 80%; WAT = 60%), followed by ceftriaxone resistance (39%) (WBT = 45%; WAT = 32%). Resistance to gentamicin and ceftazidime was found in 29% and 28% of isolates, respectively, followed by amoxicillin/clavulanic acid (24%) and ciprofloxacin (21%). Fosfomycin resistance was least common, as it was detected in only 8% of isolates. MDR was found in 27% of isolates (WBT = 32%; WAT = 21%). Prevalence of resistance to the studied antibiotics in *E. coli* isolates from WAT was lower than WBT. The differences were statistically significant for co-trimoxazole, fosfomycin, and resistance to at least one studied antibiotic (Table 1).

The distribution of MIC values for ceftazidime and ciprofloxacin susceptibility testing is presented in Figure 1. The number of *E. coli* isolates with high MIC values is large compared to the number of isolates with lower MIC values, indicating high levels as well as high proportions of resistance.

Table 1. Prevalence of resistance to studied antibiotics in *Escherichia coli* isolates found in hospital wastewater.

Studied Antibiotics	Rural Hospital (*n* = 158)				Urban Hospital (*n* = 107)				Both Hospitals (n = 265) Overall (%)
	WBT (%) *n* = 84	WAT (%) *n* = 74	*p*-Value	Overall (%)	WBT (%) *n* = 60	WAT (%) *n* = 47	*p*-Value	Overall (%)	
Amoxicillin/clavulanic acid	51	24	0.001 *	39	28	19	0.36	24	33
Ceftazidime	42	36	0.52	39	32	23	0.40	28	35
Ceftriaxone	55	42	0.11	49	45	32	0.23	39	45
Ciprofloxacin	25	35	0.22	30	23	17	0.50	21	26
Co-trimoxazole	86	53	<0.001 *	70	80	60	0.03 *	71	70
Fosfomycin	1	3	0.60	2	15	0	0.005 *	8	4
Gentamycin	51	31	0.02 *	42	33	23	0.30	29	37
Imipenem	1	0	1.00	1	0	0	N/A	0	1
At least one antibiotic	94	74	0.001 *	85	88	68	0.02 *	79	83
MDR	44	26	0.02 *	35	32	21	0.30	27	32

MDR: multidrug resistance; N/A: not available; WBT: wastewater before treatment; WAT: wastewater after treatment. * Differences in prevalence of resistant *Escherichia coli* strains isolated from WBT and WAT are significant.

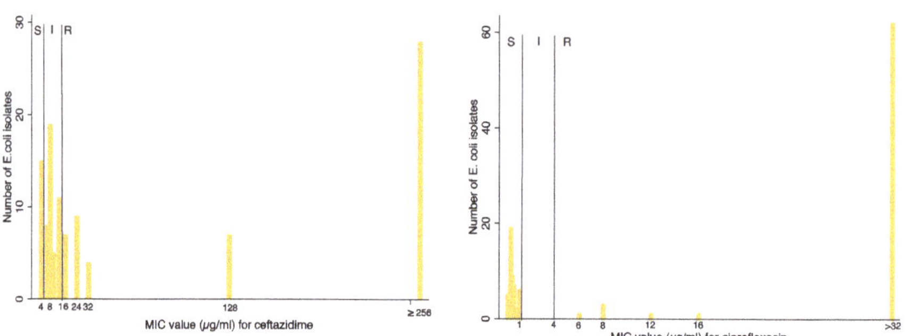

Figure 1. Distribution of minimum inhibitory concentration (MIC) values for ceftazidime and ciprofloxacin susceptibility testing.

When applying ECOFF values, we found decreased susceptibility to amoxicillin/clavulanic acid (45% of isolates), ceftazidime (39% of isolates), ceftriaxone (48% of isolates), ciprofloxacin (29% of isolates), and imipenem (2% of isolates).

MDR patterns are presented in Table 2 with identified antibiotic combinations. MDR to six out of eight studied antibiotics was found in 25 isolates (10%).

Table 2. Multidrug resistance patterns in *Escherichia coli* isolates found in hospital wastewater (the number of MDR isolates having the respective pattern).

MDR Pattern	Rural Hospital (n = 158)		Urban Hospital (n = 107)	
	WBT	WAT	WBT	WAT
CTX + CIP + SXT	0	0	0	2
CTX + GEN + SXT	7	0	3	0
AMC + GEN + SXT	0	0	1	0
GEN + CIP + SXT	0	4	7	1
CAZ + CTX + AMC + CIP	2	0	0	0
CAZ + CTX + AMC + SXT	2	0	0	2
CAZ + CTX + GEN + SXT	6	0	1	0
CAZ + CTX + CIP + SXT	0	0	0	1
CAZ + CTX + SXT + FOM	1	0	0	0
CTX + GEN + CIP + SXT	4	1	2	0
AMC + GEN + CIP + SXT	0	1	0	1
CAZ + CTX + GEN + CIP + SXT	2	2	3	0
CAZ + CTX + AMC + CIP + SXT	0	1	0	0
CTX + GEN + CIP + SXT + FOM	0	1	0	0
CTX + AMC + GEN + CIP + SXT	1	0	0	0
IMP + CAZ + CTX + AMC + SXT	1	0	0	0
CAZ + CTX + AMC + GEN + CIP + SXT	11	9	1	3
CAZ + CTX + AMC + GEN + CIP + FOM	0	0	1	0
Total (%)	37 (44%)	19 (26%)	19 (32%)	10 (21%)
	56 (35%)		29 (27%)	

MDR: multidrug resistance; WBT: wastewater before treatment; WAT: wastewater after treatment; AMC: amoxicillin/clavulanic acid; CAZ: ceftazidime; CIP: ciprofloxacin; CTX: ceftriaxone; FOM: fosfomycin; GEN: gentamicin; IMP: imipenem; SXT: trimethoprim/sulfamethoxazole (co-trimoxazole).

3.2. ESBL-Producing E. coli, ESBL, and Quinolone Resistance Genes

In the rural hospital, 76 *E. coli* isolates were ESBL-producing (48%). Among them, bla_{TEM} was detected in 97% of isolates and bla_{CTX-M} in 76%. Both bla_{CTX-M} and bla_{TEM} were detected in 75% of isolates. Quinolone-resistance gene ($qepA$) was detected in 72% of ciprofloxacin-resistant isolates. All three genes were detected in 51% of ciprofloxacin-resistant isolates (Table 3).

Table 3. Genetic analysis of extended-spectrum beta-lactamase (ESBL)-producing and ciprofloxacin-resistant *Escherichia coli* strains found in hospital wastewater.

Genetic Analysis	Rural Hospital			Urban Hospital			Both Hospitals
	WBT n (%)	WAT n (%)	Overall n (%)	WBT n (%)	WAT n (%)	Overall n (%)	Overall n (%)
ESBL-producing	45 (54)	31 (42)	76 (48)	27 (45)	12 (26)	39 (36)	115 (43)
bla_{CTX-M}	29 (64)	29 (94)	58 (76)	14 (52)	2 (17)	16 (41)	74 (64)
bla_{TEM}	44 (98)	30 (97)	74 (97)	27 (100)	10 (83)	37 (95)	111 (97)
$bla_{CTX-M} + bla_{TEM}$	29 (64)	28 (90)	57 (75)	14 (52)	2 (17)	16 (41)	73 (63)
Ciprofloxacin resistance	21 (25)	26 (35)	47 (30)	14 (23)	8 (17)	22 (21)	69 (26)
$qepA$	14 (67)	20 (77)	34 (72)	12 (86)	7 (88)	19 (86)	53 (77)
$qepA + bla_{CTX-M} + bla_{TEM}$	13 (62)	11 (42)	24 (51)	6 (43)	2 (25)	8 (36)	32 (46)

WAT: wastewater after treatment; WBT: wastewater before treatment.

In the urban hospital, 39 *E. coli* isolates were ESBL-producing strains (36%). Among them, bla_{TEM} was detected in 95% of isolates and bla_{CTX-M} in 41%. Both bla_{CTX-M} and bla_{TEM} were detected in 41% of isolates. Quinolone-resistance gene ($qepA$) was detected in 86% of ciprofloxacin-resistant isolates. All three genes, bla_{TEM}, bla_{CTX-M}, and $qepA$, were detected in 36% of ciprofloxacin-resistant isolates (Table 3).

4. Discussion

Our novel findings show that, in Vietnam, bacteria resistant to commonly used antibiotics along with genes coding for resistance are present in hospital wastewater, even after treatment. Prior to our study, Duong et al. examined *E. coli* resistance to ciprofloxacin and norfloxacin in hospital wastewater of another Hanoian hospital and reported that *E. coli* strains isolated from WAT samples were susceptible [22]. In their study, water samples were collected over two days using grab sampling and only 15 *E. coli* isolates, including three isolates from treated water samples, were tested. Conversely, in our study, water samples were collected every month over one year using continuous sampling and a total of 265 *E. coli* isolates were tested.

In both hospitals, *E. coli* isolates were most resistant to co-trimoxazole (around 70% of isolates). Previous reports show comparatively lower prevalence rates. In 2004–2005, an Indian study indicated that 55% of enteric bacteria found in hospital wastewater were resistant to co-trimoxazole [23]. A study from Poland showed that 20% of *E. coli* isolates from hospital wastewater collected before 2013 were resistant to co-trimoxazole [24]. Co-trimoxazole is a combination of sulfamethoxazole and trimethoprim. Sulfamethoxazole, one of the first antibiotics to be developed, was put into clinical use in 1935 and trimethoprim was first used in 1962. The two antibiotics first started to be used in combination in 1968. Over the past decades, its extensive use in clinical settings to treat a variety of bacterial infections, such as urinary tract infections and respiratory tract infections, which was also the case in the studied hospitals, might explain the high occurrence of co-trimoxazole resistance [25]. Moreover, both sulfamethoxazole and trimethoprim are not readily degradable and their residues found every month over the studied period in the same hospital wastewater could favour the development of co-trimoxazole resistance in the bacteria [4].

Cephalosporin resistance was also found in higher proportions than other antibiotics investigated. Similar prevalence rates of cephalosporin resistant bacteria in hospital effluent were shown by Chagas et al. [26]. Resistance mechanisms to second and third generation cephalosporins differ, with ESBLs being the most important [27]. ESBL enzymes are capable of hydrolysing and inactivating beta-lactam antibiotics and are often plasmid-mediated [28]. The plasmid genes encoding ESBLs can be transferred between different bacterial strains (horizontal gene transfer), facilitating easy spread of antibiotic resistance within as well as between species. Moreover, plasmid-encoded ESBL-producing bacteria can show co-resistance to quinolones, aminoglycosides, and sulfonamides [29]. Consequently, infections caused by ESBL-producing bacterial strains can be difficult to treat due to the restricted amount of antibiotics left for successful treatment.

ESBLs were first isolated in the 1980s [30]. In a study conducted in an urban and rural hospital in central India, Chandran et al. reported a very high prevalence of ESBL-producing *E. coli* in the hospital wastewater (96%) [31]. In our study, the prevalence was around 40%, lower than the aforementioned study but relatively higher than the figures reported by Diwan et al. (25%), Abdulhaq et al. (25%), and Korzeniewska et al. (37%) [22,32,33]. Among genes coding for ESBL production, TEM and CTX-M are the most common [34]. Our findings indicate the presence of bla_{CTX-M} and bla_{TEM} in ESBL-producing strains, with the bla_{TEM} gene being predominant. This is in accordance with the findings of Varela et al., where bla_{TEM} was found to be the most prevalent ESBL-encoding gene, followed by bla_{CTX-M} [35]. In contrast, Chandran et al. reported higher prevalence of bla_{CTX-M} than bla_{TEM} in hospital wastewater in India [31]. However, not all the bla_{TEM} genes are responsible for ESBL, and in our study, we were not able to do further sequencing to show the frequency of the ESBL bla_{TEM}. In addition, according to the PCR protocol used, the detected bla_{CTX-M} were restricted to CTX-M group 1 including CTX-M-1, CTX-M-3, and CTX-M-15. Ciprofloxacin resistance was detected in our samples, with gene *qepA* coding for high proportions of the resistant strains. Similar to ESBL-coding genes bla_{CTX-M} and bla_{TEM}, quinolone resistance gene *qepA* is plasmid-mediated and capable of horizontal gene transfer [36]. Of note is that co-existence of bla_{CTX-M}, bla_{TEM}, and *qepA* was detected, genetically proving co-resistance in the bacterial strains. In our study, the prevalence of MDR found phenotypically was around 35% with the detection of co-resistance to six out of eight of the studied antibiotics.

Fosfomycin-resistance and imipenem-resistance were also detected among the *E. coli* isolates in our study. Fosfomycin has broad activity against Gram-negative and some Gram-positive bacteria. In some countries, it is recommended as one of the first-line drugs to treat uncomplicated urinary tract infections because of increasing *E. coli* resistance towards other commonly used antibiotics, such as ciprofloxacin and co-trimoxazole [37]. Imipenem, the first carbapenem developed, is used to treat infections caused by β-lactamase-producing bacteria and should be saved to treat infections not readily treated by other antibiotics [38]. High prevalence of carbapenem-resistance in clinical isolates in Vietnamese hospitals has been reported [39]. The detection of resistance to these last-line antibiotics in bacterial isolates from hospital wastewater is of concern, since this can contribute to the spread of resistance among bacterial populations in the environment.

Although there is increasing evidence of the occurrence of antibiotic resistant bacteria in the environment, there are no standardized methods for antibiotic susceptibility testing which are directly applicable to environmental samples so far [2]. Epidemiological cut-off values developed by the European Committee on Antimicrobial Susceptibility Testing (EUCAST) can be used for the interpretation of antibiotic resistance in environmental bacteria. The ECOFF values separate bacteria with acquired resistance mechanisms (non-wild type) from the wild type population (having no resistance) [15]. We found in general decreased susceptibility in the *E. coli* isolates when using ECOFF values.

The role of hospitals in the environmental release of antibiotic-resistant bacteria and ARGs has been demonstrated and has become a growing concern for public health [40–47]. Hospital WWTPs can harbour antibiotic-resistant bacteria and ARGs [48–50]. Antibiotic residues in WWTPs can favour the development of antibiotic resistance due to the selective pressure placed on bacteria. In our previous study, we found that antibiotic concentrations in wastewater collected from hospital WWTPs were often higher than the reported predicted no-effect concentrations for resistance selection as well as the minimum selective concentrations, meaning that the selection of antibiotic-resistant bacteria can occur [4]. Published studies have reported the enrichment of antibiotic-resistant *E. coli* in WWTPs [51,52]. Our findings show significant reductions as well as the enrichment of antibiotic-resistant *E. coli* in WAT. Resistance to at least one studied antibiotic in the *E. coli* isolates from WAT was still detected in high proportions. Consequently, certain amounts of antibiotic-resistant bacteria, along with ARGs, are released into the ambient aquatic environment. They can then enter water bodies used for agriculture, irrigation, or household purposes, which poses a threat to public health. The problem can be aggravated if hospitals do not have WWTPs and the wastewater is discharged directly into the environment, which is common practice in Vietnam as well as many other low- and middle-income countries [53]. It has been shown that the prevalence of antibiotic-resistant bacteria is significantly reduced by advanced wastewater treatment processes such as ozone, UV, and ultrafiltration [54,55]. However, even in such advanced plants, resistant bacteria are not completely removed, therefore, hospitals must invest in effective WWTPs with treatment processes that completely eliminate antibiotic-resistant bacteria.

It is plausible that the dissemination of antibiotic-resistant bacteria and ARGs in the environment can result in their transmission to humans [56], however, direct evidence for this is very scarce [57]. So far, the strongest evidence available has shown the genetic similarities between human-related bacterial strains and environmental isolates collected at exposure-relevant sites [58]. Further studies are needed to identify links between the discharge of antibiotic-resistant bacteria by hospital WWTPs, their occurrence in the ambient environment, and their acquisition by humans via environmental exposure.

Our study has some limitations. Importantly, the prevalence of antibiotic resistance in *E. coli* isolates presented here might not be representative for the whole *E. coli* population in the hospital wastewater because of the limited number of *E. coli* isolates from each water sample. Moreover, due to financial constraints, we were not able to study more antibiotics and ARGs than what we have done. Screening for *bla*$_{SHV}$ gene, which is also common in ESBL-producing *E. coli,* and genes coding for imipenem resistance would make the study more comprehensive. We were also not able to do sequencing for bla_{TEM} genes to show the frequency of the genes encoding for ESBL. Another limitation is that data from the urban hospital were unavailable for three months, as sampling could not be carried out due to construction work at the hospital. Furthermore, a detailed description of the functioning of the WWTPs was not available for us.

5. Conclusions

High prevalence of antibiotic resistance and ARGs were detected in *E. coli* isolates from hospital wastewater both before and even after wastewater treatment. There is a need for inclusion and development of hospital WWTPs which are effective at eliminating antibiotic-resistant bacteria and ARGs. Further studies are needed to identify links between the discharge of antibiotic-resistant bacteria by hospital WWTPs, their occurrence in the nearby environment, and their acquisition by humans when exposed in the environment.

Acknowledgments: This work was supported by the Swedish International Development Cooperation Agency (Sida) (grant number SWE-2010-50) and the Vietnamese Government Scholarship for doctoral study of the first author.

Author Contributions: Cecilia Stålsby Lundborg, Nguyen Thi Kim Chuc, Pham Thi Lan, Nguyen Quynh Hoa, Ashok J. Tamhankar, and Vishal Diwan conceived and designed the experiments; La Thi Quynh Lien, Nguyen Thi Minh Thoa, and Pham Hong Nhung performed the experiments; La Thi Quynh Lien and Cecilia Stålsby Lundborg in collaboration with all co-authors analysed the data and contributed with reagents/materials/analysis tools; La Thi Quynh Lien wrote the first draft of the manuscript; and all authors further contributed in writing the manuscript.

Conflicts of Interest: The authors declare no conflicts of interest.

References

1. Williams, M.R.; Stedtfeld, R.D.; Guo, X.; Hashsham, S.A. Antimicrobial resistance in the environment. *Water Environ. Res.* **2016**, *88*, 1951–1967. [CrossRef] [PubMed]
2. Berendonk, T.U.; Manaia, C.M.; Merlin, C.; Fatta-Kassinos, D.; Cytryn, E.; Walsh, F.; Burgmann, H.; Sorum, H.; Norstrom, M.; Pons, M.N.; et al. Tackling antibiotic resistance: The environmental framework. *Nat. Rev. Microbiol.* **2015**, *13*, 310–317. [CrossRef] [PubMed]
3. Verlicchi, P.; Al Aukidy, M.; Galletti, A.; Petrovic, M.; Barceló, D. Hospital effluent: Investigation of the concentrations and distribution of pharmaceuticals and environmental risk assessment. *Sci. Total Environ.* **2012**, *430*, 109–118. [CrossRef] [PubMed]
4. Lien, L.T.; Hoa, N.Q.; Chuc, N.T.; Thoa, N.T.; Phuc, H.D.; Diwan, V.; Dat, N.T.; Tamhankar, A.J.; Lundborg, C.S. Antibiotics in Wastewater of a Rural and an Urban Hospital before and after Wastewater Treatment, and the Relationship with Antibiotic Use-A One Year Study from Vietnam. *Int. J. Environ. Res. Public Health* **2016**, *13*, 558. [CrossRef] [PubMed]
5. Guenther, S.; Ewers, C.; Wieler, L.H. Extended-Spectrum Beta-Lactamases Producing *E. coli* in Wildlife, yet Another Form of Environmental Pollution? *Front. Microbiol.* **2011**, *2*, 246. [CrossRef] [PubMed]
6. Drieux, L.; Haenn, S.; Moulin, L.; Jarlier, V. Quantitative evaluation of extended-spectrum beta-lactamase-producing *Escherichia coli* strains in the wastewater of a French teaching hospital and relation to patient strain. *Antimicrob. Resist. Infect. Control* **2016**, *5*, 9. [CrossRef] [PubMed]
7. World Health Organization (WHO). *Antibiotic Resistance: Global Report on Surveillance 2014*; WHO: Geneva, Switzerland, 2015.
8. Chandy, S.J.; Naik, G.S.; Balaji, V.; Jeyaseelan, V.; Thomas, K.; Lundborg, C.S. High cost burden and health consequences of antibiotic resistance: The price to pay. *J. Infect. Dev. Ctries.* **2014**, *8*, 1096–1102. [CrossRef] [PubMed]
9. Alsan, M.; Schoemaker, L.; Eggleston, K.; Kammili, N.; Kolli, P.; Bhattacharya, J. Out-of-pocket health expenditures and antimicrobial resistance in low-income and middle-income countries: An economic analysis. *Lancet Infect. Dis.* **2015**, *15*, 1203–1210. [CrossRef]
10. World Health Organization (WHO). *Global Action Plan on Antimicrobial Resistance*; WHO: Geneva, Switzeland, 2015.

11. World Health Organization (WHO). *Worldwide Country Situation Analysis: Response to Antimicrobial Resistance*; WHO: Geneva, Switzerland, 2015.

12. Velez, R.; Sloand, E. Combating antibiotic resistance, mitigating future threats and ongoing initiatives. *J. Clin. Nurs.* **2016**, *25*, 1886–1889. [CrossRef] [PubMed]

13. Roca, I.; Akova, M.; Baquero, F.; Carlet, J.; Cavaleri, M.; Coenen, S.; Cohen, J.; Findlay, D.; Gyssens, I.; Heuer, O.E.; et al. The global threat of antimicrobial resistance: Science for intervention. *New Microbes New Infect.* **2015**, *6*, 22–29. [CrossRef] [PubMed]

14. General Assembly of the United Nations. *PRESS RELEASE: High-Level Meeting on Antimicrobial Resistance*. Available online: http://www.un.org/pga/71/2016/09/21/press-release-hl-meeting-on-antimicrobial-resistance/ (accessed on 7 November 2016).

15. Hocquet, D.; Muller, A.; Bertrand, X. What happens in hospitals does not stay in hospitals: Antibiotic-resistant bacteria in hospital wastewater systems. *J. Hosp. Infect.* **2016**, *93*, 395–402. [CrossRef] [PubMed]

16. Talaro, K.P.; Talaro, A. *Foundations in Microbiology*, 4th ed.; The McGraw–Hill Companies: Boston, MA, USA, 2002; pp. 799–813.

17. Cinical and Laboratory Standards Institute (CLSI). *Performance Standards for Antimicrobial Disk Susceptibility Tests. Approved Standard-Eleventh Edition*; CLSI: Wayne, PA, USA, 2012.

18. EUCAST Antimicrobial Wild Type Distributions of Microorganisms. Available online: http://mic.eucast.org/Eucast2/SearchController/search.jsp?action=performSearch&BeginIndex=0&Micdif=dif&NumberIndex50&Antib=S-S1&Specium=162 (accessed on 2 December 2016).

19. Dallenne, C.; Da Costa, A.; Decre, D.; Favier, C.; Arlet, G. Development of a set of multiplex PCR assays for the detection of genes encoding important beta-lactamases in Enterobacteriaceae. *J. Antimicrob. Chemother.* **2010**, *65*, 490–495. [CrossRef] [PubMed]

20. Le, T.M.; Baker, S.; Le, T.P.; Cao, T.T.; Tran, T.T.; Nguyen, V.M.; Campbell, J.I.; Lam, M.Y.; Nguyen, T.H.; et al. High prevalence of plasmid-mediated quinolone resistance determinants in commensal members of the Enterobacteriaceae in Ho Chi Minh City, Vietnam. *J. Méd. Microbiol.* **2009**, *58*, 1585–1592. [PubMed]

21. Magiorakos, A.P.; Srinivasan, A.; Carey, R.B.; Carmeli, Y.; Falagas, M.E.; Giske, C.G.; Harbarth, S.; Hindler, J.F.; Kahlmeter, G.; Olsson-Liljequist, B.; et al. Multidrug-resistant, extensively drug-resistant and pandrug-resistant bacteria: An international expert proposal for interim standard definitions for acquired resistance. *Clin. Microbiol. Infect.* **2012**, *18*, 268–281. [CrossRef] [PubMed]

22. Duong, H.A.; Pham, N.H.; Nguyen, H.T.; Hoang, T.T.; Pham, H.V.; Pham, V.C.; Berg, M.; Giger, W.; Alder, A.C. Occurrence, fate and antibiotic resistance of fluoroquinolone antibacterials in hospital wastewaters in Hanoi, Vietnam. *Chemosphere* **2008**, *72*, 968–973. [CrossRef] [PubMed]

23. Alam, M.Z.; Aqil, F.; Ahmad, I.; Ahmad, S. Incidence and transferability of antibiotic resistance in the enteric bacteria isolated from hospital wastewater. *Braz. J. Microbiol.* **2013**, *44*, 799–806. [CrossRef] [PubMed]

24. Korzeniewska, E.; Korzeniewska, A.; Harnisz, M. Antibiotic resistant *Escherichia coli* in hospital and municipal sewage and their emission to the environment. *Ecotoxicol. Environ. Saf.* **2013**, *91*, 96–102. [CrossRef] [PubMed]

25. Huovinen, P. Resistance to trimethoprim-sulfamethoxazole. *Clin. Infect. Dis.* **2001**, *32*, 1608–1614. [PubMed]

26. Chagas, T.P.; Seki, L.M.; Cury, J.C.; Oliveira, J.A.; Davila, A.M.; Silva, D.M.; Asensi, M.D. Multiresistance, beta-lactamase-encoding genes and bacterial diversity in hospital wastewater in Rio de Janeiro, Brazil. *J. Appl. Microbiol.* **2011**, *111*, 572–581. [CrossRef] [PubMed]

27. Rawat, D.; Nair, D. Extended-spectrum β-lactamases in Gram Negative Bacteria. *J. Glob. Infect. Dis.* **2010**, *2*, 263–274. [CrossRef] [PubMed]

28. Bush, K. Characterization of beta-lactamases. *Antimicrob. Agents Chemother.* **1989**, *33*, 259–263. [CrossRef] [PubMed]

29. Silva, J.; Castillo, G.; Callejas, L.; Lopez, H.; Olmos, J. Frequency of transferable multiple antibiotic resistance amongst coliform bacteria isolated from a treated sewage effluent in Antofagasta, Chile. *Electron. J. Biotechnol.* **2006**, *5*, 533–540. [CrossRef]

30. Brun-Buisson, C.; Legrand, P.; Philippon, A.; Montravers, F.; Ansquer, M.; Duval, J. Transferable enzymatic resistance to third-generation cephalosporins during nosocomial outbreak of multiresistant Klebsiella pneumoniae. *Lancet* **1987**, *2*, 302–306. [CrossRef]

31. Chandran, S.P.; Diwan, V.; Tamhankar, A.J.; Joseph, B.V.; Rosales-Klintz, S.; Mundayoor, S.; Lundborg, C.S.; Macaden, R. Detection of carbapenem resistance genes and cephalosporin, and quinolone resistance genes along with oqxAB gene in *Escherichia coli* in hospital wastewater: A matter of concern. *J. Appl. Microbiol.* **2014**, *117*, 984–995. [CrossRef] [PubMed]

32. Diwan, V.; Chandran, S.P.; Tamhankar, A.J.; Lundborg, C.S.; Macaden, R. Identification of extended-spectrum -lactamase and quinolone resistance genes in *Escherichia coli* isolated from hospital wastewater from central India. *J. Antimicrob. Chemother.* **2012**, *67*, 857–859. [CrossRef] [PubMed]

33. Abdulhaq, A.; Basode, V.K. Prevalence of extended-spectrum beta-lactamase-producing bacteria in hospital and community sewage in Saudi Arabia. *Am. J. Infect. Control* **2015**, *43*, 1139–1141. [CrossRef] [PubMed]

34. Bush, K.; Jacoby, G.A. Updated functional classification of beta-lactamases. *Antimicrob. Agents Chemother.* **2010**, *54*, 969–976. [CrossRef] [PubMed]

35. Varela, A.R.; Macedo, G.N.; Nunes, O.C.; Manaia, C.M. Genetic characterization of fluoroquinolone resistant *Escherichia coli* from urban streams and municipal and hospital effluents. *FEMS Microbiol. Ecol.* **2015**, *91*. [CrossRef] [PubMed]

36. Yamane, K.; Wachino, J.; Suzuki, S.; Arakawa, Y. Plasmid-mediated qepA gene among *Escherichia coli* clinical isolates from Japan. *Antimicrob. Agents Chemother.* **2008**, *52*, 1564–1566. [CrossRef] [PubMed]

37. Alrowais, H.; McElheny, C.L.; Spychala, C.N.; Sastry, S.; Guo, Q.; Butt, A.A.; Doi, Y. Fosfomycin Resistance in *Escherichia coli*, Pennsylvania, USA. *Emerg. Infect. Dis.* **2015**, *21*, 2045–2047. [CrossRef] [PubMed]

38. Edwards, J.R.; Betts, M.J. Carbapenems: The pinnacle of the β-lactam antibiotics or room for improvement? *J. Antimicrob. Chemother.* **2000**, *45*, 1–4. [CrossRef] [PubMed]

39. Le, N.K.; Wertheim, H.F.; Vu, P.D.; Khu, D.T.; Le, H.T.; Hoang, B.T.; Vo, V.T.; Lam, Y.M.; Vu, D.T.; Nguyen, T.H.; et al. High prevalence of hospital-acquired infections caused by gram-negative carbapenem resistant strains in Vietnamese pediatric ICUs: A multi-centre point prevalence survey. *Medicine* **2016**, *95*, e4099. [CrossRef] [PubMed]

40. Laffite, A.; Kilunga, P.I.; Kayembe, J.M.; Devarajan, N.; Mulaji, C.K.; Giuliani, G.; Slaveykova, V.I.; Pote, J. Hospital Effluents Are One of Several Sources of Metal, Antibiotic Resistance Genes, and Bacterial Markers Disseminated in Sub-Saharan Urban Rivers. *Front. Microbiol.* **2016**, *7*, 1128. [CrossRef] [PubMed]

41. Mandal, S.M.; Ghosh, A.K.; Pati, B.R. Dissemination of antibiotic resistance in methicillin-resistant *Staphylococcus aureus* and vancomycin-resistant *S. aureus* strains isolated from hospital effluents. *Am. J. Infect. Control* **2015**, *43*, e87–e88. [CrossRef] [PubMed]

42. Ory, J.; Bricheux, G.; Togola, A.; Bonnet, J.L.; Donnadieu-Bernard, F.; Nakusi, L.; Forestier, C.; Traore, O. Ciprofloxacin residue and antibiotic-resistant biofilm bacteria in hospital effluent. *Environ. Pollut.* **2016**, *214*, 635–645. [CrossRef] [PubMed]

43. Roderova, M.; Sedlakova, M.H.; Pudova, V.; Hricova, K.; Silova, R.; Imwensi, P.E.; Bardon, J.; Kolar, M. Occurrence of bacteria producing broad-spectrum beta-lactamases and qnr genes in hospital and urban wastewater samples. *New Microbiol.* **2016**, *39*, 124–133. [PubMed]

44. Santoro, D.O.; Cardoso, A.M.; Coutinho, F.H.; Pinto, L.H.; Vieira, R.P.; Albano, R.M.; Clementino, M.M. Diversity and antibiotic resistance profiles of Pseudomonads from a hospital wastewater treatment plant. *J. Appl. Microbiol.* **2015**, *119*, 1527–1540. [CrossRef] [PubMed]

45. Varela, A.R.; Nunes, O.C.; Manaia, C.M. Quinolone resistant *Aeromonas* spp. as carriers and potential tracers of acquired antibiotic resistance in hospital and municipal wastewater. *Sci. Total Environ.* **2016**, *542*, 665–671. [CrossRef] [PubMed]

46. Vaz-Moreira, I.; Varela, A.R.; Pereira, T.V.; Fochat, R.C.; Manaia, C.M. Multidrug Resistance in Quinolone-Resistant Gram-Negative Bacteria Isolated from Hospital Effluent and the Municipal Wastewater Treatment Plant. *Microb. Drug Resist.* **2016**, *22*, 155–163. [CrossRef] [PubMed]

47. Anssour, L.; Messai, Y.; Estepa, V.; Torres, C.; Bakour, R. Characteristics of ciprofloxacin-resistant Enterobacteriaceae isolates recovered from wastewater of an Algerian hospital. *J. Infect. Dev. Ctries.* **2016**, *10*, 728–734. [CrossRef] [PubMed]

48. Li, C.; Lu, J.; Liu, J.; Zhang, G.; Tong, Y.; Ma, N. Exploring the correlations between antibiotics and antibiotic resistance genes in the wastewater treatment plants of hospitals in Xinjiang, China. *Environ. Sci. Pollut. Res. Int.* **2016**, *23*, 15111–15121. [CrossRef] [PubMed]

49. Harris, S.; Morris, C.; Morris, D.; Cormican, M.; Cummins, E. Antimicrobial resistant *Escherichia coli* in the municipal wastewater system: Effect of hospital effluent and environmental fate. *Sci. Total Environ.* **2014**, *468–469*, 1078–1085. [CrossRef] [PubMed]

50. Iweriebor, B.C.; Gaqavu, S.; Obi, L.C.; Nwodo, U.U.; Okoh, A.I. Antibiotic susceptibilities of enterococcus species isolated from hospital and domestic wastewater effluents in alice, eastern cape province of South Africa. *Int. J. Environ. Res. Public Health* **2015**, *12*, 4231–4246. [CrossRef] [PubMed]

51. Galvin, S.; Boyle, F.; Hickey, P.; Vellinga, A.; Morris, D.; Cormican, M. Enumeration and Characterization of Antimicrobial-Resistant *Escherichia coli* Bacteria in Effluent from Municipal, Hospital, and Secondary Treatment Facility Sources. *Appl. Environ. Microbiol.* **2010**, *76*, 4772–4779. [CrossRef] [PubMed]

52. Ferreira da Silva, M.; Vaz-Moreira, I.; Gonzalez-Pajuelo, M.; Nunes, O.C.; Manaia, C.M. Antimicrobial resistance patterns in Enterobacteriaceae isolated from an urban wastewater treatment plant. *FEMS Microbiol. Ecol.* **2007**, *60*, 166–176. [CrossRef] [PubMed]

53. Rizzo, L.; Manaia, C.; Merlin, C.; Schwartz, T.; Dagot, C.; Ploy, M.C.; Michael, I.; Fatta-Kassinos, D. Urban wastewater treatment plants as hotspots for antibiotic resistant bacteria and genes spread into the environment: A review. *Sci. Total Environ.* **2013**, *447*, 345–360. [CrossRef] [PubMed]

54. Nagulapally, S.R.; Ahmad, A.; Henry, A.; Marchin, G.L.; Zurek, L.; Bhandari, A. Occurrence of ciprofloxacin-, trimethoprim-sulfamethoxazole-, and vancomycin-resistant bacteria in a municipal wastewater treatment plant. *Water Environ. Res.* **2009**, *81*, 82–90. [CrossRef] [PubMed]

55. Rosenberg Goldstein, R.E.; Micallef, S.A.; Gibbs, S.G.; George, A.; Claye, E.; Sapkota, A.; Joseph, S.W.; Sapkota, A.R. Detection of vancomycin-resistant enterococci (VRE) at four U.S. wastewater treatment plants that provide effluent for reuse. *Sci. Total Environ.* **2014**, *466–467*, 404–411. [CrossRef] [PubMed]

56. Maheshwari, M.; Yaser, N.H.; Naz, S.; Fatima, M.; Ahmad, I. Emergence of ciprofloxacin-resistant extended-spectrum beta-lactamase-producing enteric bacteria in hospital wastewater and clinical sources. *J. Glob. Antimicrob. Resist.* **2016**, *5*, 22–25. [CrossRef] [PubMed]

57. Huijbers, P.M.; Blaak, H.; de Jong, M.C.; Graat, E.A.; Vandenbroucke-Grauls, C.M.; de Roda Husman, A.M. Role of the Environment in the Transmission of Antimicrobial Resistance to Humans: A Review. *Environ. Sci. Technol.* **2015**, *49*, 11993–12004. [CrossRef] [PubMed]

58. Hower, S.; Phillips, M.C.; Brodsky, M.; Dameron, A.; Tamargo, M.A.; Salazar, N.C.; Jackson, C.R.; Barrett, J.B.; Davidson, M.; Davis, J.; et al. Clonally related methicillin-resistant *Staphylococcus aureus* isolated from short-finned pilot whales (*Globicephala macrorhynchus*), human volunteers, and a bayfront cetacean rehabilitation facility. *Microb. Ecol.* **2013**, *65*, 1024–1038. [CrossRef] [PubMed]

International Journal of
Environmental Research and Public Health

Article

Tigecycline Resistant *Klebsiella pneumoniae* Isolated from Austrian River Water

Alexander Hladicz [1,2] , Clemens Kittinger [2] and Gernot Zarfel [2,*]

1   Center for Molecular Biology, University of Vienna, 1030 Vienna, Austria; alexander.hladicz@gmx.at
2   Institute of Hygiene, Microbiology and Environmental Medicine, Medical University of Graz,
     Neue Stiftingtalstrasse 2, 8010 Graz, Austria; clemens.kittinger@medunigraz.at
*   Correspondence: gernot.zarfel@medunigraz.at; Tel.: +43-316-385-73604

Received: 4 September 2017; Accepted: 30 September 2017; Published: 3 October 2017

Abstract: Antibiotic-resistant bacteria are spreading worldwide in medical settings but also in the environment. These resistant bacteria illustrate a major health problem in our times, and last-line antibiotics such as tigecycline represent an ultimate therapy option. Reports on tigecycline non-susceptible *Enterobacteriaceae* are presented with regard to medical settings but are rare with that for the environment. The aim of this study was to characterize two tigecycline non-susceptible *Klebsiella pneumoniae* isolates from the river Mur, and to question the resistance mechanism. The screening for chromosomal mutations revealed a deletion and a silent point mutation in one isolate and a point mutation in the other isolate all within the *ramR* allele. RamR acts as repressor and prevents overexpression of *ramA*. These mutations are likely to cause a resistant phenotype due to the overexpression of AcrAB-TolC. MLST revealed that the isolates belonged to two unrelated MLST types (ST2392 and ST2394). Both isolates only revealed resistance to tigecycline and tetracycline. This is one of the rare reports of tigecycline-resistant *Klebsiella pneumoniae* from surface water. The presence of two genetically different isolates suggests that the river water may bear substances that favor mutations that can lead to this efflux pump-driven resistance.

Keywords: ramA; efflux pump; multilocus sequence typing; surface water

1. Introduction

The emergence of antibiotic resistances is a worldwide rising phenomenon. It is not restricted to clinical settings and it reaches environmental settings and their associated ecological habitats. In particular, surface waters such as rivers, lakes or coastal waters act as reservoirs for resistant bacteria owing to anthropogenic activities and influences such as industrial or urban sewage [1–5]. The discharge of resistant bacteria in combination with antibiotics and/or other chemical compounds into the water bodies is likely to select for antibiotic resistances within microbial communities [6–8]. Therefore, effluents or insufficient water management promotes the distribution of resistant bacteria and facilitates the spread of resistance genes [9]. This trend of emerging antibiotic-resistant bacteria speeds up by the overuse of antibiotics in human and veterinary medicine, and a subsequent release of these substances into the environment [10].

The massive health problem that arises from the current situation concerns (opportunistic) pathogens that gained multidrug resistance (MDR) to a broad spectrum of antibiotics. ESBL-(extended-spectrum b-lactamase) or carbapenemase-producing *Enterobacteriaceae*, notably *Klebsiella pneumoniae*, are described not only in clinical but also in different aquatic settings all around the world, including Austria [2–4,11–13].

Last-resort antibiotics act as ultimate force to overcome those multiresistant strains. Tigecycline is such an antibiotic and is often the last or the penultimate choice (besides colistin) to treat infections caused by those pathogens [14,15]. Hence, occurrence of tigecycline resistance is a major threat to every medical institution. Cases of tigecycline non-susceptible *Klebsiella pneumoniae* in clinical settings are reported worldwide [16,17] but are rather rare regarding environmental settings.

There are different mechanisms that can lead to an acquired tigecycline resistance, most of them based on chromosomal mutations. Gene network of the efflux pump, AcrAB-TolC is associated with tigecycline non-susceptibility and its regulators has been analyzed with regard to tigecycline non-susceptibility in prior studies. In particular, mutation in the repressors RamR, MarR and SoxR of the regulators (RamA, MarA and SoxS) of the efflux pump were found to be responsible [18–22].

An additional mutation in the ribosomal RPS10 protein, which is located close to the ribosomal binding site of tigecycline is likely to influence the binding properties between the ribosome and tigecycline [21].

The aim of this study was to elucidate the resistance mechanism that causes tigecycline non-susceptibility and to question whether this mechanism is plasmid or chromosomally mediated. In order to detect a potential plasmid-encoded resistance mechanism, transformation experiments were performed.

2. Material and Methods

2.1. Sample Collection

Water samples were taken for microbiological investigations during a survey from the river Mur in the center of Graz (47°4′38″ N; 15°25′60″ E); each sample in two sterile 500 mL glass flasks, 30 cm below the river surface, 50 cm apart from the river bank.

2.2. Isolation of Bacteria

Samples were filtered using Microfil® S device (Merck, Vienna, Austria) with 0.45 μm pore filters in 4 times 250 mL portions. For each sampling, two filters were put on chromID™ ESBL Agar (bioMérieux Austria GmbH, Vienna, Austria) and two on chromID™ CARBA Agar (bioMérieux). ChromID™ agars were incubated for 24 h at 37 °C. Colonies were assessed and picked according to the manufacturer's manual. For pure cultures, colonies were transferred to blood agar and Endo agar (24 h, 37 °C) and species were finally identified with MALDI-TOF, (Vitek® MS, bioMérieux Austria GmbH, Vienna, Austria).

Thereby *Klebsiella pneumoniae* isolates could be recovered on chromID™ CARBA Agar (MurTR-KL001 on 2 February 2016; and MurTR-KL002 on 11 February 2016).

2.3. Antimicrobial Susceptibility Testing

Susceptibility testing was performed according to the Clinical and Laboratory Standards Institute (CLSI) guidelines or as recommended by the European Committee on Antimicrobial Susceptibility testing (EUCAST) using BD BBLTM, Sensi-Disc™ paper discs (Becton, Dickinson and Company, Sparks, MD, USA) [23,24].

The inhibition zone diameters were interpreted according to EUCAST guidelines with the exception for tetracycline, chloramphenicol and nalidixic acid, which were evaluated in conformity with the Clinical Laboratory Standards Institute (CLSI) guidelines. EUCAST guidelines were chosen as they are the clinical standard for Europe; whenever EUCAST criteria were not available CLSI standards were used.

The following antibiotics were used: amoxicillin/clavulanic acid (20 μg/10 μg), piperacillin/tazobactam (100 μg/10 μg), cefalexin (30 μg), cefuroxime (30 μg), cefoxitin (30 μg), cefotaxime (5 μg), ceftazidime (10 μg), cefepime (30 μg), imipenem (10 μg), meropenem (10 μg), amikacine (30 μg), gentamicin (10 μg), trimethoprim/sulfamethoxazole (1.25 μg/23.75 μg), ciprofloxacin (5 μg), moxifloxacin (5 μg), tetracycline (30 μg), nalidixic acid (30 μg) chloramphenicol (30 μg).

To determine tigecycline and colistin susceptibility, Etests® (bioMérieux Austria GmbH, Vienna, Austria) according to EUCAST guidelines for tigecycline and colistin were performed as described previously [25,26].

Escherichia coli ATCC 25922 and *Pseudomonas aeruginosa* ATCC 27853 were used as control strains in all conducted tests.

2.4. Plasmid Replicon Typing

Identification of replicon types of the 18 major plasmid incompatibility (Inc) groups present in *Enterobacteriaceae* was performed by multiplex PCR.

Standard PCR protocols and conditions were used in the following way: initial denaturation at 94 °C for 5 min; 30 cycles of 94 °C for 1 min, 60 °C for 30 s and 72 °C for 1 min; and a final incubation for 5 min at 72 °C. We used Taq DNA polymerase and dNTPs from QIAGEN (Hilden, Germany), and a T3000 Biometra thermocycler (Biometra, Gottingen, Germany).

The protocol allows detection of the following Inc groups: Hl1, Hl2, I1-Iγ, X, L/M, N, FIA, FIB, W, Y, P, FIC, A/C, T, FII$_S$, F, K, B/O [27].

2.5. Transformation by Electroporation

Preparation of Plasmid-DNA was performed with the QIAprep Spin Miniprep Kit (250) (QIAGEN).

Plasmid-DNA was desalted before electroporation, and therefore 2–3 µL of plasmid-DNA were transferred on a MF^{TM} Membrane Filter (0.025 µm VSWP, Merck), which was placed on the surface of double distilled water. Dialysis was performed for about 15 min.

Competent cells were made with two overnight cultures (each 50 mL, OD of 0.4), which were incubated on ice for 25 min, therefore reaction tubes were cooled in advance, followed by centrifugation for 10 min at 4 °C and 4.000 rpm (Eppendorf, Centrifuge 5810R). After decantation, pellets were re-suspended in 100 mL ice-cold glycerine solution (10%). After repeating this step, an additional washing step was performed, and the two pellets together were re-suspended in 10 mL glycerine solution (10%). A last washing step and resuspension were performed with 1 mL glycerine solution (10%). Aliquots of 50 µL were prepared and stored at −20 °C.

Electroporation was performed with 2 µL Plasmid-DNA and 40 µL of competent cells. Reaction tubes were cooled in advance and the DNA-cell suspension was incubated on ice for 5 min. Subsequently, the cell suspension was transferred into a sterile electro-cuvette, and transformation was performed at 2500 V using the electroporator (Eppendorf Eporator®). After the transformation, 400 µL of fresh LB liquid media were added and the cell suspension was re-transferred into the reaction tube. Incubation was performed for 40 min at 37 °C. Afterwards 100 µL of the cell suspension were plated on selection LB (lysogeny broth) plates (tetracycline 3 µg/mL or tigecycline 1 µg/mL) and a final incubation was performed over night at 37 °C.

2.6. Multilocus Sequence Typing (MLST)

MLST was performed for *Klebsiella pneumoniae* according to the Institute Pasteur MLST (http://bigsdb.web.pasteur.fr/klebsiella/klebsiella.html).

2.7. Screening for Mutations

The genes *ramR*, *marR*, *soxR* and *rpsJ* were amplified and sequenced with the primers described previously [18,21].

Standard PCR protocols and conditions were used in the following way: initial denaturation at 94 °C for 5 min; 35 cycles of 94 °C for 30 s, 52 °C for 1 min and 72 °C for 1 min; and a final incubation for 5 min at 72 °C. We used Taq DNA polymerase and dNTPs from QIAGEN (Hilden, Germany), and a T3000 Biometra thermocycler (Biometra, Germany).

Sequencing was performed with the Mix2Seq Kit (Eurofins Genomics).

Sequence analysis was performed with Serial Cloner v2.6 and BLAST (Basic Local Alignment Search Tool, https://blast.ncbi.nlm.nih.gov/Blast.cgi).

3. Results

Two *Klebsiella pneumoniae* isolates (MurTR-KL001 and MurTR-KL002) were randomly sampled from the river Mur during a study not linked to tigecycline.

3.1. Antimicrobial Susceptibility Testing

Both isolates revealed only resistance to tetracycline and tigecycline but stayed susceptible to all other tested antibiotics. Isolate MurTR-KL001 revealed a minimal inhibition concentration to tigecycline of 4 µg/mL and MurTR-KL002 of 8 µg/mL (Table 1).

3.2. Genetic Analyses

The two isolates belonged to two different unrelated MLST types: ST2392 (rpoB:1, gapA:2, mdh:172, pgi:1, phoE:9, infB:1, tonB:116) and ST2394 (rpoB:4, gapA:126, mdh:1, pgi:1, phoE:4, infB:3, tonB:351). Both MLST profiles had not been described prior to our study. Notably, *tonB* of MurTR-KL002 revealed a new allele (*tonB* 351) (Table 1).

Table 1. Characterization of tigecycline-resistant *Klebsiella pneumoniae* isolates.

Isolate	MLST	*ramR* Mutation	MIC Tetracycline	MIC Tigecycline
MurTR-KL001	ST2392	291G > A (V97V); Δ 518–521,	8 µg/mL	4 µg/mL
MurTR-KL002	ST2394	152A > C (K51T).	8 µg/mL	8 µg/mL

3.3. Determination of a Plasmid-Encoded Resistance Mechanism

The plasmid type FII$_S$ could be determined in both isolates. Transformation experiments revealed that no resistance was transferred by plasmids.

3.4. Determination of a Chromosomally-Encoded Resistance Mechanism

All alleles of *soxR*, *marR* and *rpsJ* were identical with sequences from tigecycline susceptible *Klebsiella pneumoniae* strains previously described (GenBank accession numbers: CP000647.1 [22], CP009461.1 [28], CP003999.1 [29], KC843636.1 [21]). Even though *marR* of MurTR-KL002 harbored a silent mutation (C270A), no other mutations within these genes could be observed.

With regard to the reference strain *Klebsiella pneumoniae* subsp. *pneumoniae* MGH 78578 (GenBank accession number: CP000647.1 [22]), mutations in both isolates could be observed within the *ramR* allele. *RamR* of MurTR-KL001 primarily harbored a four base pair deletion (Δ 518–521CCCG) resulting in a frameshift. Secondarily, a silent point mutation was on position 291 with a G to A mutation. *RamR* of MurTR-KL002 harbored a point mutation (152A > C), which resulted in an amino acid substitution (K51T) (Table 1).

4. Discussion

Tigecycline non-susceptible *Klebsiella pneumoniae* were recently isolated from heavily polluted coastal waters in Brazil [30] and less recently from hospital sewage in Saudi Arabia [31]. Other resistant *Enterobacteriaceae* could be recovered from drinking water samples in India [32].

However, such reports from more "decent" aquatic settings are rare in the current literature. Recent cases of tigecycline-resistant *Klebsiella pneumoniae* were reported from urban surface waters in Brazil: from a river downstream of a wastewater treatment plant, in Curitiba [1], and from an urban lake and reservoir in the city of Sao Paulo [4].

Even though the two isolates belong to two newly described and distinctly different MLST types, they seem to share the same resistance mechanism (even though owing two different mutations), which could indicate a common selective pressure. They were susceptible to all other tested antibiotics. Taking also into account that both isolates belong to two new MLST types it is very unlikely that these isolates are a contamination from a clinical source. Although low concentrations of antibiotics can cause an ecological shift towards less susceptible bacteria, it is rather unlikely that the river Mur was contaminated with tigecycline [33].

A recent study demonstrated cross-resistance to antibiotics, including tetracycline, in association with the resistance to linalool, a component of basil oil that is used as a natural preservative. The increased resistance to linalool was accompanied by the overexpression of the AcrAB efflux pump suggesting linalool as potential substrate [34]. A similar cross-resistance to antibiotics was observed in association with the resistance to pine oil and the tolerance to solvents; in both cases resistance correlated with the activity of the AcrAB efflux pump [35]. Decreased susceptibility to triclosan, a biocide, was also reported in the course of *acrAB* overexpression. Moreover, the AcrAB efflux pump extrudes dyes and detergents, and appears to play a more crucial role, as it is embedded in fundamentally physiological functions; for instance, in cell-to-cell communication and in virulence. It appears plausible that a cross-resistance to an antibiotic could easily fall within a more fundamentally microbial purpose as long as the overexpression of the efflux pump is favored within an ecological and physiological setting. In that manner, higher concentrations of any potential substrate could select for, i.e., a tigecycline resistance [36–38].

Nikaido et al. and Baucheron et al. proposed a mechanism of induction for the AcrAB locus. They suggested that indole and bile bind to RamR, thereby inhibiting its repressing effect on *ramA* transcription, and therefore promoting the induction of the *ramR* and *acrAB* locus. Yamasaki et al. further reported that different substrates can bind to RamR due to a flexible binding pocket and upon binding the DNA binding affinity of RamR decreases. Therefore, substrates could act as extracellular signals that force subsequent induction of *ramA* and *acrAB* expression, whenever the efflux system is overloaded. However, a mutation within RamR can also lead to the induction of the efflux pump resembling a permanent sensing signal. That arrangement may endure in a suitable ecological or physiological condition [39,40].

RamR represents a genetic hotspot for mutations as far as clinical Klebsiella pneumoniae isolates are concerned [18–21]. None of the reported mutations are identical, the closest mutation to the MurTR-KL001 isolate was described by Rosenblum et al. [41]. Nevertheless, reports of aquatic isolates harboring such mutations are absent in the current literature.

5. Conclusions

The presence of two genetically different isolates suggests that river water may bear substances that favor mutations that can lead to this efflux pump-driven resistance. The origin of these substances (e.g., triclosan or heavy metals) may be waste water or surface run-off after rainfall. Therefore, the occurrence and impact on human health of such mutations in bacteria in surface waters must be further investigated.

Author Contributions: Gernot Zarfel conceived and designed the study; Alexander Hladicz, Gernot Zarfel performed the experiments; Alexander Hladicz, Clemens Kittinger and Gernot Zarfel analyzed the data; Alexander Hladicz wrote the manuscript. Clemens Kittinger and Gernot Zarfel edited the manuscript.

Conflicts of Interest: The authors declare that they have no conflict of interest.

References

1. Conte, D.; Palmeiro, J.K.; da Silva Nogueira, K.; de Lima, T.M.; Cardoso, M.A.; Pontarolo, R.; Degaut Pontes, F.L.; Dalla-Costa, L.M. Characterization of CTX-M Enzymes, Quinolone Resistance Determinants, and Antimicrobial Residues from Hospital Sewage, Wastewater Treatment Plant, and River Water. *Ecotoxicol. Environ. Saf.* **2017**, *136*, 62–69. [CrossRef] [PubMed]

2. Dropa, M.; Lincopan, N.; Balsalobre, L.C.; Oliveira, D.E.; Moura, R.A.; Fernandes, M.R.; da Silva, Q.M.; Matte, G.R.; Sato, M.I.; Matte, M.H. Genetic Background of Novel Sequence Types of CTX-M-8- and CTX-M-15-Producing *Escherichia coli* and *Klebsiella pneumoniae* from Public Wastewater Treatment Plants in Sao Paulo, Brazil. *Environ. Sci. Pollut. Res. Int.* **2016**, *23*, 4953–4958. [CrossRef] [PubMed]

3. Maravic, A.; Skocibusic, M.; Cvjetan, S.; Samanic, I.; Fredotovic, Z.; Puizina, J. Prevalence and Diversity of Extended-Spectrum-Beta-Lactamase-Producing Enterobacteriaceae from Marine Beach Waters. *Mar. Pollut. Bull.* **2015**, *90*, 60–67. [CrossRef] [PubMed]

4. Nascimento, T.; Cantamessa, R.; Melo, L.; Fernandes, M.R.; Fraga, E.; Dropa, M.; Sato, M.I.Z.; Cerdeira, L.; Lincopan, N. International High-Risk Clones of *Klebsiella pneumoniae* KPC-2/CC258 and *Escherichia coli* CTX-M-15/CC10 in Urban Lake Waters. *Sci. Total Environ.* **2017**, *598*, 910–915. [CrossRef] [PubMed]

5. Piedra-Carrasco, N.; Fabrega, A.; Calero-Caceres, W.; Cornejo-Sanchez, T.; Brown-Jaque, M.; Mir-Cros, A.; Muniesa, M.; Gonzalez-Lopez, J.J. Carbapenemase-Producing Enterobacteriaceae Recovered from a Spanish River Ecosystem. *PLoS ONE* **2017**, *12*, e0175246. [CrossRef] [PubMed]

6. Baker-Austin, C.; Wright, M.S.; Stepanauskas, R.; McArthur, J.V. Co-Selection of Antibiotic and Metal Resistance. *Trends Microbiol.* **2006**, *14*, 176–182. [CrossRef] [PubMed]

7. Martinez, J.L. Environmental Pollution by Antibiotics and by Antibiotic Resistance Determinants. *Environ. Pollut.* **2009**, *157*, 2893–2902. [CrossRef] [PubMed]

8. Baquero, F.; Martinez, J.L.; Canton, R. Antibiotics and Antibiotic Resistance in Water Environments. *Curr. Opin. Biotechnol.* **2008**, *19*, 260–265. [CrossRef] [PubMed]

9. Lubbert, C.; Baars, C.; Dayakar, A.; Lippmann, N.; Rodloff, A.C.; Kinzig, M.; Sorgel, F. Environmental Pollution with Antimicrobial Agents from Bulk Drug Manufacturing Industries in Hyderabad, South India, is Associated with Dissemination of Extended-Spectrum Beta-Lactamase and Carbapenemase-Producing Pathogens. *Infection* **2017**, *45*, 479–491. [CrossRef] [PubMed]

10. Lupo, A.; Coyne, S.; Berendonk, T.U. Origin and Evolution of Antibiotic Resistance: The Common Mechanisms of Emergence and Spread in Water Bodies. *Front. Microbiol.* **2012**, *3*, 18. [CrossRef] [PubMed]

11. Zarfel, G.; Lipp, M.; Gurtl, E.; Folli, B.; Baumert, R.; Kittinger, C. Troubled Water Under the Bridge: Screening of River Mur Water Reveals Dominance of CTX-M Harboring *Escherichia coli* and for the First Time an Environmental VIM-1 Producer in Austria. *Sci. Total Environ.* **2017**, *593–594*, 399–405. [CrossRef] [PubMed]

12. Bonura, C.; Giuffre, M.; Aleo, A.; Fasciana, T.; Di Bernardo, F.; Stampone, T.; Giammanco, A.; Palma, D.M.; Mammina, C.; MDR-GN Working Group. An Update of the Evolving Epidemic of blaKPC Carrying *Klebsiella pneumoniae* in Sicily, Italy, 2014: Emergence of Multiple Non-ST258 Clones. *PLoS ONE* **2015**, *10*, e0132936. [CrossRef] [PubMed]

13. Geraci, D.M.; Bonura, C.; Giuffre, M.; Saporito, L.; Graziano, G.; Aleo, A.; Fasciana, T.; Di Bernardo, F.; Stampone, T.; Palma, D.M.; et al. Is the Monoclonal Spread of the ST258, KPC-3-Producing Clone being Replaced in Southern Italy by the Dissemination of Multiple Clones of Carbapenem-Nonsusceptible, KPC-3-Producing *Klebsiella pneumoniae*? *Clin. Microbiol. Infect.* **2015**, *21*, e15–e17. [CrossRef] [PubMed]

14. Docobo-Perez, F.; Nordmann, P.; Dominguez-Herrera, J.; Lopez-Rojas, R.; Smani, Y.; Poirel, L.; Pachon, J. Efficacies of Colistin and Tigecycline in Mice with Experimental Pneumonia due to NDM-1-Producing Strains of *Klebsiella pneumoniae* and *Escherichia coli*. *Int. J. Antimicrob. Agents* **2012**, *39*, 251–254. [CrossRef] [PubMed]

15. Kumarasamy, K.K.; Toleman, M.A.; Walsh, T.R.; Bagaria, J.; Butt, F.; Balakrishnan, R.; Chaudhary, U.; Doumith, M.; Giske, C.G.; Irfan, S.; et al. Emergence of a New Antibiotic Resistance Mechanism in India, Pakistan, and the UK: A Molecular, Biological, and Epidemiological Study. *Lancet Infect. Dis.* **2010**, *10*, 597–602. [CrossRef]

16. Pournaras, S.; Koumaki, V.; Spanakis, N.; Gennimata, V.; Tsakris, A. Current Perspectives on Tigecycline Resistance in Enterobacteriaceae: Susceptibility Testing Issues and Mechanisms of Resistance. *Int. J. Antimicrob. Agents* **2016**, *48*, 11–18. [CrossRef] [PubMed]

17. Sun, Y.; Cai, Y.; Liu, X.; Bai, N.; Liang, B.; Wang, R. The Emergence of Clinical Resistance to Tigecycline. *Int. J. Antimicrob. Agents* **2013**, *41*, 110–116. [CrossRef] [PubMed]

18. He, F.; Fu, Y.; Chen, Q.; Ruan, Z.; Hua, X.; Zhou, H.; Yu, Y. Tigecycline Susceptibility and the Role of Efflux Pumps in Tigecycline Resistance in KPC-Producing *Klebsiella pneumoniae*. *PLoS ONE* **2015**, *10*, e0119064. [CrossRef] [PubMed]

19. Sheng, Z.K.; Hu, F.; Wang, W.; Guo, Q.; Chen, Z.; Xu, X.; Zhu, D.; Wang, M. Mechanisms of Tigecycline Resistance among *Klebsiella pneumoniae* Clinical Isolates. *Antimicrob. Agents Chemother.* **2014**, *58*, 6982–6985. [CrossRef] [PubMed]

20. Wang, X.; Chen, H.; Zhang, Y.; Wang, Q.; Zhao, C.; Li, H.; He, W.; Zhang, F.; Wang, Z.; Li, S.; et al. Genetic Characterisation of Clinical *Klebsiella pneumoniae* Isolates with Reduced Susceptibility to Tigecycline: Role of the Global Regulator RamA and its Local Repressor RamR. *Int. J. Antimicrob. Agents* **2015**, *45*, 635–640. [CrossRef] [PubMed]

21. Villa, L.; Feudi, C.; Fortini, D.; Garcia-Fernandez, A.; Carattoli, A. Genomics of KPC-Producing *Klebsiella pneumoniae* Sequence Type 512 Clone Highlights the Role of RamR and Ribosomal S10 Protein Mutations in Conferring Tigecycline Resistance. *Antimicrob. Agents Chemother.* **2014**, *58*, 1707–1712. [CrossRef] [PubMed]

22. Hentschke, M.; Wolters, M.; Sobottka, I.; Rohde, H.; Aepfelbacher, M. ramR Mutations in Clinical Isolates of *Klebsiella pneumoniae* with Reduced Susceptibility to Tigecycline. *Antimicrob. Agents Chemother.* **2010**, *54*, 2720–2723. [CrossRef] [PubMed]

23. Clinical and Laboratory Standards Institute. *Performance Standards for Antimicrobial Susceptibility Testing: 18th Informational Supplement*; CLSI Document M100-S2008; Clinical and Laboratory Standards Institute: Wayne, PA, USA, 2015.

24. European Committee on Antimicrobial Susceptibility Testing (EUCAST). 2013. Available online: http://www.eucast.org/ (accessed on 4 September 2017).

25. Boyen, F.; Vangroenweghe, F.; Butaye, P.; De Graef, E.; Castryck, F.; Heylen, P.; Vanrobaeys, M.; Haesebrouck, F. Disk Prediffusion is a Reliable Method for Testing Colistin Susceptibility in Porcine *E. coli* Strains. *Vet. Microbiol.* **2010**, *144*, 359–362. [CrossRef] [PubMed]

26. Gales, A.C.; Reis, A.O.; Jones, R.N. Contemporary Assessment of Antimicrobial Susceptibility Testing Methods for Polymyxin B and Colistin: Review of Available Interpretative Criteria and Quality Control Guidelines. *J. Clin. Microbiol.* **2001**, *39*, 183–190. [CrossRef] [PubMed]

27. Carattoli, A.; Bertini, A.; Villa, L.; Falbo, V.; Hopkins, K.L.; Threlfall, E.J. Identification of Plasmids by PCR-Based Replicon Typing. *J. Microbiol. Methods* **2005**, *63*, 219–228. [CrossRef] [PubMed]

28. Hua, X.; Chen, Q.; Li, X.; Feng, Y.; Ruan, Z.; Yu, Y. Complete Genome Sequence of *Klebsiella pneumoniae* Sequence Type 17, a Multidrug-Resistant Strain Isolated during Tigecycline Treatment. *Genome Announc.* **2014**, *2*. [CrossRef] [PubMed]

29. Ramos, P.I.; Picao, R.C.; Almeida, L.G.; Lima, N.C.; Girardello, R.; Vivan, A.C.; Xavier, D.E.; Barcellos, F.G.; Pelisson, M.; Vespero, E.C.; et al. Comparative Analysis of the Complete Genome of KPC-2-Producing *Klebsiella pneumoniae* Kp13 Reveals Remarkable Genome Plasticity and a Wide Repertoire of Virulence and Resistance Mechanisms. *BMC Genom.* **2014**, *15*. [CrossRef] [PubMed]

30. Campana, E.H.; Montezzi, L.F.; Paschoal, R.P.; Picao, R.C. NDM-Producing *Klebsiella pneumoniae* ST11 Goes to the Beach. *Int. J. Antimicrob. Agents* **2017**, *49*, 119–121. [CrossRef] [PubMed]

31. Abdulhaq, A.; Basode, V.K. Prevalence of Extended-Spectrum Beta-Lactamase-Producing Bacteria in Hospital and Community Sewage in Saudi Arabia. *Am. J. Infect. Control* **2015**, *43*, 1139–1141. [CrossRef] [PubMed]

32. Tanner, W.D.; VanDerslice, J.A.; Toor, D.; Benson, L.S.; Porucznik, C.A.; Goel, R.K.; Atkinson, R.M. Development and Field Evaluation of a Method for Detecting Carbapenem-Resistant Bacteria in Drinking Water. *Syst. Appl. Microbiol.* **2015**, *38*, 351–357. [CrossRef] [PubMed]

33. Gullberg, E.; Cao, S.; Berg, O.G.; Ilback, C.; Sandegren, L.; Hughes, D.; Andersson, D.I. Selection of Resistant Bacteria at very Low Antibiotic Concentrations. *PLoS Pathog.* **2011**, *7*, e1002158. [CrossRef] [PubMed]

34. Kalily, E.; Hollander, A.; Korin, B.; Cymerman, I.; Yaron, S. Adaptation of *Salmonella enterica* Serovar Senftenberg to Linalool and its Association with Antibiotic Resistance and Environmental Persistence. *Appl. Environ. Microbiol.* **2017**, *83*. [CrossRef] [PubMed]

35. Moken, M.C.; McMurry, L.M.; Levy, S.B. Selection of Multiple-Antibiotic-Resistant (Mar) Mutants of *Escherichia coli* by using the Disinfectant Pine Oil: Roles of the Mar and AcrAB Loci. *Antimicrob. Agents Chemother.* **1997**, *41*, 2770–2772. [PubMed]

36. Sun, J.; Deng, Z.; Yan, A. Bacterial Multidrug Efflux Pumps: Mechanisms, Physiology and Pharmacological Exploitations. *Biochem. Biophys. Res. Commun.* **2014**, *453*, 254–267. [CrossRef] [PubMed]

37. Martinez, J.L.; Sanchez, M.B.; Martinez-Solano, L.; Hernandez, A.; Garmendia, L.; Fajardo, A.; Alvarez-Ortega, C. Functional Role of Bacterial Multidrug Efflux Pumps in Microbial Natural Ecosystems. *FEMS Microbiol. Rev.* **2009**, *33*, 430–449. [CrossRef] [PubMed]

38. Alvarez-Ortega, C.; Olivares, J.; Martinez, J.L. RND Multidrug Efflux Pumps: What are They Good for? *Front. Microbiol.* **2013**, *4*, 7. [CrossRef] [PubMed]

39. Nikaido, E.; Shirosaka, I.; Yamaguchi, A.; Nishino, K. Regulation of the AcrAB Multidrug Efflux Pump in *Salmonella enterica* Serovar Typhimurium in Response to Indole and Paraquat. *Microbiology* **2011**, *157*, 648–655. [CrossRef] [PubMed]

40. Yamasaki, S.; Nakashima, R.; Sakurai, K.; Yamaguchi, A.; Nishino, K. Structural Analysis and New Drug Development against Multidrug Efflux Pumps. *Yakugaku Zasshi* **2017**, *137*, 377–382. [CrossRef] [PubMed]

41. Rosenblum, R.; Khan, E.; Gonzalez, G.; Hasan, R.; Schneiders, T. Genetic Regulation of the RamA Locus and its Expression in Clinical Isolates of *Klebsiella pneumoniae*. *Int. J. Antimicrob. Agents* **2011**, *38*, 39–45. [CrossRef] [PubMed]

International Journal of
Environmental Research and Public Health

Article

Antibiotic Resistance of *Acinetobacter* spp. Isolates from the River Danube: Susceptibility Stays High

Clemens Kittinger [1], Alexander Kirschner [2,3], Michaela Lipp [1], Rita Baumert [1], Franz Mascher [1], Andreas H. Farnleitner [3,4,5] and Gernot E. Zarfel [1,*]

[1]   Institute of Hygiene, Microbiology and Environmental Medicine, Medical University Graz, Neue Stiftingtalstraße 2, 8010 Graz, Austria; Clemens.kittinger@medunigraz.at (C.K.); michaela.lipp@medunigraz.at (M.L.); rita.baumert@medunigraz.at (R.B.); franz.mascher@medunigraz.at (F.M.)

[2]   Institute for Hygiene and Applied Immunology, Water Hygiene, Medical University of Vienna, 1090 Vienna, Austria; alexander.kirschner@meduniwien.ac.at

[3]   Interuniversity Cooperation Centre for Water and Health, Vienna University of Technology, 1060 Vienna, Austria; andreas.farnleitner@wavenet.at

[4]   Institute of Chemical Engineering, Research Group Environmental Microbiology and Molecular Ecology, Vienna University of Technology, 1060 Vienna, Austria

[5]   Karl Landsteiner University for Health Sciences, 3500 Krems, Austria

*   Correspondence: gernot.zarfel@medunigraz.at; Tel.: +43-316-385-73604

Received: 29 November 2017; Accepted: 28 December 2017; Published: 30 December 2017

Abstract: *Acinetobacter* spp. occur naturally in many different habitats, including food, soil, and surface waters. In clinical settings, *Acinetobacter* poses an increasing health problem, causing infections with limited to no antibiotic therapeutic options left. The presence of human generated multidrug resistant strains is well documented but the extent to how widely they are distributed within the *Acinetobacter* population is unknown. In this study, *Acinetobacter* spp. were isolated from water samples at 14 sites of the whole course of the river Danube. Susceptibility testing was carried out for 14 clinically relevant antibiotics from six different antibiotic classes. Isolates showing a carbapenem resistance phenotype were screened with PCR and sequencing for the underlying resistance mechanism of carbapenem resistance. From the Danube river water, 262 *Acinetobacter* were isolated, the most common species was *Acinetobacter baumannii* with 135 isolates. Carbapenem and multiresistant isolates were rare but one isolate could be found which was only susceptible to colistin. The genetic background of carbapenem resistance was mostly based on typical *Acinetobacter* OXA enzymes but also on VIM-2. The population of *Acinetobacter* (*baumannii* and non-*baumannii*) revealed a significant proportion of human-generated antibiotic resistance and multiresistance, but the majority of the isolates stayed susceptible to most of the tested antibiotics.

Keywords: *Acinetobacter*; JDS3; river; water; carbapenemases

---

1. Introduction

The genus *Acinetobacter* consists of over 40 known species that can be isolated from various habitats including soil, sediment surface, and wastewater [1]. They have the ability to colonize human skin and are responsible for a growing number of nosocomial outbreaks worldwide. Although most *Acinetobacter* species have generally a low pathogenicity [2], according to Alsan et al. (2008), the intensive care unit (ICU) mortality rate is around 40% [3].

The most striking characteristic of *Acinetobacter* spp. is their natural resistance to many antibiotics and the ability to easily develop new resistances under antibiotic pressure. They overexpress efflux pumps, harbor β-lactamases, and are characterized by low membrane permeability [2]. By 2012, over 210 different β-lactamases have been identified within the genus [4]. Different *Oxacillinases* (OXA) enzyme families have their origin in *Acinetobacter*, such as OXA-21like, OXA-23like or OXA-51like [5]. These enzymes are serine hydrolases represent class D according to the Ambler classification of β-lactamases [6]. The spread of these *Acinetobacter oxaxilinases* into other species seems much more limited than, for example, the spread of CTX-M or NDM enzymes but is documented. This set of OXA enzyme enables *Acinetobacter* to adapt easily to new developed β-lactam antibiotics [4,7]. Therefore, *Acinetobacter baumannii* especially has become one of the problematic nosocomial pathogens. Infections with some of these strains, such as bloodstream infections and pneumonia, do not leave any further options for antibiotic treatment. Next to *Pseudomonas* and carbapenem-resistant *Enterobacteriaceae*, *Acinetobacter* were rated by the WHO to be the group in most urgent need of new antibiotics (http://www.who.int/medicines/publications/global-priority-list-antibiotic-resistant-bacteria/en/) [5,7,8]. In addition to the acquisition of a seemingly infinite number of resistances, such as *Pseudomonas* spp., *Acinetobacter* is characterized by a much better ability to survive hostile conditions, e.g., survival on dry surfaces. This makes *Acinetobacter* an ideal candidate for survival in clinical settings and in the environment [9–11].

Occurrence and susceptibility of *Acinetobacter* spp. in clinical settings is documented quite well, whereas their distribution and proportion of resistance in the aquatic environment remains quite unclear. Nearly all studies that investigate antibiotic resistance of *Acinetobacter* in the environment are based on selective cultivation, masking their proportion in the population, or are based on molecular methods, with all their inherent methodological weaknesses [9,12–15]. There is some evidence that environmental transport of *Acinetobacter* plays a role in the spread of clinical relevant *Acinetobacter* strains in the environment. On the other hand, there seems to be a continuous influx of novel strains into the clinical setting with the potential of new infectious features [16,17].

Participation in the Joint Danube Survey 2013 (JDS3) offered the possibility of isolating *Acinetobacter* from the total course of one of Europe's longest rivers. This chance was taken to generate an initial picture of the resistance proportion within *Acinetobacter* spp. and to get an idea of how far acquired antibiotic resistances of clinical relevance have spread in the aquatic environment.

2. .Material and Methods

2.1. Sample Collection

All samples were taken during the research expedition of the Joint Danube Survey 2013 (JDS3). The survey was organized by the International Commission for the Protection of the Danube River (ICPDR), Vienna. The water samples were taken between 12 August and 26 September 2013, from 68 sampling sites along the River Danube, starting at Böfinger Halde (Germany) downstream to the delta (Romania). At each sampling site, samples were collected at three sampling points (left, middle, right), in sterile 1 L glass flasks from 30 cm below the river surface. From each flask, duplicate volumes of 45 mL of river water were filled into sterile non-toxic 50 mL plastic vials (Techno Plastic Products AG, TPP, Trasadingen, Switzerland), containing 5 mL of glycerine (final conc. 10% v/v). The vials were completely mixed by hand and immediately stored at −20 °C on board of the cruise ship until analysis in the laboratory. After transfer to the laboratory (beginning in October 2013), the samples were stored at −80 °C. Fourteen sampling sites, four of them downstream of megacities (Vienna, Budapest, Belgrade and Bucharest), two at the beginning as well as two at the delta, four rural sampling sites, and two after confluence of two biggest tributaries (Drave, Tisa) were chosen for investigation (Table 1).

Table 1. JDS3 sampling sites chosen for isolation and their assignment to the upper-, middle-, or downstream stretches (SP = sampling point; us = upstream; ds = downstream). Country codes: Germany: DE; Austria: AT; Hungary: HU; Croatia: HR; Serbia: RS; Romania: RO; Bulgaria: BG.

SP	Name of SP	River (km)	Country
JDS2	Kelheim, gauging station	2415	DE
JDS3	Geisling power plant	2354	DE
JDS8	Oberloiben	2008	AT
JDS10	Wildungsmauer (Vienna)	1895	AT
JDS22	ds Budapest	1632	HU
JDS28	us Drava	1384	HR/RS
JDS36	ds Tisa/us Sava	1200	RS
JDS38	us Pancevo (Belgrade)	1159	RS
JDS49	Pristol/Novo Salo	834	RO/BG
JDS57	ds Ruse	488	RO/BG
JDS59	ds Arges (Bucharest)	429	RO/BG
JDS63	Siret	154	RO
JDS67	Sulina Arm	26	RO
JDS68	St. Gheorge Arm	104	RO

2.2. Isolation of Acinetobacter

The frozen samples were thawed, and 15 mL (left, middle, and right 5 mL each) were plated in 0.5 mL portions on selective agars. For the isolation of *Acinetobacter* 0.5 mL from left, middle, and right were plated on five agar-plates of CHROMagar™ (Oxoid, Germany) each. Growth conditions were 37 ± 1 °C for 18–24 h. Colonies were picked according to the manufacturer's instructions and subcultured on Columbia blood-agar (in house production). Identification of *Acinetobacter* was carried out by matrix-assisted laser desorption ionization time-of-flight mass spectrometry (MALDI-TOF-MS) as described previously [18].

2.3. Susceptibility Testing

For inoculation, colonies were picked from an overnight pure culture on Colombia blood-agar (non-selective medium) with a sterile loop and suspended in sterile saline (0.85% NaCl w/v in water) to the density of a McFarland 0.5 standard (DensiCheck, Biomerieux, Vienna, Austria). The suspension was plated on Mueller-Hinton II agar using an automatic plate rotator (Retro C80, Biomerieux, Vienna, Austria). Antibiotic test disks were stamped on the agar surface. The plates were incubated at 36 °C for 16–20 h. After incubation, inhibition zones were determined. In case of testing susceptibility with Etest®, the same procedure for preparing the plates was carried out. Interpretation of zone-diameters and Etest® was carried out according to the European Committee on Antimicrobial Susceptibility Testing (EUCAST) and if no EUCAST breakpoints were available Clinical Laboratory Standards Institute (CLSI) criteria were used for interpretation (Table 2) [19,20].

Etest for tigecycline was carried according to Altun et al. (2014) [21].

Escherichia coli ATCC 25922 and *Pseudomonas aeruginosa* ATCC 27853 were used as control strains in all performed tests.

2.4. Determination of β-Lactamase Genes

Determination of resistance genes was carried out for all *Acinetobacter* spp. isolates that revealed a resistance to at least one tested carbapenem. PCR detection and gene identification were performed for five different β-lactamases gene families, $bla_{CTX-M-1group}$, $bla_{CTX-M-2group}$, $bla_{CTX-M-9group}$, bla_{GES}, bla_{KPC}, bla_{OXA-23}, bla_{OXA-24}, bla_{OXA-51}, bla_{OXA-48}, bla_{OXA-58}, bla_{NDM}, bla_{SHV}, bla_{TEM}, and bla_{VIM}. PCR and sequencing procedures were performed as described previously [22–27].

Table 2. List of tested antibiotics, concentration on the disc (Sensi-DiscTM paper discs, BD, Vienna, Austria) or Etest® (Biomerieux) and antibiotic classes.

Antibiotic	Concentration	Antibioti Classes
piperacillin/tazobactam	100 µg/10 µg	β-lactam
cefotaxime	30 µg	β-lactam
ceftazidime	30 µg	β-lactam
cefepime	30 µg	β-lactam
imipenem	10 µg	β-lactam
meropenem	10 µg	β-lactam
amikacin	30 µg	aminoglycoside
gentamicin	10 µg	aminoglycoside
trimethoprim/sulfamethoxazole	1.25 µg/23.75 µg	folate synthesis inhibitors
ciprofloxacin	5 µg	quinolone
levofloxacin	5 µg	quinolone
tigecycline	Etest	tetracyclin
tetracycline	30 µg	tetracycline
colistin	Etest	polypeptide antibiotic

3. Results

In total, 262 *Acinetobacter* were isolated. *Acinetobacter baumannii* was the most common species with 135 isolates. *Acinetobacter johnsonii* was second most with 62 isolates; all other species were represented by less than 20 isolates; *Acinetobacter haemolyticus* 19 isolates, *Acinetobacter junii* 17 isolates, *Acinetobacter lwoffii* 16 isolates, *Acinetobacter radioresistens* four, *Acinetobacter ursingii* two and seven isolates where no distinct species identification was possible. Non-*baumannii Acinetobacter* spp. were subsumed for further analyses.

Susceptibility testing revealed that resistance to the most tested antibiotics was rare in *Acinetobacter baumannii* and non-*baumannii Acinetobacter* spp. river water isolates. The only resistance present in both groups in more than 10% of the isolates was to cefotaxime with 95.6% (129/135) of *Acinetobacter baumannii* and 43.3% (55/127) non-*baumannii Acinetobacter* spp. In addition, resistances to ceftazidime in non-*baumannii* (12.6%, 16/127) and to piperacillin/tazobactam in *Acinetobacter baumannii* (12.6%, 17/125) were present in more than 10% in one of the sample subgroups (Figure 1).

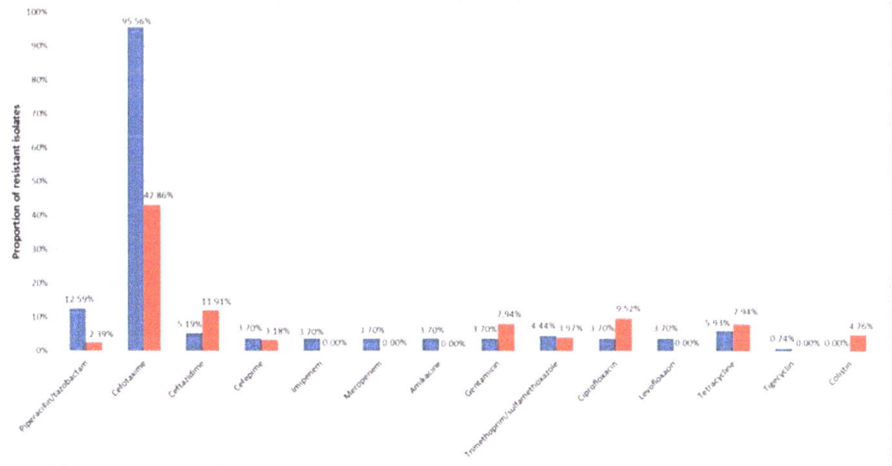

Figure 1. Percentage of resistance to tested antibiotics of isolated *Acinetobacter baumannii* (blue) and non-*baumannii Acinetobacter* spp. (red).

Resistance to fluoroquinolones showed one notable detail: all five *Acinetobacter baumannii* were resistant to ciprofloxacin and levofloxacin, whereas non-*baumannii Acinetobacter* spp. remained susceptible to the second tested fluoroquinolone. Less than 10 *Acinetobacter baumannii* isolates revealed resistance to all other tested antibiotics, all isolates were susceptible to colistin. In contrast to this, colistin resistance could be detected in six (4.7%) non-*baumannii Acinetobacter* spp., and no resistance was found to levofloxacin, imipenem, meropenem, amikacine, and tigecycline (Figure 1).

Only six (4.4%) *Acinetobacter baumannii* revealed susceptibility to all tested antibiotics, but only 16 (11.9%) were resistant to one or more tested antibiotics additionally to cefotaxime. Six (4.4%) isolates could be classified as multiresistant, with resistance to antibiotics from at least three different antibiotic classes, including one isolate only susceptible to colistin and four to colistin and tigecyline (Table 3).

Table 3. Detected resistance genes in carbapenem resistant *Acinetobacter baumannii*; us = upstream; ds = downstream.

Isolate	Site of Isolation	Susceptible Antibiotics	Detected β-Lactamases
JDS38AC017	us Pancevo (Belgrade)	colistin, tigecycline	OXA-23, OXA-51, VIM-2
JDS38AC018	us Pancevo (Belgrade)	colistin	OXA-23, OXA-51, VIM-2, TEM-1
JDS38AC020	us Pancevo (Belgrade)	colistin, tigecycline	OXA-24, OXA-51,
JDS59AC001	ds Arges (Bucharest)	colistin, tigecycline	OXA-23, OXA-51
JDS59AC007	ds Arges (Bucharest)	colistin, tigecycline	OXA-23, OXA-51

Susceptibility to all antibiotics was 10 times higher (55 isolates, 43.3%) in the non-*baumannii Acinetobacter* spp. group compared to the *Acinetobacter baumannii group*. Only 10 isolates revealed multiresistance. Two isolates with different resistance profile showed resistance to four antibiotics (JDS10AC012 to CTX, piperacillin/tazobactam, FEP, and ceftazidime; JDS38AC048 to CTX, TZP SXT and CAZ).

Five *Acinetobacter baumannii* were resistant to carbapenems (meropenem and imipenem). These isolates were analyzed for the presence of several β-lactamases genes, resulting in four different gene patterns. Classic intrinsic OXA carbapenemases were present in all isolates, but no gene was present in all isolates. JDS59AC007 and JDS59AC001 were positive for OXA-23 and OXA-51, and JDS38AC020 was positive for OXA-24 and OXA-51. In addition to OXA-23 and OXA-51, two isolates revealed β-lactamases from another Ambler class: JDS38AC018 and JDS38AC017 harbored both the gene for carbapenemase VIM-2 and JDS38AC017 harbored additionally the gene for the broad spectrum β-lactamases TEM-1 (Table 3).

4. Discussion

The presence of *Acinetobacter* with human-induced multidrug resistance phenotypes in surface water has been reported from all over the world. Their origin seems to be influenced by (treated and untreated) hospital waste water. The impact (proportion and persistence) of these strains on the *Acinetobacter* water population is not well documented [9,14,15,28]. This study shows for the first time the susceptibility phenotypes of *Acinetobacter* of a total European river system. Furthermore, our study provides a first glimpse of the anthropogenic impact on the *Acinetobacter* river population. In the *Acinetobacter* population of the River Danube, even resistance to last line antibiotics (e.g., colistin and tigecycline) is detectable, and this without using selective media by screening only a relatively small volume of water. This screening led to the detection of multidrug-resistant *Acinetobacter baumannii* isolates, whose multidrug resistance would normally be found and related only to intensive care units. The isolation of multiresistant *Acinetobacter* was limited to the area of influence of megacities, but even there the great majority of isolates remained not or only slightly influenced on their susceptibility pattern.

Looking at the resistance data for invasive *Acinetobacter* isolates for the Danube neighboring countries reveals a rather gloomy picture: In clinical isolates, the ratio for carbapenem resistance spans from 5.5% in Germany to over 80% in Romania. In our study of the river water, however, only five isolates (less than 2% of all isolates) showed resistance to carbapenems, a very low proportion of resistant *Acinetobacter* spp. compared with the clinical settings [29]. This ratio corresponds with our findings in the *Pseudomonas* population in the river Danube, where we also found a ratio of around 2% [30,31].

In a study of the Jadro River, carried out by Maravic et al., only selected multi-drug-resistant *Acinetobacter* were isolated (using selective media with supplements). Comparing these isolates with the Danube isolates, there is a remediable difference as regards aminoglycoside resistance. Maravic et al. did not detect a single isolate that was resistant to the tested aminoglycoside in contrast to 5.7% (15 isolates) from the River Danube. Furthermore, only two multiresistant Danube isolates revealed no resistance to one of the tested aminoglycoside. Carbapenem resistance in Jadro River isolates was restricted to meropenem, but the number of isolates was too low to be noteworthy. Interestingly enough, the proportion of cefotaxime resistance is nearly identical with 68% in Jadro River and 66% in the river Danube [9].

5. Conclusions

Multiresistant strains can be found in our environment, in any habitat and at any time, making chances for contact high and permanent. Further investigation will show if the spread of multiresistant *Acinetobacter* has reached its peak and if susceptible environmental *Acinetobacter* will still outnumber the clinical strains or if we have to further deal with a constant increase of non-susceptible *Acinetobacter* in the future.

Acknowledgments: Furthermore, we would like to thank Georg Reischer, Stefan Jakwerth, and Stoimir Kolarevic for their help in sampling.

Author Contributions: C.K. and G.E.Z. conceived and designed the experiments; M.L., R.B., and G.E.Z. performed the experiments; G.E.Z. and C.K. analyzed the data; K.A. and A.H.F. collected samples and data; G.E.Z., C.K., and F.M. wrote the paper.

Conflicts of Interest: The authors declare no conflict of interest.

Funding: The Joint Danube Survey was organized by the International Commission for the Protection of the Danube River (ICPDR). The study was supported by the Austrian Science Fund (FWF), project number P25817-B22.

References

1. Towner, K.J. Acinetobacter: An Old Friend, but a New Enemy. *J. Hosp. Infect.* **2009**, *73*, 355–363. [CrossRef] [PubMed]
2. Peleg, A.Y.; Seifert, H.; Paterson, D.L. *Acinetobacter baumannii*: Emergence of a Successful Pathogen. *Clin. Microbiol. Rev.* **2008**, *21*, 538–582. [CrossRef] [PubMed]
3. Alsan, M.; Klompas, M. *Acinetobacter baumannii*: An Emerging and Important Pathogen. *J. Clin. Outcomes Manag.* **2010**, *17*, 363–369. [PubMed]
4. Zhao, W.H.; Hu, Z.Q. Acinetobacter: A Potential Reservoir and Dispenser for Beta-Lactamases. *Crit. Rev. Microbiol.* **2012**, *38*, 30–51. [CrossRef] [PubMed]
5. Evans, B.A.; Amyes, S.G. OXA Beta-Lactamases. *Clin. Microbiol. Rev.* **2014**, *27*, 241–263. [CrossRef] [PubMed]
6. Ambler, R.P.; Coulson, A.F.W.; Frere, J.M.; Ghuysen, J.M.; Joris, B.; Forsman, M.; Levesque, R.C.; Tiraby, G.; Waley, S.G. A Standard Numbering Scheme for the Class-a Beta-Lactamases. *Biochem. J.* **1991**, *276*, 269–270. [CrossRef] [PubMed]
7. Poirel, L.; Nordmann, P. Carbapenem Resistance in *Acinetobacter baumannii*: Mechanisms and Epidemiology. *Clin. Microbiol. Infect.* **2006**, *12*, 826–836. [CrossRef] [PubMed]
8. Santajit, S.; Indrawattana, N. Mechanisms of Antimicrobial Resistance in ESKAPE Pathogens. *Biomed. Res. Int.* **2016**, *2016*, 2475067. [CrossRef] [PubMed]
9. Maravic, A.; Skocibusic, M.; Fredotovic, Z.; Samanic, I.; Cvjetan, S.; Knezovic, M.; Puizina, J. Urban Riverine Environment is a Source of Multidrug-Resistant and ESBL-Producing Clinically Important *Acinetobacter* spp. *Environ. Sci. Pollut. Res. Int.* **2016**, *23*, 3525–3535. [CrossRef] [PubMed]

10. Jawad, A.; Seifert, H.; Snelling, A.M.; Heritage, J.; Hawkey, P.M. Survival of *Acinetobacter baumannii* on Dry Surfaces: Comparison of Outbreak and Sporadic Isolates. *J. Clin. Microbiol.* **1998**, *36*, 1938–1941. [PubMed]

11. Fournier, P.E.; Vallenet, D.; Barbe, V.; Audic, S.; Ogata, H.; Poirel, L.; Richet, H.; Robert, C.; Mangenot, S.; Abergel, C.; et al. Comparative Genomics of Multidrug Resistance in *Acinetobacter Baumannii*. *PLoS Genet.* **2006**, *2*, e7. [CrossRef] [PubMed]

12. Lin, M.F.; Lan, C.Y. Antimicrobial Resistance in *Acinetobacter baumannii*: From Bench to Bedside. *World J. Clin. Cases* **2014**, *2*, 787–814. [CrossRef] [PubMed]

13. Deng, M.; Zhu, M.H.; Li, J.J.; Bi, S.; Sheng, Z.K.; Hu, F.S.; Zhang, J.J.; Chen, W.; Xue, X.W.; Sheng, J.F.; et al. Molecular Epidemiology and Mechanisms of Tigecycline Resistance in Clinical Isolates of *Acinetobacter baumannii* from a Chinese University Hospital. *Antimicrob. Agents Chemother.* **2014**, *58*, 297–303. [CrossRef] [PubMed]

14. Seruga Music, M.; Hrenovic, J.; Goic-Barisic, I.; Hunjak, B.; Skoric, D.; Ivankovic, T. Emission of Extensively-Drug-Resistant *Acinetobacter baumannii* from Hospital Settings to the Natural Environment. *J. Hosp. Infect.* **2017**, *96*, 323–327. [CrossRef] [PubMed]

15. Osinska, A.; Harnisz, M.; Korzeniewska, E. Prevalence of Plasmid-Mediated Multidrug Resistance Determinants in Fluoroquinolone-Resistant Bacteria Isolated from Sewage and Surface Water. *Environ. Sci. Pollut. Res. Int.* **2016**, *23*, 10818–10831. [CrossRef] [PubMed]

16. Wilharm, G.; Skiebe, E.; Higgins, P.G.; Poppel, M.T.; Blaschke, U.; Leser, S.; Heider, C.; Heindorf, M.; Brauner, P.; Jackel, U.; et al. Relatedness of Wildlife and Livestock Avian Isolates of the Nosocomial Pathogen *Acinetobacter baumannii* to Lineages Spread in Hospitals Worldwide. *Environ. Microbiol.* **2017**, *19*, 4349–4364. [CrossRef] [PubMed]

17. Rafei, R.; Hamze, M.; Pailhories, H.; Eveillard, M.; Marsollier, L.; Joly-Guillou, M.L.; Dabboussi, F.; Kempf, M. Extrahuman Epidemiology of *Acinetobacter baumannii* in Lebanon. *Appl. Environ. Microbiol.* **2015**, *81*, 2359–2367. [CrossRef] [PubMed]

18. Jamal, W.; Albert, M.J.; Rotimi, V.O. Real-Time Comparative Evaluation of bioMerieux VITEK MS Versus Bruker Microflex MS, Two Matrix-Assisted Laser Desorption-Ionization Time-of-Flight Mass Spectrometry Systems, for Identification of Clinically Significant Bacteria. *BMC Microbiol.* **2014**, *14*, 289. [CrossRef] [PubMed]

19. The European Committee on Antimicrobial Susceptibility Testing (EUCAST). Breakpoint Tables for Interpretation of MICs and Zone Diameters, Version 3.1, EU. 2013. Available online: http://www.eucast.org/fileadmin/src/media/PDFs/EUCAST_files/Breakpoint_tables/Breakpoint_table_v_3.1.xls (accessed on 29 December 2017).

20. Clinical and Laboratory Standards Institute. *Performance Standards for Antimicrobial Susceptibility Testing: 18th Informational Supplement*; CLSI Document M100-S18; Clinical and Laboratory Standards Institute: Wayne, PA, USA, 2008.

21. Altun, H.U.; Yagci, S.; Bulut, C.; Sahin, H.; Kinikli, S.; Adiloglu, A.K.; Demiroz, A.P. Antimicrobial Susceptibilities of Clinical *Acinetobacter baumannii* Isolates with Different Genotypes. *Jundishapur J. Microbiol.* **2014**, *7*, e13347. [CrossRef] [PubMed]

22. Biglari, S.; Alfizah, H.; Ramliza, R.; Rahman, M.M. Molecular Characterization of Carbapenemase and Cephalosporinase Genes among Clinical Isolates of *Acinetobacter baumannii* in a Tertiary Medical Centre in Malaysia. *J. Med. Microbiol.* **2015**, *64*, 53–58. [CrossRef] [PubMed]

23. Dallenne, C.; Da Costa, A.; Decre, D.; Favier, C.; Arlet, G. Development of a Set of Multiplex PCR Assays for the Detection of Genes Encoding Important Beta-Lactamases in Enterobacteriaceae. *J. Antimicrob. Chemother.* **2010**, *65*, 490–495. [CrossRef] [PubMed]

24. Zarfel, G.; Lipp, M.; Gurtl, E.; Folli, B.; Baumert, R.; Kittinger, C. Troubled Water Under the Bridge: Screening of River Mur Water Reveals Dominance of CTX-M Harboring *Escherichia coli* and for the First Time an Environmental VIM-1 Producer in Austria. *Sci. Total Environ.* **2017**, *593–594*, 399–405. [CrossRef] [PubMed]

25. Hornsey, M.; Phee, L.; Wareham, D.W. A Novel Variant, NDM-5, of the New Delhi Metallo-Beta-Lactamase in a Multidrug-Resistant *Escherichia coli* ST648 Isolate Recovered from a Patient in the United Kingdom. *Antimicrob. Agents Chemother.* **2011**, *55*, 5952–5954. [CrossRef] [PubMed]

26. Potron, A.; Nordmann, P.; Lafeuille, E.; Al Maskari, Z.; Al Rashdi, F.; Poirel, L. Characterization of OXA-181, a Carbapenem-Hydrolyzing Class D Beta-Lactamase from *Klebsiella pneumoniae*. *Antimicrob. Agents Chemother.* **2011**, *55*, 4896–4899. [CrossRef] [PubMed]

27. Bradford, P.A.; Bratu, S.; Urban, C.; Visalli, M.; Mariano, N.; Landman, D.; Rahal, J.J.; Brooks, S.; Cebular, S.; Quale, J. Emergence of Carbapenem-Resistant Klebsiella Species Possessing the Class A Carbapenem-Hydrolyzing KPC-2 and Inhibitor-Resistant TEM-30 Beta-Lactamases in New York City. *Clin. Infect. Dis.* **2004**, *39*, 55–60. [CrossRef] [PubMed]

28. Zhang, Y.; Marrs, C.F.; Simon, C.; Xi, C. Wastewater Treatment Contributes to Selective Increase of Antibiotic Resistance among *Acinetobacter* spp. *Sci. Total Environ.* **2009**, *407*, 3702–3706. [CrossRef] [PubMed]

29. Girlich, D.; Bonnin, R.A.; Bogaerts, P.; De Laveleye, M.; Huang, D.T.; Dortet, L.; Glaser, P.; Glupczynski, Y.; Naas, T. Chromosomal Amplification of the blaOXA-58 Carbapenemase Gene in a Proteus Mirabilis Clinical Isolate. *Antimicrob. Agents Chemother.* **2017**, *61*. [CrossRef]

30. Kittinger, C.; Lipp, M.; Baumert, R.; Folli, B.; Koraimann, G.; Toplitsch, D.; Liebmann, A.; Grisold, A.J.; Farnleitner, A.H.; Kirschner, A.; et al. Antibiotic Resistance Patterns of *Pseudomonas* Spp. Isolated from the River Danube. *Front. Microbiol.* **2016**, *7*, 586. [CrossRef] [PubMed]

31. European Centre for Disease Prevention and Control. *Antimicrobial Resistance Surveillance in Europe 2015. Annual Report of the European Antimicrobial Resistance Surveilance Network (EARS-Net)*; ECDC: Solna Municipality, Sweden, 2016.

International Journal of
Environmental Research and Public Health

Article

Antibiotic-Resistant Pathogenic *Escherichia Coli* Isolated from Rooftop Rainwater-Harvesting Tanks in the Eastern Cape, South Africa

Mokaba Shirley Malema [1,*], Akebe Luther King Abia [2], Roman Tandlich [3], Bonga Zuma [3], Jean-Marc Mwenge Kahinda [1] and Eunice Ubomba-Jaswa [4,5]

1   Council for Scientific and Industrial Research, Natural Resources and the Environment, P.O. Box 395, Pretoria 0001, South Africa; jeanmarcmk@yahoo.co.uk
2   Antimicrobial Research Unit, College of Health Sciences, University of KwaZulu-Natal, Private Bag X54001, Durban 4000, South Africa; lutherkinga@yahoo.fr
3   Faculty of Pharmacy, Pharmaceutical Chemistry Division, Rhodes University, Grahamstown 6140, South Africa; roman.tandlich@gmail.com (R.T.); b.zuma@ru.ac.za (B.Z.)
4   Department of Biotechnology, University of Johannesburg, 37 Nind Street, Doornfontein 2094, South Africa; euniceubombajaswa@yahoo.com
5   Water Research Commission, Lynnwood Bridge Office Park, Bloukrans Building, 4 Daventry Street, Lynnwood Manor, Pretoria 0081, South Africa
*   Correspondence: shirley.malema@gmail.com or smalema@csir.co.za; Tel.: +27-012-841-4630

Received: 26 March 2018; Accepted: 27 April 2018; Published: 1 May 2018

Abstract: Although many developing countries use harvested rainwater (HRW) for drinking and other household purposes, its quality is seldom monitored. Continuous assessment of the microbial quality of HRW would ensure the safety of users of such water. The current study investigated the prevalence of pathogenic *Escherichia coli* strains and their antimicrobial resistance patterns in HRW tanks in the Eastern Cape, South Africa. Rainwater samples were collected weekly between June and September 2016 from 11 tanks in various areas of the province. Enumeration of *E. coli* was performed using the Colilert®18/Quanti-Tray® 2000 method. *E. coli* isolates were obtained and screened for their virulence potentials using polymerase chain reaction (PCR), and subsequently tested for antibiotic resistance using the disc-diffusion method against 11 antibiotics. The pathotype most detected was the neonatal meningitis *E. coli* (NMEC) (*ibeA* 28%) while pathotype enteroaggregative *E. coli* (EAEC) was not detected. The highest resistance of the *E. coli* isolates was observed against Cephalothin (76%). All tested pathotypes were susceptible to Gentamicin, and 52% demonstrated multiple-antibiotic resistance (MAR). The results of the current study are of public health concern since the use of untreated harvested rainwater for potable purposes may pose a risk of transmission of pathogenic and antimicrobial-resistant *E. coli*.

Keywords: antimicrobial resistance; pathogenic *E. coli*; harvested rainwater; public health; Sub-Saharan Africa; alternative water source

---

1. Introduction

Several countries around the world, including South Africa, make use of harvested rainwater (HRW) to meet their daily water needs. However, the most significant issue relating to the use of harvested rainwater is the potential health risk associated with the presence of various pathogenic organisms in such water [1]. Indicator organisms like *E. coli* have been used to determine the microbiological safety of water meant for drinking and other human needs. Although most *E. coli* strains are non-pathogenic, certain strains may be pathogenic and carry virulence genes (VGs) [2]. Pathogenic *E. coli* strains which can cause diseases in both humans and animals are categorised as intestinal pathogenic *E. coli* (InPEC) and extraintestinal pathogenic *E. coli* (ExPEC) [3]. Intestinal strains are mostly referred to as diarrhoeagenic *Escherichia coli* (DEC) due to their ability to cause diarrhoea using diverse mechanisms [4]. The ExPEC strains have been reported to cause diseases such as urinary tract infections, neonatal meningitis, sepsis and wound infections and some examples include neonatal meningitis *Escherichia coli* (NMEC) and uropathogenic *E. coli* (UPEC) [3].

Six groups of DEC strains known to cause intestinal infections include enterotoxigenic *E. coli* (ETEC), enteropathogenic *E. coli* (EPEC), enterohemorrhagic *E. coli* (EHEC), enteroaggregative *E. coli* (EAEC), diffusely adherent *E. coli* (DAEC) and enteroinvasive *E. coli* (EIEC). Among all *E. coli* pathotypes, ETEC strains cause a cholera-like diarrhoeal disease and are the most common cause of childhood and travellers' diarrhoea in developing countries [5]. Diffusely adherent *E. coli* pathotypes were previously implicated in intestinal infections (diarrhoea in children between the ages of 18 months and 5 years) and extraintestinal infections (urinary tract infections and pregnancy complications) [6]. EIEC shows pathogenic phenotypic and genetic similarities with *Shigella* spp. and can be identified by their epithelial cell invasiveness mediated in part by the *ipaH* and *virF* genes and association with dysentery [7]. EHEC is associated with bloody diarrhoea and haemolytic uremic syndrome and expresses one or two Shiga-like toxin-encoding genes *stx1* and *stx2* [8].

Several virulence genes in these *E. coli* pathotypes are responsible for a wide array of infections such as diarrhoea or haemolytic colitis, neonatal meningitis, nosocomial septicaemia, haemolytic-uraemic syndrome and urinary tract infections [9]. Current molecular-based techniques such as polymerase chain reaction (PCR) allow for the identification of these VGs by amplifying specific target regions [10]. Virulence genes associated with these pathogenic strains have been isolated in diverse environments in South Africa. For example, the presence of DEC virulence genes in 60% of samples collected from the Apies River (water and sediments) was reported by Abia et al. [11]. In another study, a high prevalence of virulence genes associated with four pathogenic *E. coli* types (EAEC, EHEC, EPEC, and EIEC) in domestic rainwater harvesting tanks in Kleinmond, Cape Town was documented by Dobrowsky et al. [12]. Apart from being pathogenic, some of these microorganisms have developed resistance to many of the drugs designed to treat the infections they cause. For example, the antimicrobial resistance patterns of *E. coli* isolates in outpatient urinary tract infections in South Africa was studied and the results revealed that the isolated *E. coli* were resistant to trimethoprim-sulfamethoxazole (TMP-SMX; 68%), amoxicillin (65%) and ciprofloxacin (41%) [13]. Another study focused on the hospital, and community isolates of uropathogens at a tertiary hospital in South Africa and results revealed that the most isolated bacterial pathogen was *E. coli* (39%) [14]. Furthermore, levels of *E. coli* resistance to amoxicillin and co-trimoxazole ranged from 43–100% and 29–90%, respectively. The presence of such drug-resistant bacteria in human settings has placed constraints on the choice of safe, effective and inexpensive antibiotics, especially for low- and middle-income countries [15]. As such, the progression of resistant bacteria and the increasing incidence of antibiotic resistance genes (ARGs) are thus of significant public health concern [16].

Although studies have been carried out on the presence of virulence genes and antibiotic-resistant bacteria in various water sources such as wastewater effluents, taps, wells and boreholes in South Africa, very few studies have investigated their presence in harvested rainwater [12,17–19]. This study aimed at reporting on the prevalence of pathogenic *E. coli* strains and their antibiotic resistance patterns in harvested rainwater collected from tanks in the Eastern Cape Province of South Africa. Such results would highlight the need for appropriate development and implementation of effective household water treatment methods, thereby protecting the lives of populations using such water for their daily needs. Moreover, results of the current study will also add to existing research databases which report on the circulating strains of antimicrobial-resistant organisms.

2. Materials and Methods

2.1. Study Site and Sample Collection

Rooftop-harvested rainwater samples were collected from 11 rainwater-harvesting systems situated at various sites around Grahamstown west, Rhodes University campus and Kenton-on-sea in the Eastern Cape Province, South Africa. The distance between Rhodes University (33°31′36″ S, 26°51′63″ E) and Grahamstown west (33°18′36″ S; 26°31″36″ E) is approximately 4 km while the distance between Rhodes University and Kenton-on-sea (33°42′0″ S, 26°41′0″ E) is about 59.2 km. Mean annual rainfall in Grahamstown is 650 mm, with bimodal peaks in October–November and again in March–April. All the sites were selected based on the diversity in environmental conditions (e.g., presence of foliage and birds) as well as the various uses of the water stored in the tanks. A total of 110 water samples were collected from the 11 selected tanks from June 2016 to September 2016 and tested for *E. coli*. Sterile 5 L bottles were used to collect rainwater samples weekly by first rinsing the tap connected to the tanks with 70% ethanol and letting the tap run for 30 s before collection. Rainwater samples were taken from the same tanks once a week. Samples were then transported to Rhodes University laboratory on ice for microbial analysis within 6 h.

2.2. Enumeration and Isolation of E. coli

Enumeration of *E. coli* was carried out using the Colilert-18® Quanti-tray®/2000 (IDEXX Laboratories, Inc., Johannesburg, South Africa). The test was performed following the manufacturer's instructions. After incubation at 37 °C for 18–24 h, presumptive *E. coli* isolates were obtained from fluorescent quanti-tray wells as described by Abia et al. [20]. The Colilert method has a detection limit ranging from <1 MPN/100 mL to >2419.6 MPN/100 mL. *E. coli* ATCC® 25922 was used as a positive control and *Pseudomonas aeruginosa* ATCC 49189 as a negative control. One hundred (100) *E. coli* isolates were then selected from the various tanks. Of the 100 isolates selected, 66 isolates were chosen from T1-T6 (11 isolates from each tank), 20 isolates were from T7 and T8 (10 isolates from each tank) and 14 isolates from T9 and T11. T10 was excluded from further analysis due to poor growth of the selected isolates from the culture media.

2.3. Identification of Pathogenic Escherichia coli Strains Using Polymerase Chain Reaction (PCR)

DNA Extraction and Detection of Virulence Genes in *E. coli* Isolates

One hundred (100) presumptive *E. coli* isolates were randomly selected and inoculated separately into 5 mL Erlenmeyer flasks containing 2 mL nutrient broth (Merck, Johannesburg, South Africa). The flasks were incubated overnight at 37 °C on a rotary shaker at 100 rpm. DNA was extracted from 1 mL of the overnight culture using the InstaGene™ Matrix (Bio-Rad Laboratories, Johannesburg, South Africa) following the manufacturer's instruction. The template DNA was stored at −20 °C for PCR assays. All selected samples were first confirmed as *E. coli* by testing for the presence of the malate dehydrogenase (*mdh*) gene which is found in most *E. coli* strains [21]. After that, the presence of a total of eight VGs (*eaeA* (EPEC/EHEC), *eagg* (EAEC), *ipaH* (EIEC), *ST* (ETEC), *ibeA* (NMEC), *stx1* (EHEC), *stx2* (EHEC) and *flicH7* (EHEC)) were investigated. The primer sequences and the PCR-cycling conditions for the identification of the various VGs were as previously described by Abia et al. [19]. Both multiplex and singleplex PCR assays were performed for the target genes. Multiplex PCR assays were divided into 3 sets where set 1 contained *eaeA*, *eagg* and *ipaH*, set 2 contained *flicH7* and *Stx1* and finally set 3 contained *ST* and *ibeA* genes [19,22,23]. Singleplex real-time PCR assays were performed for the *mdh* and *stx2* target genes [24,25].

2.4. Screening for Antibiotic-Resistant E. coli

The remaining 1 mL from the overnight culture was used for antibiotic resistance analysis using the disk-diffusion method [26]. Briefly, 100 µL of overnight *E. coli* culture was spread on Mueller–Hinton agar (Lasec, Cape Town, South Africa) and antibiotic mastrings (Davies diagnostics, Johannesburg, South Africa) were carefully placed onto inoculated plates, incubated at 37 °C for 18–20 h. Following incubation, the diameters (in millimetres) of clear zones of growth inhibition around the antibiotic disks were measured using a ruler and compared with the Clinical Laboratory Standard Institute (CLSI) 2013 reference values. The different phenotypic profiles (resistant, intermediate or susceptible) of the isolates were then determined following the interpretation of the zones of inhibition. A total of 11 antibiotics were selected for this study (Table 1). The antibiotics were chosen for their frequent use in the treatment of bacterial infections in South Africa Both positive (*E. coli* strain ATTC 25922) and negative controls (*E. coli* strain ATTC 35218) were included in the experiments.

Table 1. Antibiotics used to determine antibiotic resistance of *E. coli* isolates.

Class	Antibiotic	Abbreviation	Concentration (µg)
β-Lactams	Ampicillin	AP	10
	Cephalothin	KF	5
Polypeptides	Colistin sulphate	CO	25
Aminoglycosides	Gentamicin	GM	10
Aminoglycosides	Streptomicin	S	10
Tetracyclines	Tetracycline	T	25
Folate pathway inhibitors	Cotrimoxazole	TS	25
Fluoroquinolones	Ciprofloxacin	CIP	5
Penicillin combination	Augmentin (amoxillin-clavulanate)	AUG	30
Sulfonamides	Trimethoprim	TM	5
Nitrofurans	Nitrofurantoin	NI	300

2.5. Data Analysis

Data were analysed using the Statistical Package for the Social Sciences (SPSS) (Version 16.0, Prentice Hall Press Company, NJ, USA) [27]. The *E. coli* counts were $\log_{10}$ transformed before computation of the means and standard deviations. A multiple antibiotic resistance (MAR) index was performed following the procedure described by Krumperman [28]. A MAR index for an isolate was calculated using the formula: $MAR = a/b$ where 'a' is the number of antibiotics from each group to which a particular isolate was resistant and 'b' is the total number of antibiotics against which the isolate was tested. A resistance index greater than 0.2 shows that *E. coli* isolates are likely to be from a high-risk source.

3. Results

3.1. Concentration of E. coli in Harvested Rainwater (HRW)

The log transformed ($\log_{10}$) *E. coli* counts and the mean *E. coli* counts in most probable number per 100 mL (MPN/100 mL) from individual tanks are shown in Table 2. The abundance of *E. coli* in the rainwater-harvesting tanks differed according to the location of the HRW system. The highest concentrations of *E. coli* were detected in tanks situated at Rhodes University (T1–T6).

Table 2. Log transformed *E. coli* (MPN/100 mL) concentrations from various rainwater tanks.

Tank ID	*n*	Minimum	Maximum	Mean ± Standard Deviation
T1	11	2.55	3.29	3.02 ± 0.21
T2	11	1.95	3.11	2.62 ± 0.35
T3	11	2.58	3.29	2.84 ± 0.25
T4	11	1.64	2.89	2.52 ± 0.42
T5	11	0.79	3.00	2.18 ± 0.82
T6	11	2.53	3.04	2.88 ± 0.19
T7	10	1.73	2.96	2.36 ± 0.37
T8	10	1.78	2.41	2.09 ± 0.22
T9	7	0.61	3.04	1.71 ± 0.82
T10	10	0.3	3.19	1.57 ± 1.04
T11	7	0.61	1.12	0.85 ± 0.26

3.2. Identification of Virulence Genes among E. coli *Isolates*

Samples which generated fluorescence from the Quanti-tray®/2000 cells were selected for the identification of the *E. coli* VGs. The most detected pathotypes were the NMEC and EHEC while the least detected pathotype was EAEC (Table 3). Of the 100 isolates tested for the VGs, 28% were identified as *ibeA* positive (Figure 1). The EAEC pathotype (*eagg* gene) was not detected among the tested isolates. Similarly, the *Stx1* gene of EHEC was not detected in any of the isolates.

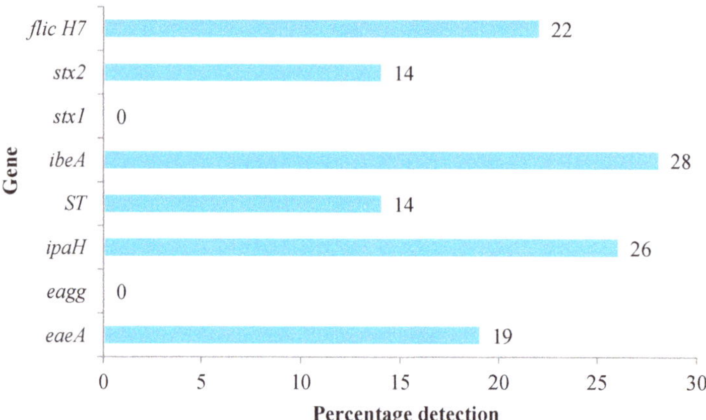

Figure 1. Overall prevalence of virulence genes in isolated *E. coli* from harvested rainwater (HRW) tanks.

Table 3. Number of virulence genes detected from rainwater-harvesting tanks.

Tank Location	Tank ID	Number of *E. coli* Isolates Tested	*EaeA* (EPEC/EHEC)	*Eagg* (EAEC)	*ipaH* (EIEC)	*ST* (ETEC)	*ibeA* (NMEC)	*Stx1* (EHEC)	*Stx2* (EHEC)	*flichH7* (EHEC)
Rhodes University	T1	11	6 (55%)	0	4 (36%)	0	4 (36%)	0	2 (18%)	4 (36%)
Rhodes University	T2	11	0	0	0	1 (9%)	1 (9%)	0	0	2 (18%)
Rhodes University	T3	11	1 (9%)	0	0	0	3 (27%)	0	0	1 (9%)
Rhodes University	T4	11	2 (18%)	0	1 (9%)	0	4 (36%)	0	0	3 (27%)
Rhodes University	T5	11	1 (9%)	0	0	0	4 (36%)	0	0	2 (18%)
Rhodes University	T6	11	1 (9%)	0	2 (18%)	2 (18%)	8 (72%)	0	0	3 (27%)
Kenton-on-sea	T7	10	2 (20%)	0	0	0	2 (20%)	0	0	2 (20%)
Kenton-on-sea	T8	10	0	0	4 (40%)	0	2 (20%)	0	1(10%)	2 (20%)
Grahamstown west	T9	7	1 (14%)	0	0	0	1 (13%)	0	0	0
Grahamstown west	T11	7	0	0	0	0	1 (13%)	0	0	0

Note: EPEC = Enteropathogenic *E. coli*, EHEC = Enterohemorrhagic *E. coli*, EAEC = Enteroaggregative *E. coli*, EIEC = Enteroinvasive *E. coli*, NMEC = Neonatal meningitis *E. coli*.

3.3. Antibiotic-Resistance Profiles of E. coli *Isolated from the Harvested-Rainwater Samples*

3.3.1. Overall Antibiotic Resistance Profiles of the *E. coli*

All the 100 *E. coli* isolates tested for the presence of VGs were further tested for antibiotic resistance. Of the 11 antibiotics tested, the highest resistance displayed by *E. coli* isolates was against Cephalothin (76%) while complete susceptibility (100%) was observed to Gentamycin. The overall percentage of antibiotic resistance found in the tested isolates is shown in Figure 2. *E. coli* isolates were resistant to 10 of the 11 antibiotics used in this study with the resistant rate ranging from 9% to 76%. Furthermore, a low percentage of the isolates showed resistance to Ciprofloxacin (15%) and Nitrofurantoin (9%).

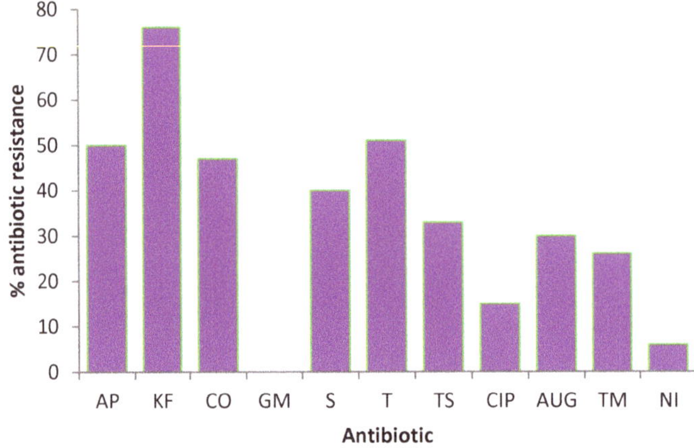

Figure 2. Percentage antibiotic resistance of *E. coli* isolates to selected antibiotics.

The bacterial resistance rate in individual tanks is shown in Table 4. Resistance to Nitrofurantoin was only observed in T1 and T2, while resistance to Augmentin was seen in all the tanks studied. Some of the selected isolates showed the presence of multiple-antibiotic resistance (MAR) where simultaneous resistance ranged from 3 to 9 antibiotics.

Table 4. Antibiotic resistance among *E. coli* strains isolated from various rainwater tanks.

Tank ID	*n*	% Resistance										
		AP	KF	CO	GM	S	T	TS	CIP	AUG	TM	NI
T1	11	72	91	63	0	45	72	39	27	27	63	9
T2	11	72	81	36	0	27	45	45	9	54	45	18
T3	11	27	36	27	0	45	18	27	18	9	18	0
T4	11	36	90	54	0	36	36	27	0	9	27	0
T5	11	36	100	54	0	0	54	36	27	18	45	0
T6	11	45	100	45	0	18	36	36	0	9	27	0
T7	10	30	30	20	0	100	20	30	20	30	30	0
T8	10	30	100	10	0	20	10	0	0	40	0	0
T9	7	75	87	75	0	62	100	87	0	75	50	0
T11	7	0	0	0	0	0	0	0	0	40	0	0

3.3.2. Prevalence of Multiple-Antibiotic Resistance

The presence of MAR was also observed for most isolates. Multiple-antibiotic resistance in this study was defined as the resistance of bacterial strains to three or more antibiotics [20]. Of the 100 isolates tested, more than half (52%) were MAR (Table 5). Ten of the 52 MAR isolates demonstrated simultaneous resistance to up to nine antibiotics. A total of 24 different MAR phenotypes were identified in this study.

Table 5. Multiple-antibiotic-resistant phenotypes of *E. coli* isolated from different rainwater tanks.

T1		T2	
MAR Phenotype	Number of Isolates	MAR Phenotype	Number of Isolates
AP-KF-CO-T	1	KF-T-NI	1
AP-KF-CO-T-TM	1	AP-KF-AUG	1
AP-KF-CO-S-T-TS-TM	1	AP-KF-NI	1
AP-KF-CO-S-T-TS-CIP-AUG-TM	1	AP-KF-S-T-TM	1
AP-KF-CO-S-T-TS-CIP-TM	1	AP-KF-CO-T-AUG	1
KF-CO-S-T-TS-AUG-TM	1	AP-KF-CO-S-T-TS-TM	1
AP-KF-CO-S-TS-CIP-TM	1	AP-KF-CO-S-T-TS-CIP-AUG-TM	1
AP-KF-CO-S-T-TS-AUG-TM	1	AP-KF-CO-T-TS-CIP-AUG-TM	1
		AP-KF-CO-S-T-TS-AUG-TM	1

T3		T4	
MAR Phenotype	Number of Isolates	MAR Phenotype	Number of Isolates
AP-KF-CO-S-TS-CIP-TM	1	KF-ST-AUG	1
AP-KF-CO-S-T-TS-TM	1	KF-T-NI	1
AP-KF-CO-S-T-TS-CIP-AUG-TM	1	AP-KF-CO-S-T-TS-TM	2
		AP-KF-CO-S-T-TS-CIP-TM	1
		AP-KF-CO-S-T-TS-CIP-AUG-TM	2

T5		T6	
MAR Phenotype	Number of Isolates	MAR Phenotype	Number of Isolates
AP-KF-CO-S-T-TS-CIP-TM	1	AP-KF-AUG	1
AP-KF-CO-T-TS-TM	1	KF-CO-S-T-TS-AUG-TM	1
AP-KF-CO-S-T-TS-CIP-AUG-TM	1	AP-KF-CO-S-TS-TM	1
KF-CO-T-TS-AUG-TM	1	AP-KF-CO-T-TS-TM-NI	1
AP-KF-CO-S-T-TS-CIP-AUG-TM	1	AP-KF-CO-S-T-TS-CIP-AUG-TM	2

T7		T8	
MAR Phenotype	Number of Isolates	MAR Phenotype	Number of Isolates
AP-KF-CO-S-T-TS-CIP-AUG-TM	1	AP-KF-T	1
AP-KF-CO-S-TS-AUG-TM	1	KF-CO-S-TS	1
AP-KF-CO-S-T-CIP-AUG-TM	1	AP-KF-CO-S-T-TS-CIP-AUG-TM	2
		AP-KF-CO-S-T-TS-AUG-TM	1

T9		T11	
MAR Phenotype	Number of Isolates	MAR Phenotype	Number of Isolates
AP-KF-CO-S-T-TS-CIP-AUG-TM	1	AP-KF-CO-S-T-TS-AUG-TM	2
AP-KF-CO-S-T-TS-AUG-TM	2	AP-KF-CO-S-T-TS-TM	1

4. Discussion

4.1. Concentration of E. coli in Harvested Rainwater

Faecal coliform bacteria such as *E. coli* have been widely used as indicator organisms to assess the possibility of pathogen presence in water [29]. Therefore, the presence of *E. coli* in roof-harvested rainwater in the Eastern Cape, South Africa, was monitored. All the 11 tanks monitored in this study were contaminated with varying concentrations of *E. coli* (0.85 ± 0.26–3.02 ± 0.21 MPN/100 mL).

Other scholars have previously reported on the high detection of *E. coli* from roof-harvested rainwater (2 to 986 CFU/100 mL; 1 to 99 MPN/100 mL and 0 to 41 CFU/100 mL) [1,30,31]. None of the tanks monitored in this study met the guidelines for drinking-water quality, as the *E. coli* amounts exceeded the South African drinking-water quality guidelines of 0 CFU/100 mL. The considerable amounts of *E. coli* in the harvested rainwater samples indicate possible faecal contamination.

The variations in the number of *E. coli* contamination in different HRW systems could be attributed to the fact that some of the HRW systems (Rhodes University) had a constant presence of birds which could have landed and dropped faecal matter on the roof, thereby contaminating tank water. Bird faecal droppings may negatively impact roof-harvested rainwater quality due to the presence of zoonotic pathogens [32]. A study conducted in South Africa investigated antibiotic resistance in *E. coli* isolates from roof-harvested rainwater tanks and urban pigeon faeces as the likely source of contamination and concluded that urban pigeons, the most likely source of HRW contamination, are also reservoirs of multiple antibiotic-resistant bacteria [33]. The findings of the South African study on bird faeces and antibiotic-resistant *E. coli* have a similar conclusion to our study where bird faecal matter was suspected to contribute to the contamination of HRW. In cases where the sources of faecal pollution in rainwater tanks are suspected to be from birds, the application of bird faecal markers may have the potential to confirm the sources of faecal contamination in a rainwater tank [32]. In another study to identify the likely sources of potential clinically significant *E. coli* in rainwater tanks, a source-tracking approach was used where a biochemical-fingerprinting method for typing of *E. coli* strains revealed that of the 43 strains from rainwater tank samples, 14 (from 7 tanks) and 9 (from 6 tanks) had identical biochemical phenotypes to those found in bird and possum faecal samples, respectively [34]. Furthermore, five strains from 4 rainwater tanks were identical to those isolated from both bird and possum faecal samples [34].

The rainwater tanks in the current study are used for various purposes such drinking and toilet flushing (for tanks situated at Rhodes University). Tanks situated at Grahamstown west were mainly used for gardening and sometimes drinking, depending on the availability of the municipal supply, while Kenton-on-sea tanks were used for indoor potable uses such dish-washing and laundry. In order to reduce or limit the risk of pathogenic and antimicrobial resistant *E. coli*, constant cleaning and maintenance of the catchment area may significantly improve the quality of the HRW, as the catchment area is suspected to contribute largely to the deterioration of the HRW in the Eastern Cape due to birds landing on the roof. Installation of first flush diverters may also help to improve the quality of the HRW. A study conducted in South Africa on the quality of HRW reported that 100% of the samples tested for *E. coli* exceeded the recommended standard of 0 CFU/100 mL [12]. Their results were similar to the ones observed in this study where all of the samples showed high levels of *E. coli*. In the Eastern Cape, where harvested rainwater is used for various household purposes including drinking, the presence of *E. coli* in the rainwater tanks is a major health concern as the presence of *E. coli* could imply the presence of other bacterial pathogens which may be detrimental to the health of rainwater users. The findings of the current study are of significant health concern as antibiotic-resistant pathogenic *E. coli* isolates may cause diseases if the users of the HRW consume the water without treatment. Furthermore, resistance of the isolated pathogenic *E. coli* to commonly used antibiotics in South Africa may lead to antibiotic treatment failure with serious public health implications for the population and the country.

4.2. Identification of Virulence Genes among E. coli Isolates

Pathogenic *E. coli* strains are a major cause of infections worldwide, the most common of which are diarrhoeal diseases. All the 100 *E. coli* isolates from the tanks tested positive for one or more VGs. The most detected pathotype was the NMEC (*ibeA*; 28%) which is responsible for neonatal meningitis and endothelial cell invasion [35]. The *ibeA* gene is also reportedly found in avian pathogenic *E. coli* (APEC) and causes avian colibacillosis, which is the most significant infectious bacterial disease of poultry worldwide [35]. The detection of the *ibeA*-positive strains in this study possibly indicates that the observed pathotype may be due to the presence of birds around the HRW systems. Although the present study did not investigate whether the *ibeA* gene detected was of human or avian origin, the presence of *ibeA*-positive isolates in the HRW systems is still of health concern given that there could be a possibility of zoonotic infections arising from the consumption of untreated rainwater containing these strains. Genes pertaining to other pathotypes of public health concern were also detected in the present study. For example, the *flicH7* (22%) and *Stx2* (14%) genes of EHEC were also detected in the isolates. Members of the EHEC group have been involved in many diarrhoeal disease outbreaks around the world, and they are known to cause hemorrhagic colitis and hemolytic uremic syndrome in humans [36].

The EHEC pathotype showed high prevalence across all the sampling sites except for the sites located in Grahamstown west. Both T1 and T6 which yielded a high percentage in VGs detection were situated at Rhodes University. Prevalence of the virulence gene *ipaH* (26%) (pathotype EIEC) was also noticeable in 4 tanks; 3 of the tanks were situated on campus and 1 in Kenton-on-sea. A previous study conducted in Cape Town, South Africa, reported that EPEC and EHEC (3% each) were detected in lower numbers, whereas EIEC was not identified in any of the rainwater tanks tested in their study [12]. The results differ from the findings of the current study where EIEC (26%) was the second most detected pathotype. This shows that the location of the tank could affect the pathotypes detected. Due to the detection of *E. coli* pathotypes in the current study, there is a great need to create awareness on household treatment technologies among users of HRW. Available treatment options which have proven to be successful in the treatment of HRW such as boiling, closed-couple solar pasteurizer, and solar disinfection (SODIS) can be used to decontaminate HRW [37–39]. In this study, all the rainwater tanks did not have any treatment option fitted, such as first-flush diverters and filters, except for T5 which had a chlorinator. However, due to limited maintenance of the rainwater-harvesting systems, the chlorinator in T5 was clogged in the middle of the sampling season and the *E. coli* counts increased going forward. The interruption of the treatment option observed in this study is also a clear indication of lack of proper maintenance of the HRW systems.

4.3. Detection of Antibiotic-Resistant E. coli *in Harvested Rainwater*

Results of the antibiotic-resistance profiling of the isolates from harvested rainwater analysed in the current study revealed that most of the *E. coli* isolates were resistant to the commonly prescribed antibiotics in South Africa. In areas such as the Eastern Cape where most of the population rely on harvested rainwater, exposure to antibiotic-resistant bacteria can further increase the health risk, particularly to children, the elderly and immune-compromised individuals. Antibiotic resistance is on the increase worldwide as most microorganisms now exhibit resistance to a large number of known antibiotics. The *E. coli* isolates from harvested rainwater in this study revealed resistance to Cephalothin (76%), Tetracyclines (51%), Colistin sulphate (47%), Ampicillin (50%) and Streptomycin (40%). The antibiotics most used in South Africa are the penicillins (Cephalothin) and fluoroquinolones, (Ciprofloxacin and glycopeptides) [40]. Tetracyclines and trimethoprim are also extensively used in the treatment of bacterial infections in both human and animals [41].

Cephalothin belongs to the β-Lactam class of antibiotics which are characterised by a β-lactam ring in their molecular structure [42]. Resistance to beta-lactam antibiotics has been highly documented as bacterial strains that produce extended-spectrum beta-lactamases have become more common [43]. Extended-spectrum beta-lactamase (ESBL)-producing *E. coli* are highly resistant to an array of antibiotics and infections by these strains are difficult to treat [43]. Furthermore, genes for ESBLs are most often encoded on plasmids, which can readily be transferred between bacteria [44]. Given that most of the isolates carrying virulence genes, especially the *ibeA* gene, were also resistant to Cephalothin, this could suggest that most of the isolated *E. coli* strains may carry the ESBL genes with the possibility of transfer to related organisms within the rainwater tanks. However, it is necessary to conduct further studies to ascertain such ARGs' transfer within harvested-rainwater systems. Results of such studies would highlight the need for implementation of appropriate treatment options and better policies for the safe use of harvested rainwater, especially where such water is the main source of water for personal and household uses, thus protecting the lives of users of harvested rainwater.

In the current study, the tested *E. coli* isolates showed resistance to one or more antibiotics with the highest *E. coli* resistance recorded against Cephalothin, Ampicillin and Tetracyclines. Also, there was evidence of MAR *E. coli* in almost all the HRW systems with some isolates showing simultaneous resistance to a panel of up to nine antibiotics. These results indicate that in the case of infections occurring due to the consumption of contaminated harvested rainwater, treatment may fail because of the persistent resistance of the *E. coli* isolates detected in the HRW systems. A similar study carried out in Pretoria and Johannesburg, South Africa, showed that the resistances most encountered were against Ampicillin, Gentamicin, Amikacin and Tetracyclines [34]. These results were not in agreement with our findings, where *E. coli* isolates were resistant to Cephalothin and 100% susceptible to Gentamicin, although the same method and concentration was used for Gentamicin in both studies. The difference in antibiotic resistance results from the two studies could be attributed to the fact that roof-harvested rainwater samples were collected from different locations (Gauteng and Eastern Cape). Our findings were, however, similar to the those of Chidamba and Korsten [34] in that the authors also reported a substantial prevalence of MAR. All the isolates tested in this study showed a MAR index greater than 0.2, suggesting that a greater proportion of the isolates were likely to be from a high-risk source such as faecal material. These results and the differences observed with other studies could inform those implementing antibiotic-resistance surveillance schemes that would address different geographical locations. Also, the presence of MAR *E. coli* in harvested rainwater could pose a severe health risk to the public in general, as antibiotic resistance decreases the efficiency of antibiotics used in the treatment of infections. These findings are of major concern, as more households are now reported to be using harvested rainwater for their daily water needs.

5. Conclusions

Rainwater samples tested in this study showed contamination with varying concentrations of pathogenic *E. coli* strains. The outcome of the study further demonstrates that HRW tanks could serve as reservoirs for not only pathogenic but also antibiotic-resistant *E. coli* strains including MAR strains. These findings suggest that the tested harvested rainwater was not fit for human consumption and, therefore, should not be used for potable purposes without appropriate treatment. Furthermore, routine monitoring and treatment are essential to ensure that harvested rainwater is fit for intended use as well as to stimulate the need for strategies (e.g., maintenance of HRW systems, constant cleaning of the roof, and installation of first-flush diverters to minimise faecal contamination) that would prevent the spread of antibiotic-resistant bacteria.

Author Contributions: A.L.K.A. and E.U.-J. conceived and designed the experiments as well as editing the manuscript; M.S.M. performed the experiments, analyzed the data and wrote the manuscript; R.T. and B.Z. contributed reagents/materials/analysis tools, sample collection as well as laboratory analysis and input on manuscript write-up; and J.-M.M.K. acquired the financial support for the project leading to this publication, supervised the project, and edited the manuscript.

Acknowledgments: This study was supported by funds from the Parliamentary Grant of the Council for Scientific and Industrial Research (CSIR Project ECHS043) and the Water Research Commission (WRC project K5/2593). Rhodes University provided access to monitoring sites and laboratory space for sample analysis. CSIR colleague Lisa Schaefer is thanked for advice on laboratory PCR analysis.

Conflicts of Interest: The authors declare no conflict of interest.

References

1. Ahmed, W.; Hodgers, L.; Masters, N.; Sidhu, J.P.S.; Katouli, M.; Toze, S. Occurrence of intestinal and extraintestinal virulence genes in *Escherichia coli* isolates from rainwater tanks in Southeast Queensland, Australia. *Appl. Environ. Microbiol.* **2011**, *77*, 7394–7400. [CrossRef] [PubMed]
2. Masters, N.; Wiegand, A.; Ahmed, W.; Katouli, M. *Escherichia coli* virulence genes profile of surface waters as an indicator of water quality. *Water Res.* **2011**, *45*, 6321–6333. [CrossRef] [PubMed]

3. Russo, T.A.; Johnson, J.R. Proposal for a new inclusive designation for extraintestinal pathogenic isolates of *Escherichia coli*: ExPEC. *J. Infect. Dis.* **2000**, *181*, 1753–1754. [CrossRef] [PubMed]

4. Canizalez-Roman, A.; Gonzalez-Nuñez, E.; Vidal, J.E.; Flores-Villaseñor, H.; León-Sicairos, N. Prevalence and antibiotic resistance profiles of diarrheagenic *Escherichia coli* strains isolated from food items in northwestern Mexico. *Int. J. Food Microbiol.* **2013**, *164*, 36–45. [CrossRef] [PubMed]

5. Jafari, A.; Aslani, M.M.; Bouzari, S. *Escherichia coli*: A brief review of diarrheagenic pathotypes and their role in diarrheal diseases in Iran. *Iran. J. Microbiol.* **2012**, *4*, 102–117. [PubMed]

6. Servin, A.L. Pathogenesis of human diffusely adhering *Escherichia coli* expressing Afa/Dr adhesins (Afa/Dr DAEC): Current insights and future challenges. *Clin. Microbiol. Rev.* **2014**, *27*, 823–869. [CrossRef] [PubMed]

7. Karmali, M.A.; Gannon, V.; Sargeant, J.M. Verocytotoxin-producing *Escherichia coli* (VTEC). *Vet. Microbiol.* **2010**, *140*, 360–370. [CrossRef] [PubMed]

8. Viazis, S.; Diez-Gonzalez, F. Enterohemorrhagic *Escherichia coli*. In *The Twentieth Century's Emerging Foodborne Pathogen: A Review*, 1st ed.; Elsevier Inc.: New York, NY, USA, 2011; Volume 111.

9. Anastasi, E.M.; Matthews, B.; Gundogdu, A.; Vollmerhausen, T.L.; Ramos, N.L.; Stratton, H.; Ahmed, W.; Katouli, M. Prevalence and persistence of *Escherichia coli* strains with uropathogenic virulence characteristics in sewage treatment plants. *Appl. Environ. Microbiol.* **2010**, *76*, 5882–5886. [CrossRef] [PubMed]

10. Costa, C.F.M.; Monteiro Neto, V.; Santos, B.R.d.C.; Costa, B.R.R.; Azevedo, A.; Serra, J.L.; Mendes, H.B.R.; Nascimento, A.R.; Mendes, M.B.P.; Kuppinger, O. Enterobacteria identification and detection of diarrheagenic *Escherichia coli* in a Port Complex. *Braz. J. Microbiol.* **2014**, *45*, 945–952. [CrossRef] [PubMed]

11. Abia, A.L.K.; Ubomba-Jaswa, E.; Momba, M.N.B. Occurrence of diarrhoeagenic *Escherichia coli* virulence genes in water and bed sediments of a river used by communities in Gauteng, South Africa. *Environ. Sci. Pollut. Res.* **2016**, *23*, 15665–15674. [CrossRef] [PubMed]

12. Dobrowsky, P.H.; van Deventer, A.; De Kwaadsteniet, M.; Ndlovu, T.; Khan, S.; Cloete, T.E.; Khan, W. Prevalence of virulence genes associated with pathogenic *Escherichia coli* strains isolated from domestically harvested rainwater during low- and high-rainfall periods. *Appl. Environ. Microbiol.* **2014**, *80*, 1633–1638. [CrossRef] [PubMed]

13. Bosch, F.J.; van Vuuren, C.; Joubert, G. Antimicrobial resistance patterns in outpatient urinary tractinfections—The constant need to revise prescribing habits. *S. Afr. Med. J.* **2011**, *101*, 328–331. [CrossRef] [PubMed]

14. Habte, T.M.; Dube, S.; Ismail, N.; Hoosen, A.A. Hospital and community isolates of uropathogens at a tertiary hospital in South Africa. *S. Afr. Med. J.* **2009**, *99*, 584–587. [PubMed]

15. Detels, R.; Gulliford, M.; Karim, Q.A.; Tan, C.C.; Press, O.U. *Oxford Textbook of Global Public Health: The Practice of Public Health*, 6th ed.; Detels, R., Beaglehole, R., Lansang, M., Gulliford, M., Eds.; Oxford University: New York, NY, USA, 2015; Volume 3.

16. Xu, Y.; Guo, C.; Luo, Y.; Lv, J.; Zhang, Y.; Lin, H.; Wang, L.; Xu, J. Occurrence and distribution of antibiotics, antibiotic resistance genes in the urban rivers in Beijing, China. *Environ. Pollut.* **2016**, *213*, 833–840. [CrossRef] [PubMed]

17. Kinge, W.C.N.; Ateba, C.N.; Kawadza, D.T. Antibiotic resistance profiles of *Escherichia coli* isolated from different water sources in the Mmabatho locality, Northwest Province, South Africa. *S. Afr. J. Sci.* **2010**, *106*, 44–49.

18. Adefisoye, M.A.; Okoh, A.I. Identification and antimicrobial resistance prevalence of pathogenic *Escherichia coli* strains from treated wastewater effluents in Eastern Cape, South Africa. *Microbiologyopen* **2016**, *5*, 143–151. [CrossRef] [PubMed]

19. Abia, A.L.K.; Schaefer, L.; Ubomba-Jaswa, E.; Le Roux, W. Abundance of pathogenic *Escherichia coli* virulence-associated genes in well and borehole water used for domestic purposes in a peri-urban community of South Africa. *Int. J. Environ. Res. Public Health* **2017**, *14*, 320. [CrossRef] [PubMed]

20. Abia, A.L.K.; Ubomba-Jaswa, E.; Momba, M.N.B. High prevalence of multiple-antibiotic-resistant (MAR) *Escherichia coli* in river bed sediments of the Apies River, South Africa. *Environ. Monit. Assess.* **2015**, *187*, 652. [CrossRef] [PubMed]

21. Hsu, S.C.; Tsen, H.Y. PCR primers designed from malic acid dehydrogenase gene and their use for detection of *Escherichia coli* in water and milk samples. *Int. J. Food Microbiol.* **2001**, *64*, 1–11. [CrossRef]

22. Caine, L.; Nwodo, U.; Okoh, A.; Ndip, R.; Green, E. Occurrence of virulence genes associated with diarrheagenic *Escherichia coli* isolated from raw cow's milk from two commercial dairy farms in the Eastern Cape Province, South Africa. *Int. J. Environ. Res. Public Health* **2014**, *11*, 11950–11963. [CrossRef] [PubMed]

23. Titilawo, Y.; Obi, L.; Okoh, A. Occurrence of virulence gene signatures associated with diarrhoeagenic and non-diarrhoeagenic pathovars of Escherichia coli isolates from some selected rivers in South-Western Nigeria. *BMC Microbiol.* **2015**, *15*, 204. [CrossRef] [PubMed]

24. Omar, K.B.; Barnard, T.G. Detection of diarrhoeagenic *Escherichia coli* in clinical and environmental water sources in South Africa using single-step 11-gene m-PCR. *World J. Microbiol. Biotechnol.* **2014**, *30*, 2663–2671. [CrossRef] [PubMed]

25. Omar, K.B.; Potgieter, N.; Barnard, T.G. Development of a rapid screening method for the detection of pathogenic *Escherichia coli* using a combination of Colilert® Quanti-Trays/2000 and PCR. *Water Sci. Technol. Water Supply* **2010**, *10*, 7–13. [CrossRef]

26. Stange, C.; Sidhu, J.P.S.; Tiehm, A.; Toze, S. Antibiotic resistance and virulence genes in coliform water isolates. *Int. J. Hyg. Environ. Health* **2016**, *219*, 823–831. [CrossRef] [PubMed]

27. Norusis, M. *SPSS 16.0 Statistical Procedures Companion*; Prentice Hall Press: Upper Saddle River, NJ, USA, 2008.

28. Krumperman, P.H. Multiple antibiotic resistance indexing of *Escherichia coli* to indentify high-risk sources of fecal contamination of foods. *Appl. Environ. Microbiol.* **1983**, *46*, 165–170. [PubMed]

29. Hachich, E.M.; Di Bari, M.; Christ, A.P.G.; Lamparelli, C.C.; Ramos, S.S.; Sato, M.I.Z. Comparison of thermotolerant coliforms and *Escherichia coli* densities in freshwater bodies. *Braz. J. Microbiol.* **2012**, *43*, 675–681. [CrossRef] [PubMed]

30. Spinks, J.; Phillips, S.; Robinson, P.; Van Buynder, P. Bushfires and tank rainwater quality: A cause for concern? *J. Water Health* **2006**, *4*, 21–28. [PubMed]

31. Sazakli, E.; Alexopoulos, A.; Leotsinidis, M. Rainwater harvesting, quality assessment and utilization in Kefalonia Island, Greece. *Water Res.* **2007**, *41*, 2039–2047. [CrossRef] [PubMed]

32. Ahmed, W.; Sidhu, J.P.S.; Toze, S. An attempt to identify the likely sources of *Escherichia coli* harboring toxin genes in rainwater tanks. *Environ. Sci. Technol.* **2012**, *46*, 5193–5197. [CrossRef] [PubMed]

33. Chidamba, L.; Korsten, L. Antibiotic resistance in *Escherichia coli* isolates from roof-harvested rainwater tanks and urban pigeon faeces as the likely source of contamination. *Environ. Monit. Assess.* **2015**, *187*, 405. [CrossRef] [PubMed]

34. Ahmed, W.; Hamilton, K.; Gyawali, P.; Toze, S.; Haas, C. Evidence of avian and possum fecal contamination in rainwater tanks as determined by microbial source tracking approaches. *Appl. Environ. Microbiol.* **2016**. [CrossRef] [PubMed]

35. Johnson, T.J.; Wannemuehler, Y.; Kariyawasam, S.; Johnson, J.R.; Logue, C.M.; Nolan, L.K. Prevalence of avian-pathogenic *Escherichia coli* strain O1 genomic islands among extraintestinal and commensal *E. coli* isolates. *J. Bacteriol.* **2012**, *194*, 2846–2853. [CrossRef] [PubMed]

36. Hamilton, M.J.; Hadi, A.Z.; Griffith, J.F.; Ishii, S.; Sadowsky, M.J. Large scale analysis of virulence genes in *Escherichia coli* strains isolated from Avalon Bay, CA. *Water Res.* **2010**, *44*, 5463–5473. [CrossRef] [PubMed]

37. Spinks, A.T.; Dunstan, R.H.; Coombes, P.; Kuczera, G. Thermal destruction analyses of water related pathogens at domestic hot water system temperatures. In Proceedings of the 28th International Hydrology and Water Resources Symposium, Wollongong, Australia, 10–13 November 2003.

38. Dobrowsky, P.H.; Carstens, M.; De Villiers, J.; Cloete, T.E.; Khan, W. Efficiency of a closed-coupled solar pasteurization system in treating roof harvested rainwater. *Sci. Total Environ.* **2015**, *536*, 206–214. [CrossRef] [PubMed]

39. Amin, M.T.; Nawaz, M.; Amin, M.N.; Han, M. Solar disinfection of *Pseudomonas aeruginosa* in harvested rainwater: A step towards potability of rainwater. *PLoS ONE* **2014**, *9*, e90743. [CrossRef] [PubMed]

40. Essack, S.Y.; Schellack, N.; Pople, T.; van der Merwe, L.; Suleman, F.; Meyer, J.C.; Gous, A.G.S.; Benjamin, D. Part III. GARP: Antibiotic supply chain and management in human health. *S. Afr. Med. J.* **2011**, *101*, 562–566. [PubMed]

41. Ruhe, J.J.; Menon, A. Tetracyclines as an oral treatment option for patients with community onset skin and soft tissue infections caused by methicillin-resistant Staphylococcus aureus. *Antimicrob. Agents Chemother.* **2007**, *51*, 3298–3303. [CrossRef] [PubMed]

42. Beceiro, A.; Tomás, M.; Bou, G. Antimicrobial resistance and virulence: A successful or deleterious association in the bacterial world? *Clin. Microbiol. Rev.* **2013**, *26*, 185–230. [CrossRef] [PubMed]
43. Paterson, D.L.; Bonomo, R.A. Extended-Spectrum beta-Lactamases: A clinical update. *Clin. Microbiol. Rev.* **2005**, *18*, 657–686. [CrossRef] [PubMed]
44. Haenni, M.; Châtre, P.; Madec, J.Y. Emergence of *Escherichia coli* producing extended-spectrum AmpC β-lactamases (ESAC) in animals. *Front. Microbiol.* **2014**, *5*, 1–7. [CrossRef] [PubMed]

International Journal of
Environmental Research and Public Health

MDPI

Article

Pathogenic *Escherichia coli* Strains Recovered from Selected Aquatic Resources in the Eastern Cape, South Africa, and Its Significance to Public Health

Kingsley Ehi Ebomah [1,2,*], Martins Ajibade Adefisoye [1,2] and Anthony Ifeanyi Okoh [1,2]

[1] SAMRC Microbial Water Quality Monitoring Centre, University of Fort Hare, Alice 5700, South Africa;
MAdefisoye@ufh.ac.za (M.A.A.); AOkoh@ufh.ac.za (A.I.O.)
[2] Applied and Environmental Microbiology Research Group (AEMREG), Department of Biochemistry
and Microbiology, University of Fort Hare, Alice 5700, South Africa
* Correspondence: 201411619@ufh.ac.za; Tel.: +27-769-541-783

Received: 15 June 2018; Accepted: 10 July 2018; Published: 17 July 2018

Abstract: The prevalence of pathogenic microorganisms, as well as the proliferation of antimicrobial resistance, pose a significant threat to public health. However, the magnitude of the impact of aquatic environs concerning the advent and propagation of resistance genes remains vague. *Escherichia coli* (*E. coli*) are widespread and encompass a variety of strains, ranging from non-pathogenic to highly pathogenic. This study reports on the incidence and antibiotic susceptibility profiles of *E. coli* isolates recovered from the Nahoon beach and its canal waters in South Africa. A total of 73 out of 107 (68.2%) Polymerase chain reaction confirmed *E. coli* isolates were found to be affirmative for at least one virulence factor. These comprised of enteropathogenic *E. coli* 11 (10.3%), enteroinvasive *E. coli* 14 (13.1%), and neonatal meningitis *E. coli* 48 (44.9%). The phenotypic antibiogram profiles of the confirmed isolates revealed that all 73 (100%) were resistant to ampicillin, whereas 67 (91.8%) of the pathotypes were resistant to amikacin, gentamicin, and ceftazidime. About 61 (83.6%) and 51 (69.9%) were resistant to tetracycline and ciprofloxacin, respectively, and about 21.9% (16) demonstrated multiple instances of antibiotic resistance, with 100% exhibiting resistance to eight antibiotics. The conclusion from our findings is that the Nahoon beach and its canal waters are reservoirs of potentially virulent and antibiotic-resistant *E. coli* strains, which thus constitute a potent public health risk.

Keywords: *E. coli*; surface water; antibiotic-resistance gene; MARI; MARP; multidrug resistance

1. Introduction

Water has a function in numerous metabolic activities and is hence an essential ingredient for hydration which sustains health and sanitation, while also having industrial and agricultural applications. Thus, poor water quality has a demoralizing impact on public health, and polluted water sources can lead to waterborne disease outbreaks [1]. According to Gorde and Jadhav [2], the human population is most likely to suffer from waterborne diseases due to the use of contaminated or polluted water.

Irrespective of enormous developments in therapeutic treatment options as well as wastewater treatment facilities, waterborne infections still pose a major threat to public health worldwide [3]. These infections, caused by contaminations of surface water bodies by pathogenic microorganisms transmitted via contact with polluted water, are responsible for the illness of millions of people each year, while also causing numerous deaths [4]. The majority of these infections occur in developing nations which, in comparison with developed nations, often have less than desirable levels of sanitation, socioeconomic conditions, and public health awareness [5].

As beaches are typical spots for human recreation, they can gain a lot of patronage from both domestic and international tourists. Such recreational centers fortify development and prove to be a significant economic contribution to tourism in coastal areas [6]. Unfortunately, many beaches have been subjected to high levels of contamination in recent years [7], which is why this phenomenon has become a matter of urgency [8]. This study thus outlines the importance of maintaining a clean environment in the coastal areas and reports the discovery of pathogenic strains of bacteria exhibiting multidrug resistance.

Fecal contamination of water bodies presents severe public health issues in many countries [9] and owes the source of its threat to microbial pathogens. These are often shed by diseased humans and animals, and may be conveyed via the sewer system and agricultural run-offs [10]. In a study conducted by Okoh et al. [11], it was found that the release of ineffectively-treated effluents were the major source of enteric pathogens in aquatic environs. Due to the low monitoring of health risk that could be associated with beach water, literature has shown that potential risks may be associated with nonhuman fecal contamination [12]. *E. coli* is one of the bacteria used as an indicator organism for the monitoring of water bodies, and different strains of these bacteria are pathogenic. The pathogenicity of a specific *E. coli* pathotype is primarily determined by explicit virulent influences [13]. Globally, *E. coli* strains have been associated with human and animal diseases by means of pathogens, on the basis of their virulent elements and clinical symptoms. According to Mellata [14] and Titilawo et al. [15], *E. coli* strains can be categorized into two groups: extra-intestinal pathogenic *E. coli* (ExPEC) and intestinal pathogenic *E. coli* (InPEC). However, InPEC can also be subdivided into enteroinvasive *E. coli* (EIEC), enteroaggregative *E. coli* (EAEC), diffusely adherent *E. coli* (DAEC), enterotoxigenic *E. coli* (ETEC), enteropathogenic *E. coli* (EPEC), and enterohemorrhagic *E. coli* (EHEC). ExPEC can also be classified into neonatal meningitis *E. coli* (NMEC), uropathogenic *E. coli* (UPEC), and avian pathogenic *E. coli* (APEC) [16]. A further class known as diarrhoeagenic *E. coli* pathotypes has been proposed, such as cell-detaching *E. coli* (CDEC) although their significance remains unclear [16]. The majority of infections caused by *E. coli* are treated by using antimicrobial agents. However, the effects of some of these agents have been compromised by some types of bacteria [17]. Evidently, antimicrobial-resistant bacteria (ARB) can be released into the environment via the discarding of human and animal waste [18]. Moreover, the use of antibiotics for the treatment of infections in humans and farm animals has also been reported to cause an increase in ARB [19], and numerous antibiotics have become ineffective against their targets due to the frequent exposure of pathogens to antimicrobial agents [20,21]. The aim of this study was thus to identify and characterize the *E. coli* isolates into various pathotypes, while also determining the phenotypic resistance pattern of the confirmed isolates.

2. Materials and Methods

2.1. Description of the Study Site and Sampling Points

Nahoon beach and its canal are located in East London, Eastern Cape, South Africa on the coast of the Indian Ocean (geographical coordinates: 32.99° S and 27.95° E). As shown in Figure 1 below, the study area was in Buffalo City Metropolitan Municipality in the Eastern Cape, as highlighted in the map. The Nahoon canal is observed as an extension of the Nahoon River, which flows into the beach. A total of six sampling points for the beach and the canal (three points each) were mapped along the sea shore. Nahoon canal point 3 flowed into the Nahoon beach at point 1, and Nahoon canal point 2 had some domestic effluent flowing into it. There was a release of final effluent from the East Bank Reclamation Works (sited in East London, South Africa) into the Indian Ocean at Bats Cave, which is represented by sampling point 2 on the beach site in this study.

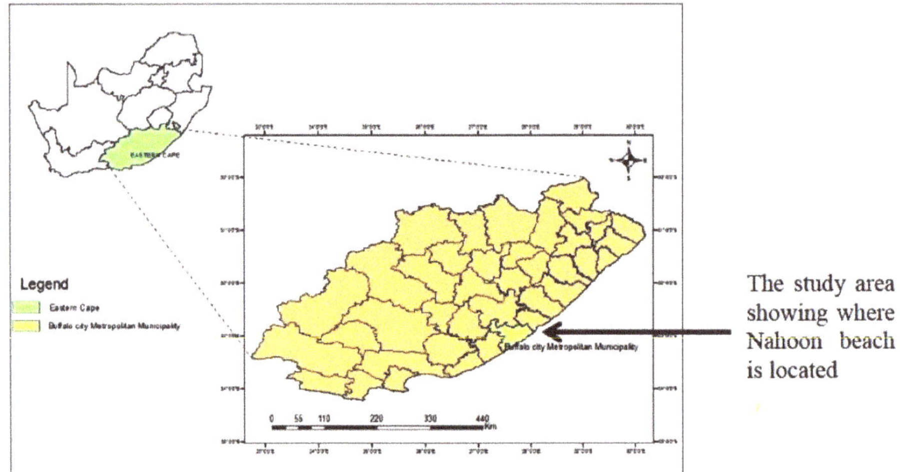

Figure 1. A map showing the location of the study area.

2.2. Sample Collection

Water samples were collected bi-weekly using Nalgene sterile bottles from 6 different sampling points along the Nahoon beach and canal for a period of twelve months, between 8:00 a.m. and 11:00 a.m. The sampling points where the water samples were collected are: canal point 1, canal point 2 (where domestic effluents flowed into canal), canal point 3 (where canal flowed into beach), beach point 1, beach point 2 (where final effluent was released into beach), and beach point 3. The samples were then transported on ice to the Applied and Environmental Microbiology Research Group (AEMREG) laboratory, University of Fort Hare, Alice within 6 h for analyses.

2.3. Isolation and DNA Extraction

The water samples were filtered using the membrane filtration technique, after which the filter papers were aseptically picked, placed onto *E. coli* chromogenic agar, and incubated at 37 °C for 18–24 h. After incubation, the isolates were re-streaked onto nutrient agar (NA) plates and incubated at 37 °C for 24 h. A total of 260 presumptive *E. coli* isolate colonies were picked from the NA plates, inoculated into nutrient broth, and incubated at 37 °C for 24 h. Thereafter, glycerol stock was prepared from the cultured broths, and DNA was extracted following the method of Torres et al. [22], and stored at −80 °C for further analyses.

2.4. Molecular Identification and Characterization of the Recovered E. coli Isolates

Molecular identification of the presumptive *E. coli* isolates targeting the *uidA* gene and the various genes of the *E. coli* pathotypes screened were determined by following the method of Titilawo et al. [15] as shown in Table 1. The PCR products were resolved in a 2% (*w/v*) agarose gel in 1 × TAE buffer (40 mM Tris–HCl, 20 mM Sodium Acetate, 1 mM EDTA, pH 8.5), stained with 0.5 mg mL ethidium bromide (EtBr), and visualized under the Alliance BioDoc-It System (UFH, Alice 5700, South Africa) [23–25].

Table 1. Primer sequences of target genes and their respective amplicon sizes and PCR (polymerase chain reaction techniques) cycling conditions.

Target Strain	Target Gene	Primer Sequence (5′→3′)	Amplicon Size (bp)	PCR Cycling Condition
E. coli	*uidA*	F: AAA ACG GCA AGA AAA AGC AG R: ACG CGT GGT TAA CAG TCT TGC G	147	Initial denaturation of 5 min at 94 °C followed by 35 cycles, denaturation at 95 °C for 30 s, annealing at 58 °C for 1 min, extension at 72 °C for 1 min and final extension at 72 °C for 8 min
EPEC	*eae*	F: TCA ATG CAG TTC CGT TAT CAG TT R: GTA AAG TCC GTT ACC CCA ACC TG R: GGA ATC AGA CGC AGA CTG GTA GT	482	Initial denaturation of 15 min at 95 °C followed by 35 cycles, denaturation at 94 °C for 45 s, annealing at 55 °C for 45 s, extension at 68 °C for 2 min and final extension at 72 °C for 5 min
ETEC	*lt*	F: GGC GAC AGA TTA TAC CGT GC R: CGG TCT CTA TAT TCC CTG TT	450	Initial denaturation of 2 min at 94 °C followed by 35 cycles, denaturation at 94 °C for 1 min, annealing at 55 °C for 1 min, extension at 72 °C for 1 min and final extension at 72 °C for 5 min
EAEC	*eagg*	F: AGA CTC TGG CGA AAG ACT GTA TC R: ATG GCT GTC TGT AAT AGA TGA GAA C	194	Initial denaturation of 15 min at 95 °C followed by 35 cycles, denaturation at 94 °C for 45 s, annealing at 55 °C for 45 s, extension at 68 °C for 2 min and final extension at 72 °C for 5 min
EIEC	*ipaH*	F: CTC GGC ACG TTT TAA TAG TCT GG R: GTG GAG AGC TGA AGT TTC TCT GC	933	Initial denaturation of 2 min at 94 °C followed by 35 cycles, denaturation at 94 °C for 1 min, annealing at 55 °C for 1 min, extension at 72 °C for 1 min and final extension at 72 °C for 5 min
DAEC	*daaE*	F: GAA CGT TGG TTA ATG TGG GGT AA R: TAT TCA CCG GTC GGT TAT CAG T	542	Initial denaturation of 2 min at 94 °C followed by 40 cycles, denaturation at 92 °C for 30 s, annealing at 59 °C for 30 s, extension at 72 °C for 30 s and final extension at 72 °C for 5 min
EHEC	*stx1*	F: CAG TTA ATG TGG TGG CGA AGG R: CAC CAG ACA ATG TAA CCG CTG	384	Initial denaturation of 15 min at 95 °C followed by 35 cycles, denaturation at 94 °C for 45 s, annealing at 55 °C for 45 s, extension at 68 °C for 2 min and final extension at 72 °C for 5 min
NMEC	*ibeA*	F: TGG AAC CCC GCT CGT AAT ATA C R: CTG CCT GTT CAA GCA TTG CA	342	Initial denaturation of 2 min at 94 °C followed by 30 cycles, denaturation at 94 °C for 1 min, annealing at 55 °C for 1 min, extension at 72 °C for 1 min and final extension at 72 °C for 5 min
UPEC	*papC*	F: GAC GGC TGT ACT GCA GGG TGT GGC G R: ATA TCC TTT CTG CAG GGA TGC AAT A	328	Initial denaturation of 2 min at 94 °C followed by 30 cycles, denaturation at 94 °C for 1 min, annealing at 55 °C for 1 min, extension at 72 °C for 1 min and final extension at 72 °C for 5 min

Source: Titilawo et al. [15].

2.5. Antimicrobial Susceptibility Pattern of the Confirmed E. coli Strains

The antimicrobial susceptibility test of the confirmed *E. coli* isolates was determined by the disc diffusion technique on Mueller Hinton agar (MHA) plates, following Clinical and Laboratory Standards Institute (CLSI) [26] guidelines. Fresh culture from the glycerol stock was streaked onto nutrient agar plates and incubated at 37 °C for 24 h. Colonies were transferred into a test tube of 5 mL of normal sterile saline, and adjusted to attain turbidity matching the 0.5 McFarland standard. The isolates were then streaked onto MHA plates, and disks infused with antimicrobial agents were dispensed onto the inoculated plates and incubated for 18 to 24 h at 37 °C. After incubation, the zones of inhibition were measured, and isolates were categorized as resistant or susceptible to the antimicrobial agents used, while those that were intermediate were considered resistant. The following eight commercial antibiotic discs: Amikacin (30 µg), ampicillin (10 µg), ceftazidime (30 µg), ciprofloxacin (5 µg), gentamicin (10 µg), norflaxacin (10 µg), tetracycline (30 µg), and trimethoprim (10 µg) were tested against the confirmed isolates.

2.6. Interpretation of Multiple Antibiotic-Resistance Index (MARI)

The multiple antibiotic resistance index (MAR Index) of isolates that exhibited resistance against the actions of three or more antibiotics which were tested was expressed as x/y, where x indicates the sum of antibiotics to which the isolate was resistant to and y indicates the number of antibiotics tested against the isolate. Multidrug resistance was interpreted as the display of resistance to three or more antibiotics used, whereas the MARI (multidrug antibiotic-resisted indices) of the isolates was approximated, as previously described by Krumperman [27]. The multiple antibiotic resistance index (MARI) = $w/(u \times v)$, where: w is the summation of antibiotics resistance scores of the isolates; u is the sum of antibiotics used; and v is the sum of isolates which resisted the antibiotics employed.

3. Results

3.1. Molecular Identification and Characterization of the Recovered E. coli Isolates

A total of 260 presumptive *E. coli* isolates were obtained from the water samples following microbiological analysis. The presumptive isolates were confirmed by polymerase chain reaction techniques (PCR) targeting the *uidA* gene. Results showed that 41.2% (107/260) of the *E. coli* isolates were positive, as shown in Figure 2. The confirmed *E. coli* isolates were further characterized into different pathotypes using specific primers for each pathotype, and the result is shown in Table 2. A total number of 26 isolates belonging to the three pathotypes identified were isolated from the 3 sampling points (canal), while 47 isolates belonging to the three pathotypes identified were recovered from the 3 sampling points (beach).

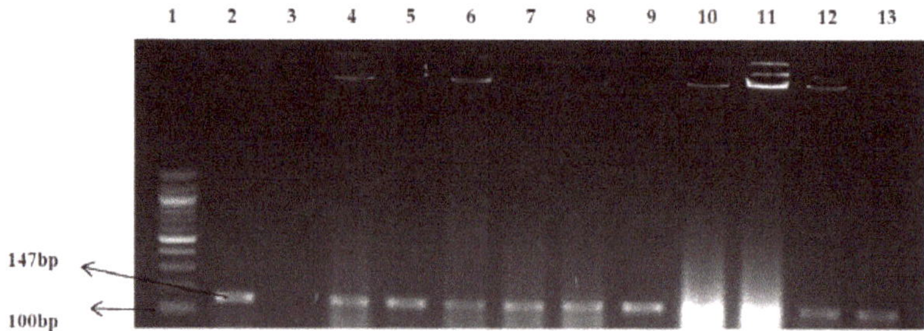

Figure 2. PCR products of the amplification of the *uidA* gene (*E. coli*) Lane 1: 100 bp molecular weight marker; Lane 2: positive control (*E. coli* ATCC 25922); Lane 3: negative control; Lanes 4–13: positive isolates.

Table 2. Results of *E. coli* pathotypes.

No. of Isolates Screened	Pathotype/Target Gene	No. of Positive Isolates (%)
107	EPEC/*eae*	11 (10.3%)
107	ETEC/*lt*	0
107	EAEC/*eagg*	0
107	EIEC/*ipaH*	14 (13.1%)
107	DAEC/*daaE*	0
107	EHEC/*stx1*	0
107	NMEC/*ibeA*	48 (44.9%)
107	UPEC/*papC*	0

3.2. Antimicrobial Susceptibility Pattern of the Confirmed E. coli Pathotypes

Of the 8 test antimicrobial agents which were selected, ampicillin had the highest resistance frequency (100%). Nevertheless, amikacin and gentamycin both had quite high frequencies of 98.6% (72/73), while 70 of the strains were resistant to ceftazidime, with a frequency of 95.9% (Figure 3). About 45 strains (93% of the NMEC strains) exhibited resistance to each of ampicillin, amikacin, gentamycin, tetracycline, and ceftazidime, while 9.1% (1/11) and 91% (10/11) of the EPEC strains displayed resistance to ciprofloxacin and tetracycline respectively. For NMEC, 26 strains showed a resistance frequency of 54.2% against ciprofloxacin. Similarly, the EIEC strains demonstrated resistance ranging between 7% (1/14) and 50% (7/14) to amikacin, ampicillin, ceftazidime, ciprofloxacin, gentamycin, norflaxacin tetracycline, and trimethoprim. The results from the *E. coli* isolates which were subjected to the selected antimicrobial agents are summarized in Figure 3, which highlights all the sensitivity percentages of the isolates. 23 isolates of the various strains identified showed resistance to 8 antibiotics (19 NMEC, 3 EPEC and 1 EIEC), while 19 strains showed resistance to 7 antibiotics (5 NMEC, 4 EPEC and 10 EIEC).

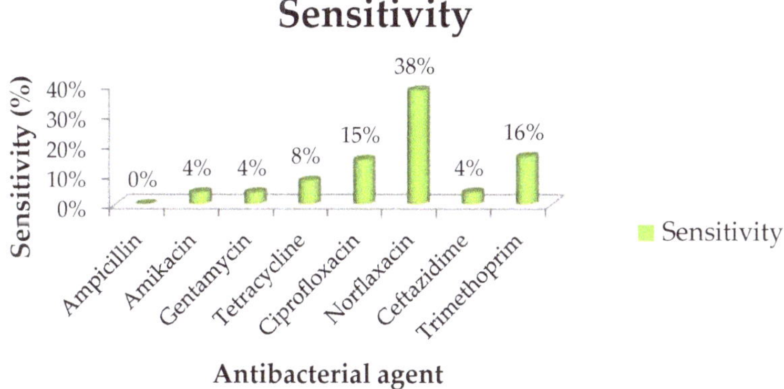

Figure 3. Sensitivity percentages of *E. coli* isolates to 8 antibacterial agents. The antibiotic susceptibility patterns of the isolates of the several antibiotics tested following the CLSI (Clinical and Laboratory Standards Institute) guideline [26] showed that the isolates displayed highest resistance to ampicillin (100%). The following is the order of the level of resistance exhibited against the remaining antibiotics; amikacin (96%), gentamycin (96%), ceftazidime (96%), tetracycline (92%), ciprofloxacin (85%), trimethoprim (84%), norflaxacin (62%). However, the isolates were mostly susceptible to norflaxacin.

3.3. Multiple Antibiotic-Resistance Index (MARI)

MARI of the isolates were expressed using the formula MARI = $w/(uxv)$, as explained above. For the sampling site, the multiple antibiotic resistance index (MARI) is estimated at 0.0514.

The summation score was obtained from the total sum of MAR Index isolates from each sampling point, and the MARI value was calculated for the six sampling points.

4. Discussion

This study investigated the occurrence of potentially pathogenic strains of *E. coli* recovered from beach water samples. Among the total number of presumptive *E. coli* isolates screened, 73 (68%) were confirmed positive by molecular techniques, in accordance with the report of Whitman et al. [28]. The presence of these bacteria in beach water poses a high risk with regard to human contact with this water, and there are certain factors which may be responsible for the fewer number of confirmed pathogens. There is the tendency of a low survival rate due to the depth of the beach water and rapid movement of the sea waves, with a possibly high level of dilution involved [29]. Moreover, there appears to be a higher level of fecal contamination near the sea shore around the sampling points, due to the high turbidity [30,31]. In general, the presence of pathogenic *E. coli* obtained from all sampling points of this recreational facility can pose serious health risks to both tourists and bathers. A study by Tsai et al. [32] suggested that certain pedigrees of *E. coli* have adapted and become accustomed to the different aquatic milieu, and this corroborates with our results. It was observed that bacterial counts from the sampling points where wastewater was being discharged into the beach had the highest number of positive isolates during the spring season and festive period, and our findings support the report of de Carvalho and Neto [33]. Although there are many probable sources of contamination, sewage treatment plants (STP) have become a constant source of beach pollution in respect of the quality of the final effluents that are released into receiving waters [34].

The result from the PCR products of the 260 presumptive *E. coli* isolates screened is; 107 isolates were positive, and our result is in agreement with the report of da Costa Andrade et al. [35]. Another study by Partyka et al. [36] also identified *E. coli* from beach water, and this is also in line with our result. From the eight different *E. coli* pathotypes screened for, three groups of *E. coli* pathotypes were identified as belonging to the two categories, InPEC (EPEC and EIEC) and ExPEC (NMEC) and the frequencies of detection ranged between 10% [InPEC] and 45% [ExPEC]. The molecular identification of *E. coli* pathogens in beach water poses high risk to the people in that area who use the beach for recreational activity. This study showed that 11 (10.3%) of the 73 positive strains of *E. coli* belonged to enteropathogenic *E. coli*. A study by Byappanahalli et al. [37] has also reported the presence of EPEC strains in beach water, and this is also in line with our result. Another study by Maloo [38], carried out in India, also identified various pathotypes of *E. coli* recovered from beach water, and our report is in line with their findings. The order of the percentage of phenotypic resistance levels exhibited by the isolates against the antibiotics is as follows: ampicillin (100%), amikacin (96%), gentamycin (96%), ceftazidime (96%), tetracycline (92%), ciprofloxacin (85%), trimethoprim (84%), and norflaxacin (62%). However, the isolates were mostly susceptible to norflaxacin. A study conducted by Stoll et al. [39] in Germany and Australia revealed a high resistance rate in *E. coli* isolates recovered from surface water samples that were resistant against ampicillin and tetracycline, and our result is in accordance with their report. A high percentage of the phenotypic resistance observed in the *E. coli* isolated could either be from the origin of WWTP or agricultural waste (poultry droppings), as most of the final effluents have been discharged into water bodies [40,41].

A multiple antibiotic resistance index (MARI) was carried out in order to evaluate or assess health risks that were concomitant with the rise and spread of multidrug resistance in the environment. The MARI value of 0.2 (arbitrary) was utilized to distinguish between low and high risk to public health. In addition, a MARI value above 0.2 proposed that the pathogenic strain of bacteria originated from an environment which was highly contaminated or which had high levels of antibiotics usage [19,26]. From our study, the MARI value (0.05) obtained for the isolates was less than 0.2, signifying that the isolates originated from environments with minimal antimicrobial use. The low MARI value estimated in this study provides an opportunity for further research in this area. This could be as a result of unsuitable use of antibiotics among the populace in the study area, and any greater MARI value obtained will suggest exposure to antimicrobial pressure, which may perhaps eventually lead to an increase in multidrug resistance.

5. Conclusions

This research demonstrates that the aquatic environs of the Nahoon beach are potential reservoirs of pathogenic *E. coli* strains that may probably combine a high level of antimicrobial resistance. This is an indication of the pressure mount by antimicrobial usage and poses a serious public health risk to humans upon exposure, consequently, presenting a public health hazard to the people around where the beach is located.

Author Contributions: A.I.O. and K.E.E. conceived and designed the experiments; K.E.E. performed the experiments; K.E.E. and M.A.A. analyzed the data; A.I.O. and M.A.A. contributed reagents/materials/analysis tools; K.E.E. and A.I.O. wrote the paper.

Funding: This research received funding from the South Africa Medical Research Council (SAMRC).

Acknowledgments: The authors would like to express sincere appreciation to the South Africa Medical Research Council (SAMRC) and the University of Fort Hare for their financial support.

Conflicts of Interest: The authors declare no conflict of interest.

References

1. Hlavsa, M.C.; Roberts, V.A.; Anderson, A.R.; Hill, V.R.; Kahler, A.M.; Orr, M.; Garrison, L.E.; Hicks, L.A.; Newton, A.; Hilborn, E.D.; et al. Surveillance for waterborne disease outbreaks and other health events associated with recreational water—United States, 2007–2008. *MMWR Surveill. Summ.* **2011**, *60*, 1–32. [PubMed]

2. Gorde, S.P.; Jadhav, M.V. Assessment of water quality parameters: A review. *J. Eng. Res. Appl.* **2013**, *6*, 2029–2035.

3. Okoh, A.I.; Odjadjare, E.E.; Igbinosa, E.O.; Osode, A.N. Wastewater treatment plants as a source of microbial pathogens in receiving watersheds. *Afr. J. Biotechnol.* **2007**, *6*. [CrossRef]

4. Lebaron, P.; Cournoyer, B.; Lemarchand, K.; Nazaret, S.; Servais, P. Environmental and human pathogenic microorganisms. In *Environmental Microbiology: Fundamentals and Applications*; Springer: Dordrecht, The Netherlands, 2015; pp. 619–658.

5. Borowy, I. Global health and development: Conceptualizing health between economic growth and environmental sustainability. *J. Hist. Med. Allied Sci.* **2012**, *68*, 451–485. [CrossRef] [PubMed]

6. DuBois, B.B. Beaches, People, and Change: A Political Ecology of Rockaway Beach after Hurricane Sandy. Ph.D. Thesis, The City University of New York, New York, NY, USA, 2016.

7. Brown, A.C.; McLachlan, A. Sandy shore ecosystems and the threats facing them: Some predictions for the year 2025. *Environ. Conserv.* **2002**, *29*, 62–77. [CrossRef]

8. Cloutier, D.D. Microbial Communities and the Diverse Ecology of Fecal Indicators at Lake Michigan Beaches. Ph.D. Thesis, The University of Wisconsin-Milwaukee, Milwaukee, WI, USA, 2017.

9. Massoud, M.A.; Tarhini, A.; Nasr, J.A. Decentralized approaches to wastewater treatment and management: Applicability in developing countries. *J. Environ. Manag.* **2009**, *90*, 652–659. [CrossRef] [PubMed]

10. Johannessen, G.S.; Wennberg, A.C.; Nesheim, I.; Tryland, I. Diverse land use and the impact on (irrigation) water quality and need for measures—A case study of a Norwegian river. *Int. J. Environ. Res. Public Health* **2015**, *12*, 6979–7001. [CrossRef] [PubMed]

11. Okoh, A.I.; Sibanda, T.; Gusha, S.S. Inadequately Treated Wastewater as a Source of Human Enteric Viruses in the Environment. *Int. J. Environ. Res. Public Health* **2010**, *7*, 2620–2637. [CrossRef] [PubMed]

12. Soller, J.A.; Schoen, M.E.; Bartrand, T.; Ravenscroft, J.E.; Ashbolt, N.J. Estimated human health risks from exposure to recreational waters impacted by human and non-human sources of faecal contamination. *Water Res.* **2010**, *44*, 4674–4691. [CrossRef] [PubMed]

13. Watkins, E.R.; Maiden, M.C.; Gupta, S. Metabolic competition as a driver of bacterial population structure. *Future Microbiol.* **2016**, *11*, 1339–1357. [CrossRef] [PubMed]

14. Mellata, M. Human and avian extraintestinal pathogenic *Escherichia coli*: Infections, zoonotic risks, and antibiotic resistance trends. *Foodborne Pathog. Dis.* **2013**, *10*, 916–932. [CrossRef] [PubMed]

15. Titilawo, Y.; Obi, L.; Okoh, A. Occurrence of virulence gene signatures associated with diarrhoeagenic and non-diarrhoeagenic pathovars of *Escherichia coli* isolates from some selected rivers in South-Western Nigeria. *BMC Microbiol.* **2015**, *15*, 204. [CrossRef] [PubMed]

16. Richards, V.P.; Lefébure, T.; Bitar, P.D.P.; Dogan, B.; Simpson, K.W.; Schukken, Y.H.; Stanhope, M.J. Genome based phylogeny and comparative genomic analysis of intra-mammary pathogenic *Escherichia coli*. *PLoS ONE* **2015**, *10*, e0119799. [CrossRef] [PubMed]

17. Thangamani, S.; Younis, W.; Seleem, M.N. Repurposing celecoxib as a topical antimicrobial agent. *Front. Microbiol.* **2015**, *6*, 750. [CrossRef] [PubMed]

18. Agga, G.E.; Arthur, T.M.; Durso, L.M.; Harhay, D.M.; Schmidt, J.W. Antimicrobial-resistant bacterial populations and antimicrobial resistance genes obtained from environments impacted by livestock and municipal waste. *PLoS ONE* **2015**, *10*, e0132586. [CrossRef] [PubMed]

19. Adefisoye, M.A.; Okoh, A.I. Identification and antimicrobial resistance prevalence of pathogenic *Escherichia coli* strains from treated wastewater effluents in Eastern Cape, South Africa. *Microbiol. Open* **2016**, *5*, 143–151. [CrossRef] [PubMed]

20. Wellington, E.M.; Boxall, A.B.; Cross, P.; Feil, E.J.; Gaze, W.H.; Hawkey, P.M.; Johnson-Rollings, A.S.; Jones, D.L.; Lee, N.M.; Otten, W.; et al. The role of the natural environment in the emergence of antibiotic resistance in gram-negative bacteria. *Lancet Infect. Dis.* **2013**, *13*, 155–165. [CrossRef]

21. Rizzo, L.; Manaia, C.; Merlin, C.; Schwartz, T.; Dagot, C.; Ploy, M.C.; Michael, I.; Fatta-Kassinos, D. Urban wastewater treatment plants as hotspots for antibiotic resistant bacteria and genes spread into the environment: A review. *Sci. Total Environ.* **2013**, *447*, 345–360. [CrossRef] [PubMed]

22. Torres, B.; Jaenecke, S.; Timmis, K.N.; García, J.L.; Díaz, E. A dual lethal system to enhance containment of recombinant micro-organisms. *Microbiology* **2003**, *149*, 3595–3601. [CrossRef] [PubMed]

23. Finley, R.L.; Collignon, P.; Larsson, D.J.; McEwen, S.A.; Li, X.Z.; Gaze, W.H.; Reid-Smith, R.; Timinouni, M.; Graham, D.W.; Topp, E. The scourge of antibiotic resistance: The important role of the environment. *Clin. Infect. Dis.* **2013**, *57*, 704–710. [CrossRef] [PubMed]

24. Ventola, C. The antibiotic resistance crisis: Part 1: Causes and threats. *Pharm. Ther.* **2015**, *40*, 277.

25. Cagney, C.; Crowley, H.; Duffy, G.; Sheridan, J.J.; O'Brien, S.; Carney, E.; Anderson, W.; McDowell, D.A.; Blair, I.S.; Bishop, R.H. Prevalence and numbers of *Escherichia coli* O157: H7 in minced beef and beef burgers from butcher shops and supermarkets in the Republic of Ireland. *Food Microbiol.* **2004**, *21*, 203–212. [CrossRef]

26. Clinical and Laboratory Standards Institute. *Performance Standards for Antimicrobial Susceptibility Testing; Twenty-Fourth Informational Supplement*; Clinical and Laboratory Standards Institute: Wayne, NY, USA, 2014.

27. Krumperman, P.H. Multiple antibiotic resistance indexing of *Escherichia coli* to identify high-risk sources of fecal contamination of foods. *Appl. Environ. Microbiol.* **1983**, *46*, 165–170. [PubMed]

28. Whitman, R.L.; Nevers, M.B.; Korinek, G.C.; Byappanahalli, M.N. Solar and temporal effects on *Escherichia coli* concentration at a Lake Michigan swimming beach. *Appl. Environ. Microbiol.* **2004**, *70*, 4276–4285. [CrossRef] [PubMed]

29. Wu, M.Z.; O'Carroll, D.M.; Vogel, L.J.; Robinson, C.E. Effect of low energy waves on the accumulation and transport of fecal indicator bacteria in sand and pore water at freshwater beaches. *Environ. Sci. Technol.* **2017**, *51*, 2786–2794. [CrossRef] [PubMed]

30. Rao, G.; Eisenberg, J.N.; Kleinbaum, D.G.; Cevallos, W.; Trueba, G.; Levy, K. Spatial variability of *Escherichia coli* in rivers of Northern Coastal Ecuador. *Water* **2015**, *7*, 818–832. [CrossRef] [PubMed]

31. Vogel, L.J.; O'Carroll, D.M.; Edge, T.A.; Robinson, C.E. Release of *Escherichia coli* from foreshore sand and pore water during intensified wave conditions at a recreational beach. *Environ. Sci. Technol.* **2016**, *50*, 5676–5684. [CrossRef] [PubMed]

32. Tsai, Y.L.; Palmer, C.J.; Sangermano, L.R. Detection of *Escherichia coli* in sewage and sludge by polymerase chain reaction. *Appl. Environ. Microbiol.* **1993**, *59*, 353–357. [PubMed]

33. De Carvalho, D.G.; Neto, J.A.B. Microplastic pollution of the beaches of Guanabara Bay, Southeast Brazil. *Ocean Coast. Manag.* **2016**, *128*, 10–17. [CrossRef]

34. Schoen, M.E.; Ashbolt, N.J. Assessing pathogen risk to swimmers at non-sewage impacted recreational beaches. *Environ. Sci. Technol.* **2010**, *44*, 2286–2291. [CrossRef] [PubMed]

35. Da Costa Andrade, V.; Zampieri, B.D.B.; Ballesteros, E.R.; Pinto, A.B.; De Oliveira, A.J.F.C. Densities and antimicrobial resistance of *Escherichia coli* isolated from marine waters and beach sands. *Environ. Monit. Assess.* **2015**, *187*, 342. [CrossRef] [PubMed]

36. Partyka, M.L.; Bond, R.F.; Chase, J.A.; Atwill, E.R. Monitoring bacterial indicators of water quality in a tidally influenced delta: A Sisyphean pursuit. *Sci. Total Environ.* **2017**, *578*, 346–356. [CrossRef] [PubMed]
37. Byappanahalli, M.N.; Nevers, M.B.; Whitman, R.L.; Ishii, S. Application of a microfluidic quantitative polymerase chain reaction technique to monitor bacterial pathogens in beach water and complex environmental matrices. *Environ. Sci. Technol. Lett.* **2015**, *2*, 347–351. [CrossRef]
38. Maloo, A.; Fulke, A.B.; Mulani, N.; Sukumaran, S.; Ram, A. Pathogenic multiple antimicrobial resistant *Escherichia coli* serotypes in recreational waters of Mumbai, India: A potential public health risk. *Environ. Sci. Pollut. Res.* **2017**, *24*, 11504–11517. [CrossRef] [PubMed]
39. Stoll, C.; Sidhu, J.P.S.; Tiehm, A.; Toze, S. Prevalence of clinically relevant antibiotic resistance genes in surface water samples collected from Germany and Australia. *Environ. Sci. Technol.* **2012**, *46*, 9716–9726. [CrossRef] [PubMed]
40. Laroche-Ajzenberg, E.; Flores Ribeiro, A.; Bodilis, J.; Riah, W.; Buquet, S.; Chaftar, N.; Pawlak, B. Conjugative multiple-antibiotic resistance plasmids in *Escherichia coli* isolated from environmental waters contaminated by human faecal wastes. *J. Appl. Microbiol.* **2015**, *118*, 399–411. [CrossRef] [PubMed]
41. Corsi, S.R.; Borchardt, M.A.; Carvin, R.B.; Burch, T.R.; Spencer, S.K.; Lutz, M.A.; McDermott, C.M.; Busse, K.M.; Kleinheinz, G.T.; Feng, X.; et al. Human and bovine viruses and bacteria at three Great Lakes beaches: Environmental variable associations and health risk. *Environ. Sci. Technol.* **2015**, *50*, 987–995. [CrossRef] [PubMed]

International Journal of
*Environmental Research
and Public Health*

Article

Susceptibility of Multidrug-Resistant Bacteria, Isolated from Water and Plants in Nigeria, to Ceragenins

Marjan M. Hashemi [1], Augusta O. Mmuoegbulam [2], Brett S. Holden [1], Jordan Coburn [1], John Wilson [1], Maddison F. Taylor [1], Joseph Reiley [1], Darius Baradaran [1], Tania Stenquist [1], Shenglou Deng [1] and Paul B. Savage [1,*]

[1] Department of Chemistry and Biochemistry, Brigham Young University, Provo, UT 84602, USA; marjan.mhashemi@gmail.com (M.M.H.); bholden222@yahoo.com (B.S.H.); jccoburn2@gmail.com (J.C.); johnmartinwilson@gmail.com (J.W.); maddifaytaylor@gmail.com (M.F.T.); joseph.reiley@hotmail.com (J.R.); darius.baradaran@gmail.com (D.B.); tnance99@aol.com (T.S.); shengloudeng@hotmail.com (S.D.)

[2] Department of Microbiology, University of Calabar, Calabar PMB 1115, Nigeria; oluchidinma@yahoo.com

* Correspondence: paul_savage@byu.edu; Tel.: +1-801-422-4020

Received: 12 October 2018; Accepted: 3 December 2018; Published: 6 December 2018

Abstract: The continuous emergence of multidrug resistant pathogens is a major global health concern. Although antimicrobial peptides (AMPs) have shown promise as a possible means of combatting multidrug resistant strains without readily engendering resistance, costs of production and targeting by proteases limit their utility. Ceragenins are non-peptide AMP mimics that overcome these shortcomings while retaining broad-spectrum antimicrobial activity. To further characterize the antibacterial activities of ceragenins, their activities against a collection of environmental isolates of bacteria were determined. These isolates were isolated in Nigeria from plants and water. Minimum inhibitory concentrations (MICs) and minimum bactericidal concentrations (MBCs) of selected ceragenins and currently available antimicrobials against these isolates were measured to determine resistance patterns. Using scanning electron microscopy (SEM), we examined the morphological changes in bacterial membranes following treatment with ceragenins. Finally, we investigated the effectiveness of ceragenins in inhibiting biofilm formation and destroying established biofilms. We found that, despite high resistance to many currently available antimicrobials, including colistin, environmental isolates in planktonic and biofilm forms remain susceptible to ceragenins. Additionally, SEM and confocal images of ceragenin-treated cells confirmed the effective antibacterial and antibiofilm activity of ceragenins.

Keywords: ceragenin; multidrug-resistant bacteria; biofilm; antimicrobial peptides; colistin

---

1. Introduction

The discovery and widespread use of antibiotics was one of the most important advances in medicine. These drugs were heralded for their effectiveness, and, as a result, began to be prescribed across the world. However, widespread use of antibiotics has resulted in the generation of mutational resistance in bacteria as well as identification of adaptational resistance mechanisms. These have led to the rise of hyper-resistant bacteria, often called superbugs [1,2]. Today, the phenomenon of antibiotic resistance has become a global public health concern, with 700,000 deaths across the globe each year attributed to antimicrobial resistance. This count is expected to reach 10 million by 2050 as the decreasing effectiveness of available market drugs continues to compound this problem [3]. Of particular concern is the widespread use of antimicrobial agents in food animals, which may be a major source of antimicrobial resistance that can spread drug-resistant pathogens to humans directly or through the environmental pollution of farm effluents [4].

Endogenous antimicrobial peptides (AMPs) are a key component of the body's innate immune system, which is critical in fighting bacteria, fungi, and lipid-enveloped viruses. AMPs are typically cationic and amphiphilic in nature, which facilitates targeted association with negatively-charged pathogenic membranes, causing membrane disruption and cell death [5,6]. Interestingly, evidence has shown that bacteria are unable to achieve high levels of resistance to AMPs, making this an important area of antimicrobial research. However, AMPs can be expensive to manufacture synthetically and can be degraded in the presence of bacterial and host proteases [7,8]. In order to circumvent these challenges, ceragenins were developed from a common bile acid as non-peptide mimics of AMPs. Structure of ceragenins are shown in Figure 1. Ceragenins are cationic and amphiphilic, giving them analogous antimicrobial properties to AMPs. They are relatively inexpensive to produce and have shown potent activity against a broad spectrum of organisms. Of particular note is that ceragenins are active against methicillin-resistant *Staphylococcus aureus* [9], colistin-resistant *Klebsiella pneumoniae* [10], and fluconazole-resistant *Candida albicans* [11] and *Candida auris* [12]. To date, no bacteria have been shown to achieve high levels of resistance to ceragenins [13,14]. Ceragenins appear to be well tolerated in tissues and exhibit both the antimicrobial and secondary properties that are characteristic of many AMPs. Because of their promising therapeutic properties, ease of production, and possible synergistic effects, ceragenins represent an important target of study for further clinical development [15–17]. In this study, the antimicrobial resistance patterns of ten Nigerian bacterial strains isolated from the environment were determined by selected ceragenins and compared to commonly used antibiotics. The effects of ceragenins on the cell membranes of these isolates were observed by scanning electron microscopy (SEM). Additionally, we assessed the potential of selected ceragenins to eradicate biofilms formed by multidrug-resistant environmental isolates.

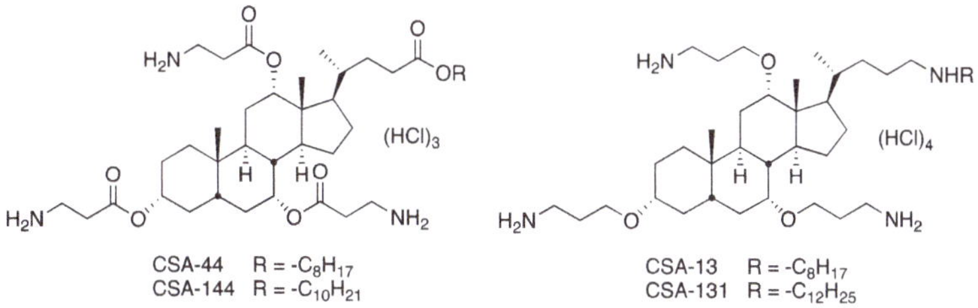

Figure 1. Structures of ceragenins CSA-44, CSA-144, CSA-13 and CSA-131.

2. Materials and Methods

Ceragenins CSA-13, CSA-131, CSA-44, and CSA-144 were synthesized from a cholic acid scaffolding as previously described [18]. Colistin, chlorhexidine, kanamycin, polymyxin B, erythromycin, tetracycline, vancomycin, and ampicillin were purchased from Sigma–Aldrich (St Louis, MO, USA).

2.1. Isolation and Maintenance of Bacterial Isolates

Bacteriological analyses of water samples, including heterotrophic plate count (HPC), fecal coliform (FC) and total coliform count (TCC), were determined using both the direct pour plate method and membrane filtration techniques. No serial dilution was carried out. Nutrient agar medium was used for heterotrophic bacteria plate counts. MacConkey agar (MCA) was used for total coliform counts, and membrane fecal coliform (MF-C) agar medium was used for fecal coliform counts. The plates were inoculated in triplicate. Inoculations for HPC and TCC were conducted by adding 1 mL of sample to each plate. For membrane filtration, 100 mL of each water sample was filtered through a 0.45 μm membrane filter before aseptic transfer of the membrane onto MF-C agar or MCA for FC and TCC respectively. No dilution was used in either method. MCA was used for isolation of lactose fermenters (coliforms).

For isolates taken from plants, natural rubber latex (1 mL) was added to sterile distilled water (9 mL) and then serially diluted using a sterile micropipette. One gram of deteriorated rubber latex was added to sterile distilled water (100 mL) in a conical flask. The combination was mixed well, and an aliquot from this mixture was serially diluted in water (10^{-1} to 10^{-10} dilution). Selected dilutions were then used for the inoculation of agar plates. Each sample was plated in triplicate using the pour plate method. Inoculated plates were incubated for 24 h at 35–37 °C. Colonies were enumerated using a colony counter for total heterotrophic bacteria and total coliform counts. Discrete colonies were sub-cultured onto fresh nutrient agar plates aseptically to obtain pure cultures of the isolates and were stored in a refrigerator at 4 °C for further identification.

2.2. Identification of Isolates

Bacterial isolates were characterized based on microscopic appearance, colony morphology, gram staining reactions, and appropriate biochemical tests based on Bergey's Manual of Determinative Bacteriology and as described by Cheesbrough [19]. The isolates were identified by comparing their characteristics with those of known taxa, as described by Cruickshank et al. [20] and Holt [21]. Table 1 shows the source of ten isolates used in this study.

Table 1. Isolation source of bacteria used in this study.

	Strains	**Isolation Source**
1	*Enterococcus* spp.	Rubber plant
2	*Actinomyces odontolyticus*	Rubber plant
3	*Actinomyces israelii*	Rubber plant
4	*Acetobacter* spp.	Rubber plant
5	*Moraxella* spp.	Rubber plant
6	*Enterobacter dissolvens*	Water
7	*Pseudomonas* spp.	Rubber plant
8	*Klebsiella pneumoniae*	Water
9	*Legionella pneumophila*	Water
10	*Erwinia Stewartii*	Water

2.3. Susceptibility Testing

Minimum inhibitory concentrations (MICs) were determined using a broth microdilution method in a 96-well microdilution plate according to the Clinical Laboratory Standards Institute protocol [22]. Briefly, 96-well plates were prepared with individual wells containing doubling concentrations of selected ceragenins, including CSA-13, CSA-44, CSA-131, and CSA-144 in the appropriate culture medium for a total volume of 100 µL. A selection of commercial antimicrobials including chlorhexidine, kanamaycin, colistin, polymyxin B, erythromycin, tetracycline, vancomycin and ampicillin was also used for comparison. An inoculation of 100 µL at 10^6 CFU/mL was added to each well. Each of the 10 isolates was tested in duplicate and each plate contained positive and negative controls. Plates were incubated for 24 h at 37 °C. After incubation, results were obtained by examining wells for turbidity. Minimum bactericidal concentrations (MBCs) were determined by taking 10 µL from each well and plating on agar media. The MBC was defined as the lowest concentration of an antibacterial agent giving no visible colonies after 24 h incubation at 37 °C [10].

2.4. Scanning Electron Microscopy (SEM)

To observe the effect of ceragenins on cell membranes, selected isolates were cultured to mid-log phase and washed three times with PBS. Bacteria were re-suspended in PBS (OD_{600} = 0.2). CSA-131 (25 µg/mL) was then added and the mixtures were incubated at 37 °C for 1 h. A control was prepared by incubating the bacterial suspension without adding CSA-131. After collection via centrifugation, cells were washed with PBS three times. Gluteraldehyde (2.5% (w/v)) was added to fix the cells at 4 °C overnight. Resulting material was washed five times with PBS at 5000 rpm for 10 min using a microhematocrit centrifuge (Hettich Mikro 20, Hettich, Tuttlingen, Germany) to remove the glutaraldehyde. Osmium tetroxide (0.5 mL) was used as a second fixative reagent, and samples were stored at room temperature under a protective laboratory hood system for 2–3 h. Cells were washed with PBS five times at 14,000 rpm for 8 min. A graded ethanol series including 10%, 30%, 50%, 70%, 90% (1 time), 100% (3 times) and HMDS (2 times) for 15 min each was used to dehydrate the cells. Samples were collected by centrifuge each time and the supernatant was discarded after each centrifugation. Finally, dried bacterial specimens were sputter-coated with 5–10 nm of a Gold-Palladium alloy and visualized under a scanning electron microscope (FEI Helios NanoLab 600 SEM/FIB, Hillsboro, Oregon, USA) [12].

2.5. Biofilm Study Using XTT Assay

Biofilms of each isolate were grown for 48 h in separate wells in 96-well plates. Planktonic cells were then removed by washing three times with PBS. The biofilm-containing wells were treated with CSA-131 (100 μg/mL) and incubated for another 24 h. After another PBS wash, 100 μL of a mixed solution of 0.5 mg/mL 2,3-bis (2-methoxy-4-nitro-5-sulfophenyl)-5-[(phenylamino) carbonyl]-2H-tetrazolium hydroxide (XTT) and 10 mM menadione in acetone was added to each well. Plates were covered in aluminum foil and incubated for 2 to 3 h at 37 °C. Remaining solution was removed from each well and remaining dye in the wells was then quantified with a microtiter plate reader at 490 nm. Optical density results of test wells were compared with controls to determine the percent of biofilm remaining in each [23].

2.6. Confocal Laser Scanning Microscopy

Biofilms of *Acetobacter* spp. were formed on glass slides by complete submersion of slides in inoculated media (50 mL) and incubation for 48 h. Selected slides were treated with CSA-131 (100 μg/mL) and slides were further incubated at 37 °C for 24 h. Following incubation, glass slides were carefully removed from the solutions and rinsed three times with PBS. Using protocols of a BacLight Live/Dead Viability Kit (L13152, Molecular Probes, Inc), biofilms were stained and further imaged by a confocal laser scanning microscope (Olympus FluoView FV1000) at ×60 magnification [12].

3. Results and Discussion

3.1. Susceptibility of Isolated Bacteria

In an initial set of experiments, the MICs of several common antimicrobials against ten bacterial isolates were determined (Table 2). Gram-positive bacteria, including *Actinomycins* species, were relatively susceptible to chlorhexidine, vancomycin and ampicillin; however, all tested Gram-negative isolates showed very high MICs with selected antimicrobials, including chlorhexidine, kanamycin, colistin, polymyxin B, erythromycin and tetracyclin. For example, the MIC of *Enterococcus* species with chlorhexidine, kanamycin, colistin, polymyxin B, erythromycin and tetracyclin was 64, 100, >100, 100, 8 and 32 μg/mL, respectively. Of particular note is the MIC of colistin, which was more than 100 μg/mL with most of isolates tested. Colistin is generally considered the therapeutic of last resort for multidrug-resistant Gram-negative bacterial infections, so the presence of highly colistin-resistant isolates in this study is alarming.

Table 2. Minimum inhibitory concentrations (MICs) (µg/mL) of ten isolates with common antibiotics.

Strains	Chl	Kan	Col	Pol B	Ery	Tet	Van	Amp
Enterobacter dissolvens	8	64	>100	100	2	2	nm	nm
Erwinia stewartii	8	2	100	100	1	2	nm	nm
Enterococcus spp.	64	100	>100	100	8	32	nm	nm
Pseudomonas spp.	32	16	>100	>100	16	8	nm	nm
Klebsiella pneumoniae	32	32	>100	100	16	1	nm	nm
Acetobacter spp.	32	32	16	8	32	16	nm	nm
Moraxella spp.	64	64	>100	100	32	32	nm	nm
Legionella pneumophila	64	64	32	16	8	4	nm	nm
Actinomyces odontolyticus	4	nm	nm	nm	nm	nm	1	1
Actinomyces israelii	4	nm	nm	nm	nm	nm	2	2

Chl: chlorhexidine; Kan: kanamycin; Col: colistin; Pol B: polymyxin B; Ery: erythromycin; Tet: tetracycline; Van: vancomycin; Amp: ampicillin. nm: not measured.

To determine the activity of ceragenins against multidrug-resistant isolates, the MICs of selected ceragenins CSA-13, CSA-44, CSA-131 and CSA-144 were measured. These results are shown in Table 3. Ceragenins retained activity against all multidrug-resistant strains and showed low MICs compared to the commonly used antimicrobials. The MIC of CSA-13 and CSA-131 of 1–2 µg/mL with all of the highly multidrug-resistant isolates was of particular note. This result is consistent with our previous studies showing that ceragenins are highly active against methicillin-resistant *Staphylococcus aureus* [9] and colistin-resistant *Klebsiella pneumoniae* [10].

Table 3. Comparison of the MIC (minimum bactericidal concentrations (MBC)) (µg/mL) of ten isolates to selected ceragenins.

Strains	CSA-13	CSA-44	CSA-131	CSA-144
Enterobacter dissolvens	1(8)	2(10)	2(8)	2(10)
Erwinia stewartii	2(8)	4(10)	2(8)	4(10)
Actinomyces odontolyticus	1(1–2)	1(1)	1(1)	2(2)
Actinomyces israelii	2(4)	1(2)	2(4)	2(4)
Pseudomonas spp.	1(1)	4(4)	1(1)	4(4)
Legionella pneumophila	8(8)	4(8)	4(4)	16(32)
Enterococcus spp.	16(32)	8(8)	4(32)	32(100)
Moraxella spp.	10(32)	4(16)	4(16)	24(100)
Acetobacter spp.	2(32)	4(64)	2(32)	4(64)
Klebsiella pneumoniae	2(4)	4(8)	2(4)	4(4)

To confirm that the activity of ceragenins is bactericidal, MBCs of the same ceragenins were measured with the multidrug-resistant isolates. All tested ceragenins were found to be bactericidal at a range of 1–100 µg/mL, exhibiting bactericidal activity against strains such as *Pseudomonas* and *Actinomyces* spp. at the same concentrations as the corresponding MICs, suggesting that antibacterial activity of ceragenins are likely bactericidal rather than bacteriostatic.

3.2. Scanning Electron Microscopy (SEM)

To visualize the effect of ceragenins on the cell membrane, selected isolates were treated with a lead ceragenin, CSA-131, and their morphology was studied via SEM. Scanning electron photomicrographs of *Klebsiella pneumoniae*, *Moraxella* and *Legionella pneumophila* are shown in Figure 2. In the control without ceragenin treatment (Figure 2A,C,E), cells maintained normal morphology. In contrast, there were significant alterations in the morphology of cells treated with CSA-131 (Figure 2B,D,F). Treated cells are characterized by disruptions in the cell membrane along with increased roughness and wrinkling on the cell surface, confirming the membrane activity of ceragenins. Importantly, reported morphological changes in this study are consistent with previous reports describing changes in the structure of bacteria and fungi after treatment with ceragenins [24,25].

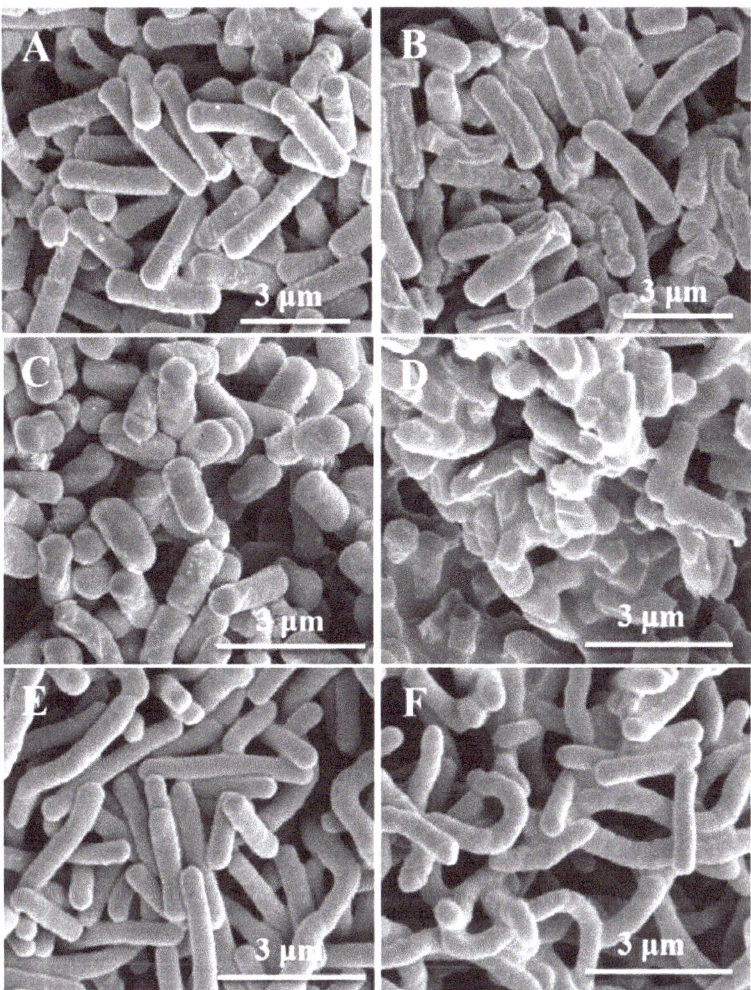

Figure 2. Scanning electron photomicrograph of untreated (**A**) and treated (**B**) *Klebsiella pneumoniae*, untreated (**C**) and treated (**D**) *Moraxella* spp., untreated (**E**) and treated (**F**) *Legionella pneumophila* with 25 µg/mL CSA-131.

3.3. Determination of Susceptibility Profiles of Bacterial Biofilms

It is well established that biofilms have greater resistance to antimicrobials than planktonic cells [26]. There are several explanations for this increase in resistance. The extracellular matrix surrounding the cells in the biofilm prevents targeting and subsequent penetration by antimicrobials. The reduced growth rate of cells in biofilms, compared to planktonic cells, increases resistance to antimicrobials that target growth-specific factors. Other mechanisms include the inactivation or degradation of antimicrobials and efflux pumps that remove antimicrobials from the cells [27]. Despite these challenges, ceragenins have been shown to permeate the biofilm extracellular matrix, due to their relatively small size, and eradicate biofilms at relatively low concentrations. This activity is likely due to the mechanism of action of ceragenins, which is not dependent on the metabolic state of their targets [28].

To quantify the impact of ceragenins on biofilm formation by multidrug-resistant isolates, an XTT assay was performed. The XTT assay measures metabolic activity of cells in the biofilm following a change in color. Corresponding biofilm growth reduction for each strain was calculated compared to a negative control that was not treated with any drugs. As shown in Figure 3, all representative ceragenins demonstrated strong antibiofilm activity against both Gram-positive and Gram-negative multidrug-resistant isolates and caused a substantial reduction of growth. Treatment of *Moraxella* spp., *K. pneumoniae* and *L. pneumophila* decreased biofilm mass by more than 96% compared to the negative control. A previous study showed that in a comparison of CSA-13 with ciproflaxicin, CSA-13 was shown to have greater activity against established biofilms formed by methicillin-resistant *S. aureus* [29].

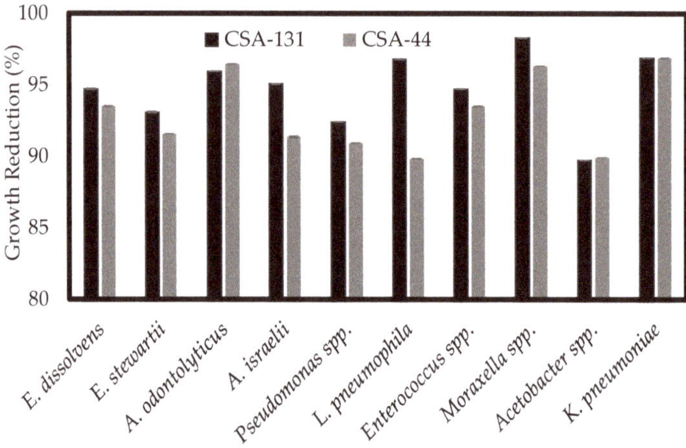

Figure 3. Reduction of established biofilms of ten isolates after 48 h incubation with CSA-131 or/and CSA-44 (100 µg/mL). Using the 2,3-bis (2-methoxy-4-nitro-5-sulfophenyl)-5-[(phenylamino) carbonyl]-2H-tetrazolium hydroxide (XTT) colorimetric based assay, metabolic activity of ceragenin-treated biofilms was measured and the percent of growth reduction was calculated in comparison to an untreated biofilm (control).

3.4. Confocal Laser Scanning of Biofilms

To visualize antibiofilm properties of ceragenins, biofilms of *Acetobacter* spp. were treated with a lead ceragenin, CSA-131, and prepared for confocal microscopy. Confocal images are shown in Figure 4. In the images, a lack of biofilm is seen in some areas, which could be due to sample preparation in which slides were rinsed prior to staining to remove loosely adhered and planktonic organisms. Overall, as expected, untreated biofilms showed expected aggregates of live cells (green dye, Figure 4A), while ceragenin-treated biofilms exhibited comparable aggregates of dead cells (red dye, Figure 4B). Lack of biofilm was observed more often in the treated than in the untreated cells, which highlights ceragenins' ability to destabilize established biofilms, facilitating their detachment from slide surfaces. Nagent, et al., [9] conducted a biofilm study using confocal microscopy and their images revealed that ceragenins efficiently penetrated established biofilms and led to cell death without significant alterations to the extracellular matrix. Additionally, a recent study demonstrated prolonged inhibition of biofilm formation on endotracheal tube surfaces when the tubes were coated with a CSA-131-containing hydrogel [30].

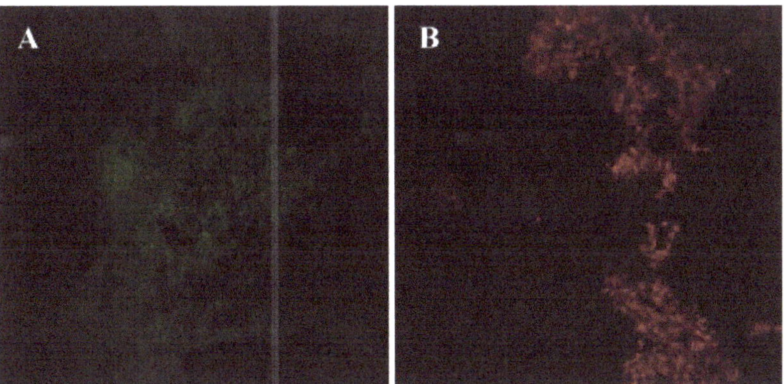

Figure 4. Confocal laser scanning micrographs (×60 magnification) of stained bacterial biofilms. Green: live cells; red: dead cells. (**A**) Untreated *Acetobacter* spp. (**B**) treated with CSA-131 (100 µg/mL).

4. Conclusions

Obstacles to the development of novel antimicrobial agents include concerns that generation of resistance to one antimicrobial agent may result in cross-resistance to other antimicrobials. Since higher organisms have co-evolved with bacteria, the mechanisms by which they control bacterial growth may provide guidance for development of antimicrobial agents to which bacteria do not readily generate resistance. AMPs represent one of the key means by which higher organisms control bacterial growth. Ceragenins mimic key AMP structural features, specifically, multiple cationic (positive) charges juxtaposed with hydrophobic structure. The studies presented herein demonstrate that even highly multidrug-resistant environmental isolates largely remain susceptible to ceragenins. Additionally, previous studies showed that CSA-13 toxicity is comparable to LL-37 in tested human keratinocytes and it is not toxic to HatCat cells at bactericidal concentrations [31].

The ceragenins tested in this study gave MICs in the single µg/mL range in spite of the high MICs of commonly used antimicrobials, including the last resort antibiotic colistin. SEM images gave results comparable to earlier studies, demonstrating that ceragenins interact with bacterial membranes. Morphological changes to Gram-negative bacterial membranes are a hallmark of the activity of many AMPs, and we have shown, via transmission electron microscopy (TEM) and atomic force microscopy (AFM), that similar changes occur in bacterial membranes upon treatment with a ceragenin [24,32]. Further characterization of the lead ceragenins, CSA-131 and CSA-44, demonstrated that reduction of growth in a preformed biofilm was also successful; however, the extent of reduction was much less compared to the inhibition activity against planktonic cells. Confocal images verified the antibiofilm activity of ceragenins that occurs through penetration of the compound into the extracellular matrix of the biofilm.

Multidrug resistance in Nigeria is on the rise [33]. The highly resistant nature of the Nigerian environmental isolates analyzed in this study suggests that the careful designing and adoption of a multi-sectoral antimicrobial resistance surveillance plan for research and diagnostic purposes should be implemented. Relevant ministries and governmental agencies should consider the following: registration and observation of production premises, particularly where food-producing animals are concerned; improved biosecurity compliance in food-animal environments; banning antibiotic use for animal growth promotion or prophylactic treatment in animal husbandry; implementation of a drug withdrawal period for food animals.

Author Contributions: M.M.H. and A.O.M. designed the experiments and wrote the manuscript. B.S.H., J.C., J.W., M.F.T., J.R., D.B., T.S., and S.D. performed the experiments and data analysis. P.B.S. supervised and edited the manuscript. P.B.S. is a paid consultant for N8 Medical and CSA Biotech.

Acknowledgments: Generous funding is acknowledged from N8 Medical, Inc., CSA Biotech and Brigham Young University.

Funding: This research received no external funding.

Conflicts of Interest: The authors declare no conflict of interest.

References

1. Padiyara, P.; Inoue, H.; Sprenger, M. Global governance mechanisms to address antimicrobial resistance. *Infect. Dis.* **2018**, *11*, 1–4. [CrossRef] [PubMed]

2. Aoki, W.; Kuroda, K.; Ueda, M. Next generation of antimicrobial peptides as molecular targeted medicines. *J. Biosci. Bioeng.* **2012**, *114*, 365–370. [CrossRef] [PubMed]

3. Simpkin, V.L.; Renwick, M.J.; Kelly, R.; Mossialos, E. Incentivising innovation in antibiotic drug discovery and development: Progress, challenges and next steps. *J. Antibiot.* **2017**, *70*, 1087–1096. [CrossRef] [PubMed]

4. Marshall, B.M.; Levy, S.B. Food animals and antimicrobials: Impacts on human health. *Clin. Microbiol. Rev.* **2011**, *24*, 718–733. [CrossRef]

5. Zhang, L.-J.; Gallo, R.L. Antimicrobial peptides. *Curr. Biol.* **2016**, *26*, R14–R19. [CrossRef]

6. Ganz, T. The role of antimicrobial peptides in innate immunity. *Integr. Comp. Biol.* **2003**, *43*, 300–304. [CrossRef]

7. Mangoni, M.L.; McDermott, A.M.; Zasloff, M. Antimicrobial peptides and wound healing: Biological and therapeutic considerations. *Exp. Dermatol.* **2016**, *25*, 167–173. [CrossRef]

8. Bahar, A.A.; Ren, D. Antimicrobial peptides. *Pharmaceuticals* **2013**, *6*, 1543–1575. [CrossRef]

9. Nagant, C.; Pitts, B.; Stewart, P.S.; Feng, Y.; Savage, P.B.; Dehaye, J.-P. Study of the effect of antimicrobial peptide mimic, CSA-13, on an established biofilm formed by Pseudomonas aeruginosa. *Microbiologyopen* **2013**, *2*, 318–325. [CrossRef]

10. Hashemi, M.M.; Rovig, J.; Weber, S.; Hilton, B.; Forouzan, M.M.; Savage, P.B. Susceptibility of colistin-resistant, Gram-negative bacteria to antimicrobial peptides and ceragenins. *Antimicrob. Agents Chemother.* **2017**, *61*. [CrossRef]

11. Durnaś, B.; Wnorowska, U.; Pogoda, K.; Deptuła, P.; Wątek, M.; Piktel, E.; Głuszek, S.; Gu, X.; Savage, P.B.; Niemirowicz, K.; et al. Candidacidal activity of selected ceragenins and human cathelicidin LL-37 in experimental settings mimicking infection sites. *PLoS ONE* **2016**, *11*, e0157242. [CrossRef]

12. Hashemi, M.M.; Rovig, J.; Holden, B.S.; Taylor, M.F.; Weber, S.; Wilson, J.; Hilton, B.; Zaugg, A.L.; Ellis, S.W.; Yost, C.D.; et al. Ceragenins are active against drug-resistant Candida auris clinical isolates in planktonic and biofilm forms. *J. Antimicrob. Chemother.* **2018**, *73*, 1537–1545. [CrossRef] [PubMed]

13. Pollard, J.E.; Snarr, J.; Chaudhary, V.; Jennings, J.D.; Shaw, H.; Christiansen, B.; Wright, J.; Jia, W.; Bishop, R.E.; Savage, P.B. In vitro evaluation of the potential for resistance development to ceragenin CSA-13. *J. Antimicrob. Chemother.* **2012**, *67*, 2665–2672. [CrossRef] [PubMed]

14. Hashemi, M.M.; Holden, B.S.; Savage, P.B. Ceragenins as non-peptide mimics of endogenous antimicrobial peptides. In *Fighting Antimicrobial Resistance*, 1st ed.; Budimir, A., Ed.; IAPC Publishing: Zagreb, Croatia, 2018; pp. 139–169, ISBN 978-953-56942-6-7.

15. Sinclair, K.; Pham, T.; Farnsworth, R.; Williams, D.L.; Loc-Carrillo, C.; Horne, L.A.; Ingebretsen, S.H.; Bloebaum, R.D. Development of a broad spectrum polymer-released antimicrobial coating for the prevention of resistant strain bacterial infections. *J. Biomed. Mater. Res. A* **2012**, *100*, 2732–2738. [CrossRef] [PubMed]

16. Bucki, R.; Niemirowicz, K.; Wnorowska, U.; Byfield, F.J.; Piktel, E.; Wątek, M.; Janmey, P.A.; Savage, P.B. Bactericidal activity of ceragenin CSA-13 in cell culture and in an animal model of peritoneal infection. *Antimicrob. Agents Chemother.* **2015**, *59*, 6274–6282. [CrossRef]

17. Olekson, M.; You, T.; Savage, P.B.; Leung, K.P. Ceragenin peptide-mimics inhibit biofilms and affect mammalian cell viability and migration in vitro. *FEBS Open Bio* **2017**, *7*, 953–967. [CrossRef]

18. Li, C.; Peters, A.S.; Meredith, E.L.; Allman, G.W.; Savage, P.B. Design and synthesis of potent sensitizers of Gram-negative bacteria based on a cholic acid scaffolding. *J. Am. Chem. Soc.* **1998**, *120*, 2961–2962. [CrossRef]

19. Cheesbrough, M. *District Laboratory Practice in Tropical Countries*; Cambridge University Press: Cambridge, UK, 2006; pp. 143–157, ISBN 978-0-521-67631-1.

20. Cruickshank, R.; Duguid, J.O.; Marmon, B.P.; Swain, R.H.A. *Medical Microbiology: The Practice of Medical Microbiology*, 12th ed.; Churchill Livingstone: Edinburg, NY, USA, 1980; p. 585, ISBN 978-0443011115.

21. Holt, J.G. *Bergey's Manual of Determinative Bacteriology*, 9th ed.; Williams & Wilkins: Baltimore, MD, USA, 2000; pp. 532–551, ISBN 978-0683006032.

22. Wikler, M.A. *Methods for Dilution Antimicrobial Susceptibility Tests for Bacteria that Grow Aerobically: Approved Standard*; Clinical and Laboratory Standards Institute: Wayne, PA, USA, 2006; ISBN 1-56238-783-9.

23. Hashemi, M.M.; Holden, B.S.; Taylor, M.F.; Wilson, J.; Coburn, J.; Hilton, B.; Nance, T.; Gubler, S.; Genberg, C.; Deng, S.; et al. Antibacterial and antifungal activities of poloxamer micelles containing ceragenin CSA-131 on ciliated tissues. *Molecules* **2018**, *23*, 596. [CrossRef]

24. Bucki, R.; Sostarecz, A.G.; Byfield, F.J.; Savage, P.B.; Janmey, P.A. Resistance of the antibacterial agent ceragenin CSA-13 to inactivation by DNA or F-actin and its activity in cystic fibrosis sputum. *J. Antimicrob. Chemother.* **2007**, *60*, 535–545. [CrossRef]

25. Ding, B.; Guan, Q.; Walsh, J.P.; Boswell, J.S.; Winter, T.W.; Winter, E.S.; Boyd, S.S.; Li, C.; Savage, P.B. Correlation of the antibacterial activities of cationic peptide antibiotics and cationic ateroid antibiotics. *J. Med. Chem.* **2002**, *45*, 663–669. [CrossRef]

26. Stewart, P.S. Antimicrobial tolerance in biofilms. *Microbiol. Spectr.* **2015**, *3*. [CrossRef]

27. Sandasi, M.; Leonard, C.; Viljoen, A. The in vitro antibiofilm activity of selected culinary herbs and medicinal plants against Listeria monocytogenes. *Lett. Appl. Microbiol.* **2010**, *50*, 30–35. [CrossRef]

28. Hashemi, M.M.; Holden, B.S.; Durnaś, B.; Bucki, R.; Savage, P.B. Ceragenins as mimics of endogenous antimicrobial peptides. *J. Antimicrob. Agents* **2017**, *3*, 141. [CrossRef]

29. Pollard, J.; Wright, J.; Feng, Y.; Geng, D.; Genberg, C.; Savage, P.B. Activities of ceragenin CSA-13 against established biofilms in an in vitro model of catheter decolonization. *Antiinfect. Agents Med. Chem.* **2009**, *8*, 290–294. [CrossRef]

30. Hashemi, M.M.; Rovig, J.; Bateman, J.; Holden, B.S.; Modelzelewski, T.; Gueorguieva, I.; von Dyck, M.; Bracken, R.; Genberg, C.; Deng, S.; et al. Preclinical testing of a broad-spectrum antimicrobial endotracheal tube coated with an innate immune synthetic mimic. *J. Antimicrob. Chemother.* **2018**, *73*, 143–150. [CrossRef] [PubMed]

31. Leszczyńska, K.; Namiot, D.; Byfield, F.J.; Cruz, K.; Zendzian-Piotrowska, M.; Fein, D.E.; Savage, P.B.; Diamond, S.; McCulloch, C.A.; Janmey, P.A. Antibacterial activity of the human host defence peptide LL-37 and selected synthetic cationic lipids against bacteria associated with oral and upper respiratory tract infections. *J. Antimicrob. Chemother.* **2013**, *68*, 610–618. [CrossRef] [PubMed]

32. Lai, X.-Z.; Feng, Y.; Pollard, J.; Chin, J.N.; Rybak, M.J.; Bucki, R.; Epand, R.F.; Epand, R.M.; Savage, P.B. Ceragenins: Cholic acid-based mimics of antimicrobial peptides. *Acc. Chem. Res.* **2008**, *41*, 1233–1240. [CrossRef] [PubMed]

33. Oloso, N.O.; Fagbo, S.; Garbati, M.; Olonitola, S.O.; Awosanya, E.J.; Aworh, M.K.; Adamu, H.; Odetokun, I.A.; Fasina, F.O. Antimicrobial resistance in food animals and the environment in Nigeria: A review. *Int. J. Environ. Res. Public health* **2018**, *15*, 1284. [CrossRef]

International Journal of
***Environmental Research
and Public Health***

Article

Antimicrobial Susceptibility of *Staphylococcus aureus* Isolated from Recreational Waters and Beach Sand in Eastern Cape Province of South Africa

Olufemi Emmanuel Akanbi *, Henry Akum Njom, Justine Fri, Anthony C. Otigbu and Anna M. Clarke

Microbial Pathogenicity and Molecular Epidemiology Research Group (MPMERG), Department of Biochemistry and Microbiology, University of Fort Hare, Private Bag X1314, Alice 5700, South Africa; hnjom@ufh.ac.za (H.A.N.); jfri@ufh.ac.za (J.F.); aotigbu@gmail.com (A.C.O.); aclarke@ufh.ac.za (A.M.C.)

*   Correspondence: femo.emman@gmail.com; Tel.: +27-733-302-440

Received: 20 June 2017; Accepted: 30 August 2017; Published: 1 September 2017

Abstract: *Background*: Resistance of *Staphylococcus aureus* to commonly used antibiotics is linked to their ability to acquire and disseminate antimicrobial-resistant determinants in nature, and the marine environment may serve as a reservoir for antibiotic-resistant bacteria. This study determined the antibiotic sensitivity profile of *S. aureus* isolated from selected beach water and intertidal beach sand in the Eastern Cape Province of South Africa. *Methods*: Two hundred and forty-nine beach sand and water samples were obtained from 10 beaches from April 2015 to April 2016. *Staphylococcus aureus* was isolated from the samples using standard microbiological methods and subjected to susceptibility testing to 15 antibiotics. Methicillin-resistant *Staphylococcus aureus* (MRSA) was detected by susceptibility to oxacillin and growth on Brilliance MRSA II agar. Antibiotic resistance genes including *mec*A, *fem*A *rpo*B, *bla*Z, *erm*B, *erm*A, *erm*C, *van*A, *van*B, *tet*K and *tet*M were screened. *Results*: Thirty isolates (12.3%) were positive for *S. aureus* by PCR with over 50% showing phenotypic resistance to methicillin. Resistance of *S. aureus* to antibiotics varied considerably with the highest resistance recorded to ampicillin and penicillin (96.7%), rifampicin and clindamycin (80%), oxacillin (73.3%) and erythromycin (70%). *S. aureus* revealed varying susceptibility to imipenem (96.7%), levofloxacin (86.7%), chloramphenicol (83.3%), cefoxitin (76.7%), ciprofloxacin (66.7%), gentamycin (63.3%), tetracycline and sulfamethoxazole-trimethoprim (56.7%), and vancomycin and doxycycline (50%). All 30 (100%) *S. aureus* isolates showed multiple antibiotic-resistant patterns (resistant to three or more antibiotics). The *mec*A, *fem*A, *rpo*B, *bla*Z, *erm*B and *tet*M genes were detected in 5 (22.7%), 16 (53.3%), 11 (45.8%), 16 (55.2%), 15 (71.4%), and 8 (72.7%) isolates respectively; *Conclusions*: Results from this study indicate that beach water and sand from the Eastern Cape Province of South Africa may be potential reservoirs of antibiotic-resistant *S. aureus* which could be transmitted to exposed humans and animals.

Keywords: *S. aureus*; antibiotic resistance; beaches; multiple-antibiotic resistance

---

1. Introduction

Staphylococcus aureus are Gram-positive cocci ranging from 0.5 to 1.5 μm in diameter, which may or may not contain a polysaccharide capsule. They are non-motile, non-spore forming facultative anaerobes that produce catalase and coagulase enzymes [1–3]. Yearly, microbial contamination of marine waters is predicted to be responsible for millions of gastrointestinal and acute respiratory infections (ARIs) [4], in addition to several skin infections [5]. Although *S. aureus* is typically a commensal organism, it has been known to be opportunistic. Invasive infections due to wound invasion can lead to numerous diseases, including scalded skin syndrome, abscesses, septicaemia, pneumonia, food poisoning, and toxic shock syndrome [6,7].

S. aureus is a potentially harmful human pathogen associated with both nosocomial and community-acquired infections, and it is increasingly becoming resistant to most antibiotics. Previous studies of *S. aureus* in marine environments have linked swimmers to the dissemination of *S. aureus* in marine water [8], via the shedding of the bacterium from their nose, skin, and respiratory tract [9]. On recreational beaches, *S. aureus* has occasionally been found in high abundance in both water and sand, which can be directly associated with bather density and human activities around the beach [9–11].

The human skin is directly exposed to infectious agents during swimming [12], and this exposure can lead to the colonization of *S. aureus* with the potential to invade the immune system and cause infections. There is a relationship between seawater exposure and *S. aureus* infection rates which suggests that recreational waters are potential sources of community-acquired *S. aureus* infections [9]. There is also a positive correlation between the concentrations of *S. aureus* and total staphylococci to skin, eye, and ear infections among bathers [13–15].

Methicillin-resistant *Staphylococcus aureus* (MRSA) is defined as any strain of *S. aureus* that has acquired resistance to methicillin and other beta lactam antibiotics [16] and it is responsible for several intractable infections in humans [17]. *S. aureus* and MRSA are both shed by swimmers [18,19] and have been reported in beach seawater and sand [18,20–24].

The resistance of *S. aureus* to methicillin is due to the production of penicillin-binding protein 2a (PBP2a), which is encoded by the *mec*A gene located on the mobile gene element (MGE) of the staphylococcal chromosome cassette *mec* (SCC*mec*), which has a low affinity for beta-lactam antibiotics [25,26].

The fact that *S. aureus* is resistant to multiple classes of antimicrobial agents in the hospital environment is a challenge currently facing clinicians when treating *S. aureus* infections [27]. This resistance stems from a history of over 50 years of recurrent adaptation of *S. aureus* to different antibiotics introduced into clinical practice over the years. Abuse of as well as indiscriminate use of antimicrobials are contributing factors to the spread of resistance [27]. Antibiotic-resistance genes are carried on plasmids and transposons, and can be transferred from one staphylococcal species to another and among other Gram-positive bacteria [28].

Antimicrobials act by targeting important bacterial functions such as cell wall synthesis (beta-lactams and glycopeptides), protein synthesis (aminoglycosides, tetracyclines, macrolides, lincosamides, chloramphenicol, mupirocin and fusidic acid), nucleic acid synthesis (quinolones), RNA synthesis (rifampin), and metabolic pathways such as folic acid metabolism (sulphonamides and trimethoprim) [29–31]. The overuse of antimicrobials elicits resistance either by the emergence of point mutations or by the acquisition of foreign resistance genes, which leads to alteration of the antimicrobial target and the degradation of the antimicrobial or reduction of the cell's internal antimicrobial concentration [27,29–31].

This study was carried out to determine the antimicrobial resistance pattern of *Staphylococcus aureus* and methicillin-resistant *Staphylococcus aureus* isolated from seawater and sand from selected beaches in the Eastern Cape Province of South Africa. We also determined whether isolates carried any antibiotic-resistance gene markers for methicillin, beta-lactams, tetracycline, vancomycin, erythromycin and rifampicin.

2. Materials and Methods

2.1. Study Site

Sea water and sand samples were obtained from ten beaches in four major cities in the Eastern Cape Province; Nahoon beach (32°59′20.09″ S 27°57′1.30″ E), Eastern beach (33°0′32.00″ S 27°55′31.02″ E), East beach (33°36′6.07″ S 26°54′4.94″ E), West beach (33°36′18.80″ S 26°53′56.53″ E),

Kelly's beach (33°36′37.20″ S 26°53′25.86″ E), Kariega beach (33°41′1.05″ S 26°40′59.28″ E), Middle beach (33°41′21.16″ S 26°40′36.09″ E), King's beach (33°58′16.92″ S 25°38′49.87″ E), Hobie beach (33°58′49.75″ S 25°39′35.18″ E), and Pollock beach (33°59′6.59″ S 25°40′21.92″ E) (Figure 1).

Figure 1. Aerial view of sampling sites [32].

2.2. Sample Collection

A total of 245 (178 marine water, 67 marine sand) samples were collected monthly from 10 selected beaches in the Eastern Cape Province of South Africa between April 2015 and April 2016. Water samples were collected in 2 L sterile containers against an incoming wave. Beach sand was also collected in sterile 100 mL containers. Samples were transported at 4 °C and processed within 24 h.

Isolation and Molecular Confirmation of *S. aureus*

Sand samples were vigorously hand shaken in Phosphate Buffered Saline (PBS), where a ratio of 2 g of sand to 80 mL of PBS was used [20,33]. Both sand and water samples were enriched in tryptone soy broth and incubated at 37 °C for 24 h, followed by sub-culturing on mannitol salt agar (MSA), and further incubated at 37 °C for 24 h. Presumptive *S. aureus*, identified by the fermentation of mannitol (yellow colonies) were purified on nutrient agar. Presumptive isolates were stored in 25% glycerol at −80 °C.

Polymerase chain reaction (PCR) was used for confirmation of *S. aureus* as previously described [20]. DNA was extracted using the boiling method where 2 mL of overnight pure Nutrient broth cultures were transferred to sterile eppendorf tubes and centrifuged at 13,000 rpm for 3 min. The supernatant was discarded and cells re-suspended in 200 µL sterile distilled water. The cell solution was then heated at 100 °C in an Accu dri-block (Lasec, SA) for 10 min, followed by centrifugation at 13,000 rpm for 2 min to pellet the cells [34]. The supernatants were transferred to clean, sterile tubes and used directly as templates for PCR assay or stored at −20 °C for subsequent use.

A method previously described by Maes [35] was used for identification of *S. aureus*, based on the detection of a specie-specific nuc-gene. *S. aureus* ATCC 25923 was used as a positive control. Each 25 µL PCR reaction mix constituted 12.5 µL of 2X PCR master mix, 0.5 µL each of both reverse and forward primers (Table 1), 6.5 µL nuclease-free water and 5 µL of template DNA. PCR was conducted in a T1000 Touch Thermal Cycler (Bio-Rad, Hercules, CA, USA). Cycling conditions are shown in Table 1. The PCR products were separated by agarose gel electrophoresis in 1% agarose, stained with ethidium bromide. A 100 bp DNA ladder was included in each run.

2.3. Antimicrobial Susceptibility Testing

Isolates confirmed by PCR as *S. aureus* were subjected to antimicrobial susceptibility testing to 15 antibiotics. Profiling was performed by the Kirby-Bauer disk diffusion method on Mueller-Hinton agar according to Clinical and Laboratory Standards Institute guidelines [36,37]. An inoculum for each isolate was prepared by emulsifying colonies from an overnight pure culture in sterile normal saline (0.85%) in test tubes with the turbidity adjusted to 0.5 McFarland standard (0.5 mL of 1% *w/v* BaCl$_2$ and 99.5 mL of 1% *v/v* H$_2$SO$_4$), equivalent to 1.0×10^8 cfu/mL. The bacterial suspension was uniformly streaked on Mueller Hinton agar plates using sterile swabs and left for 3 min prior to introduction of the antibiotics. Antibiotics commonly used for treatment of *S. aureus* infections were selected for this assay, namely penicillin, ampicillin, gentamycin, erythromycin, levofloxacin, ciprofloxacin, tetracycline, doxycycline, vancomycin, cefoxitin, imipenem, sulfamethoxazole-trimethoprim, clindamycin, rifampicin and chloramphenicol. Plates were incubated at 35 °C for 24 h, and the diameters of zone of inhibition were measured and results interpreted according to Clinical Laboratory Standards institute [37].

2.4. Detection of MRSA

2.4.1. Phenotypic

All isolates confirmed to be *S. aureus* by PCR were subjected to antibiotic susceptibility testing to oxacillin (5 µg) [37] by disc diffusion test as well as growth on Brilliance MRSA II agar [38], to determine phenotypic resistance to methicillin. Inoculated plates were incubated at 37 °C for 24 h [38]. All isolates that tested positive on Brilliance MRSA II agar (blue to violet colonies) or resistant by oxacillin disc were considered to be presumptive MRSA.

2.4.2. Molecular Confirmation of MRSA

Presumptive isolates from Brilliance MRSA II agar, as well as isolates that were phenotypically resistant to oxacillin, were further confirmed by PCR detection of the *mec*A gene (responsible for methicillin resistance) using specific primers (Table 1) as earlier described [20,39]. The *fem*A gene, a factor essential for methicillin resistance, was also evaluated [40] by PCR using specific primers (Table 1). A 25 uL reaction was set up consisting of 12.5 µL master mix, 0.5 µL forward primer, 0.5 µL reverse primer, 6.5 µL nuclease-free water and 5 µL of DNA. PCR was conducted using a T1000 Touch Thermal Cycler (Bio-Rad, Johannesburg, SA, USA). The cycling conditions used for confirmation of the *mec*A and *fem*A gene are shown in Table 1. The amplicons were separated using 1.5% agarose stained with ethidium bromide and visualized under a transilluminator (UVITEC Alliance 4.7, Bio-Active., Ltd., Bangkok, Thailand).

Table 1. Oligonucleotide primers and cycling conditions used in the molecular confirmation of *S. aureus* and antibiotic-resistance genes.

Primer	Sequence (5′-3′)	Product Size (bp)	Cycling Conditions	Reference
nuc-F nuc-R	GCGATTGATGGTGGATACGGT AGCCAAGCCTTGACGAACTAAAGC	279	Initial denaturation at 94 °C for 5 min, followed by 40 cycles of 94 °C for 45 s, 58 °C for 45 s and 72 °C for 90 s. Final extension at 72 °C for 10 min	[35]
mecA-F mecA-R	TCCAGGAATGCAGAAAGACCAAAGC GACACGATAGCCATCTTCATGTTGG	499	Initial denaturation at 94 °C for 3 min, followed by 40 cycles of 94 °C for 30 s, 59 °C for 30 s and 72 °C for 1 min. Final extension at 72 °C for 8 min.	[41]
ermA-F ermA-R	TATCTTATCGTTGAGAAGGGATT CTACACTTGGCTTAGGATGAAA	139	Initial denaturation at 94 °C for 5 min, followed by 40 cycles of 94 °C for 40 s, 48 °C for 40 s and 72 °C for 90 s. Final extension at 72 °C for 8 min.	[42]
ermB-F ermB-R	CTATCTGATTGTTGAAGAAGGATT GTTTACTCTTGGTTTAGGATGAAA	142	Initial denaturation at 94 °C for 5 min, followed by 40 cycles of 94 °C for 40 s, 47 °C for 40 s and 72 °C for 90 s. Final extension at 72 °C for 8 min.	[42]
ermC-F ermC-R	CTTGTTGATCACGATAATTTCC ATCTTTTAGCAAACCCGTATTC	190	Initial denaturation at 94 °C for 5 min, followed by 40 cycles of 94 °C for 40 s, 49 °C for 40 s and 72 °C for 90 s. Final extension at 72 °C for 8 min.	[42]
blaZ-F blaZ-R	ACTTCAACACCTGCTGCTTTC TGACCACTTTTATCAGCAACC	173	Initial denaturation at 94 °C for 3 min, followed by 35 cycles of 94 °C for 30 s, 49 °C for 30 s and 72 °C for 1 min. Final extension at 72 °C for 8 min.	[42]
rpoB1-F rpoB2-R	ACCGTCGTTTACGTTCTGTA TCAGTGATAGCATGTGTATC	460	Initial denaturation at 94 °C for 5 min, followed by 40 cycles of 94 °C for 40 s, 45.5 °C for 40 s and 72 °C for 90 s. Final extension at 72 °C for 8 min.	[43]
tetM-F tetM-R	AGTGGAGCGATTACAGAA CATATGTCCTGGCGTGTCTA	158	Initial denaturation at 94 °C for 3 min, followed by 40 cycles of 94 °C for 30 s, 45 °C for 30 s and 72 °C for 1 min. Final extension at 72 °C for 8 min.	[44]
tetK-F tetK-R	GTAGCGACAATAGGTAATAGT GTAGTGACAATAAACCTCCTA	360	Initial denaturation at 94 °C for 3 min, followed by 40 cycles of 94 °C for 30 s, 47 °C for 30 s and 72 °C for 1 min. Final extension at 72 °C for 8 min.	[44]
vanA vanA	GCGCGGTCCACTTGTAGATA TGAGCAACCCCCAAACAGTA	314	Initial denaturation at 94 °C for 3 min, followed by 35 cycles of 94 °C for 1 min, 56.5 °C for 1 min and 72 °C for 1 min. Final extension at 72 °C for 10 min.	[45]
vanB vanB	AGACATTCCGGTCGAGGAAC GCTGTCAATTAGTGCGGGAA	220	Initial denaturation at 94 °C for 3 min, followed by 35 cycles of 94 °C for 1 min, 56.5 °C for 1 min and 72 °C for 1 min. Final extension at 72 °C for 10 min.	[45]
femA-F femA-R	AAAAAAGCACATAACAAGCG GATAAAGAAGAAACCAGCAG	132	Initial denaturation at 94 °C for 5 min, followed by 40 cycles of 94 °C for 40 s, 45.5 °C for 40 s and 72 °C for 90 s. Final extension at 72 °C for 8 min.	[40]

2.5. PCR Detection of Antibiotic Resistance Genes

Based on the phenotypic antibiotic resistance profiles, (29/30, 24/30, 22/30, 21/30, 17/30, 15/30) isolates showing resistance to β-lactam, rifampicin, methicillin, erythromycin, tetracycline, vancomycin respectively, were investigated for the presence of associated antibiotic-resistance genes (ARGs). These were *blaZ*, *rpoB*, *mecA ermB*, *ermA*, *ermC*, *tetK*, *tetM*, *vanA* and *vanB* genes respectively. The reactions were performed as singleplex PCRs in a total volume of 25 μL consisting of 12.5 μL 2X PCR master mix, 0.5 μL each of the forward and reverse primer, 6.5 μL nuclease-free water and 5 μL of template DNA performed in a T1000 Touch Thermal Cycler (Bio-Rad, Johannesburg, SA, USA). The amplicons were separated on 1.5% agarose stained with ethidium bromide, visualized and photographed using a transilluminator (UVITEC Alliance 4.7, Bio-Active., Ltd., Bangkok, Thailand). Table 1 shows the primer sequences used, and cycling conditions for PCR detection of *S. aureus* and antibiotic resistance gene markers.

3. Results

3.1. Molecular Identification of Staphylococcus aureus in Recreational Beach Water and Sand Samples

A total of 245 samples were screened; beach water (n = 178) and sand samples (n = 67) of which 143 isolates (one isolate from each sample) were presumptive by culture on MSA. A 12.3% (30/245) of the isolates were confirmed by PCR as *S. aureus*, with 12.4% (22/178) of isolates from seawater, and 11.9% (8/67) from marine sand. Of the 22 confirmed *S. aureus* isolates from seawater, 6 isolates each were from Middle beach and Eastern beach, 5 isolates from Nahoon beach, 2 each from Kariega beach and East beach and 1 isolate from West beach. Of the 8 confirmed *S. aureus* isolates from sand, 4 isolates were from Middle beach and 2 each were from East beach and Kariega beach respectively.

3.2. Antimicrobial Susceptibility Test (AST)

Antibiotic susceptibility of 30 *S. aureus* isolates revealed varying degrees of susceptibility patterns against the antimicrobial agents. Generally, cefoxitin 76.7% (23/30), chloramphenicol 83.3% (25/30), levofloxacin 86.7% (26/30), and imipenem 96.7% (29/30) were the most effective antibiotics to *S. aureus*. A low, ≥50% susceptibility was recorded to vancomycin and doxycycline (50%; 15/30), tetracycline and sulfamethoxazole-trimethoprim (56.7%; 17/30), gentamycin (63.3%; 19/30), and ciprofloxacin (66.7%; 20/30). A higher resistance to erythromycin (70%; 21/30) and clindamycin and rifampicin (80%, 24/30) was identified, with resistance to penicillin G and ampicillin the highest (each recording 96.7%; 29/30). The percentage of antimicrobial resistance of *S. aureus* isolates are shown in Figure 2.

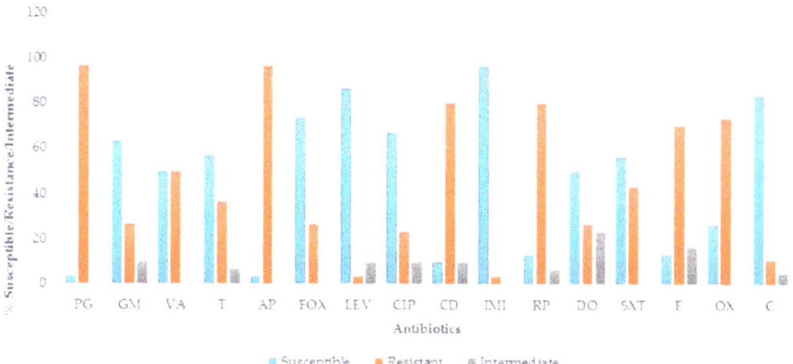

Figure 2. The percentage of antimicrobial resistance profiles of *S. aureus* isolates. PG = penicillin, GM = gentamicin, VA = vancomycin, T = tetracycline, AP = ampicillin, FOX = cefoxitin, LEV = levofloxacin, CIP = ciprofloxacin, CD = clindamycin, IMI = imipenem, RP = rifampicin, DO = doxycycline, SXT = sulfamethoxazole-trimethoprim, E = erythromycin, OX = oxacillin, C = chloramphenicol.

3.3. Phenotypic Detection of MRSA

A methicillin-resistant *S. aureus* isolate was defined as resistant by any of the two methods tested. Fifteen (50%) isolates showed phenotypic resistance to methicillin after culturing on selective media (Brilliance MRSA II agar) while 73.3% (22/30) of the isolates showed phenotypic resistance to oxacillin (Figure 2), which could be used as a proxy for determining methicillin resistance [46]. All those that were positive on Brilliance MRSA II agar were also positive for the oxacillin disc diffusion test.

3.4. Multiple Antibiotic Resistance (MAR)/MAR Phenotypes of S. aureus

All isolates tested were multi-drug resistant, (100%; 30/30) (resistant to three or more antimicrobials), with 3 isolates resistant to 12 of the 15 antibiotics tested. Resistance to 8 antibiotics was the most common, shown by 5 (16.7%) isolates, followed by resistance to 4 and 5 antibiotics recorded by 4 (13.3%) isolates each. Twenty-three different MAR patterns were observed from the 30 isolates. The most common of these were PG-GM-VA-T-AP-FOX-CIP-CD-RP-DO-SXT-E-OX, PG-VA-T-AP-FOX-CD-RP-DO-SXT-E-OX and PG-VA-AP-FOX-CD-RP-SXT-E-OX, observed in 3 (10%), 2 (6.7%), and 2 (6.7%) isolates, respectively.

3.5. Prevalence of Antibiotic Resistance Genes

Generally, a total of five of 10 ARGs tested were detected in one or more resistant isolates, with higher frequencies recorded in isolates recovered from seawater. Of the ten ARGs tested (*blaZ*, *mecA*, *rpoB*, *ermB*, *ermA*, *ermC*, *tetK*, *tetM*, *vanA* and *vanB*), the *blaZ* gene, coding for resistance to beta-lactam antibiotics (penicillin & ampicillin), was detected in 16 (55.2%, $n = 29$) of the isolates, the *mecA* gene, coding for methicillin resistance was detected in 5 (22.7%, $n = 22$), the *rpoB* gene, coding for rifampicin resistance, was detected in 11 (45.8%, $n = 24$), the *ermB* gene, coding for erythromycin resistance, in 15 (71.4%, $n = 21$) and the *tetM* gene, coding for tetracycline resistance, was detected in 8 (72.7%, $n = 11$) of the isolates. However, other ARGs such as *ermA*, *ermC*, *tetK*, *vanA* and *vanB* investigated were absent in the isolates. Table 2 shows the various ARGs detected in beach sand and water while Figure 3 shows a representative gel of the PCR amplified products for these genes.

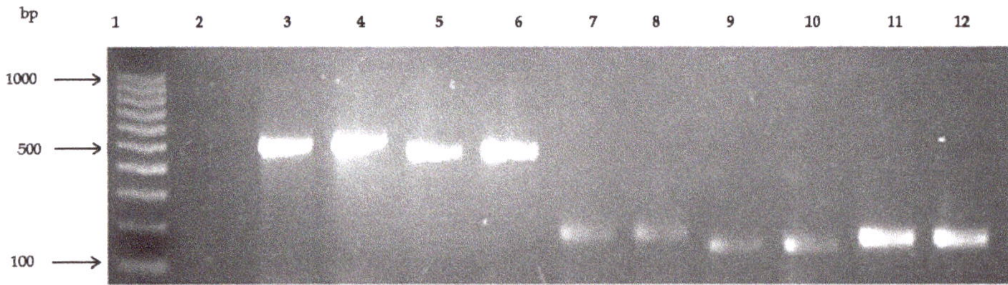

Figure 3. Representative gel showing PCR amplified products of antibiotic resistance genes of *mec*A, *rpo*B, *bla*Z, *erm*B and *tet*M separated on 1.5% agarose. Lane 1: 100 bp DNA ladder (Fermentas Life Sciences, Vilnius, Lithuania), Lane 2: negative control, Lane 3, 4: *mec*A (499 bp) positive isolates, Lane 5, 6: *rpo*B (460 bp) positive isolates, Lane 7, 8: *bla*Z (173 bp) positive isolates, Lane 9, 10: *erm*B (142 bp) positive isolates and Lane 11, 12: *tet*M (142 bp) positive isolates.

The *fem*A gene, a factor also responsible for methicillin resistance [40], was identified in 53.3% (16/30) of the isolates. Figure 4 shows the gel electrophoresis of PCR amplified products for the *fem*A gene.

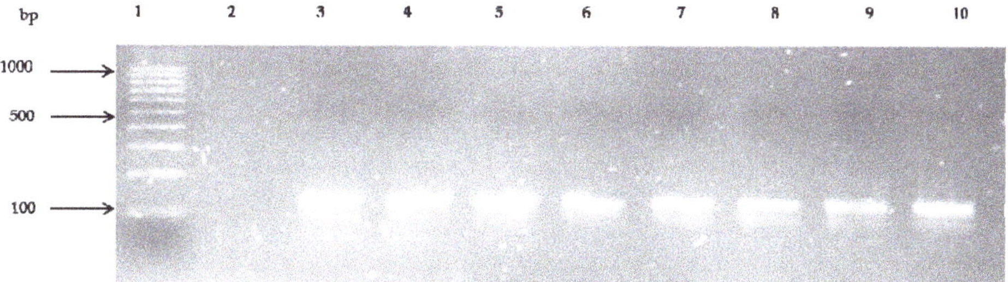

Figure 4. Representative gel showing PCR amplified products of *fem*A gene separated on 1.5% agarose. Lane 1: 100 bp DNA ladder (Fermentas Life Sciences, Vilnius, Lithuania), Lane 2: negative control, Lane 3–10: *fem*A (132 bp) positive isolate.

Table 2. Antibiotic resistance genes detected in *S. aureus* isolates from beach sand and seawater.

No. Resistant by Disc Diffusion	Associated ARG Tested	ARG Detected		
		Sand (%)	Water (%)	Total (%)
Ampicillin & Penicillin (*n* = 29)	*bla*Z	4 (25%)	12 (75%)	16 (55.2%)
Methicillin (*n* = 22)	*mec*A	1 (20%)	4 (80%)	5 (22.7%)
Rifampicin (*n* = 24)	*rpo*B	2 (18.2%)	9 (81.8%)	11 (45.8%)
Erythromycin (*n* = 21)	*erm*B	3 (20%)	12 (80%)	15 (71.4%)
Tetracycline (*n* = 11)	*tet*M	1 (12.5%)	7 (87.5%)	8 (72.7%)

4. Discussion

Humans and animals have been reported as sources of antibiotic-resistant organisms in water environments and can transfer antibiotic resistance genes to other pathogens and naturally occurring water microbes through transposons, plasmids and integrons [47,48]. Bacteria isolated from beach sand, seawater and sediments have recorded resistance to various antimicrobials [48–52].

The occurrence of *S. aureus* and MRSA is on the rise, resulting in increased incidences of hospital-acquired and community-acquired infections worldwide, posing a major public health concern [53–55]. Moreover, microbial ecosystems can also be potentially altered by the presence of varying antibiotics of industrial origin, circulating in water environs [47]. *S. aureus* is one of the most successful and adaptable human pathogens due to its proficiency in acquiring antibiotic-resistant mechanisms and pathogenic determinants, leading to its emergence in both nosocomial and community settings [54]. Nosocomial colonisation of *S. aureus* and MRSA can go undetected, and signs of infection may only appear months after a patient is exposed to the infection. Infected patients may then serve as reservoirs for further transmission, especially as most of these strains carry SCC*mec* types coding for resistance to methicillin and other beta lactams [56].

To the best of our knowledge, this is the first study which has used a mixture of phenotypic and genotypic approaches simultaneously to determine the occurrence and antibiotic resistance profiles of *S. aureus* strains from beach water and sand in the study area. In this study, *S. aureus* was isolated from beach water and sand samples. Other studies have also reported this organism in marine water and/or sand [18,20–22,24,57–59], however, the frequency (12.2%) of isolation was lower in our study than observed in other studies [20,21,24]. This study only analyzed a single isolate for every sample, which could account for the lower detection frequency.

Of the isolates evaluated in this study, individual resistances of *S. aureus* to penicillin G and ampicillin was high (96.7%; 29/30). High resistance to these β-lactam antibiotics was not surprising, as ampicillin is one of the most commonly used antibiotics for treatment of infections in humans and animals [60], with penicillin developing resistance to *S. aureus* since the 1960s [61].

In addition, ampicillin-resistant isolates may cross-select for resistance to other beta-lactams [62]. Resistance to ampicillin may therefore indicate resistance of the isolates to other β-lactam antibiotics. This was observed in our study, as resistance to both ampicillin and penicillin occurred in equal proportion. Resistances observed to erythromycin, chloramphenicol, sulfamethoxazole-trimethoprim, and tetracycline were similar to that previously reported [21].

Given the relatively small number of isolates evaluated, a 50% vancomycin resistance was of concern, as this antibiotic is historically regarded as the antibiotic of final resort and the highest quality level antimicrobial for the treatment of genuine MRSA diseases [30]. The first case of a fully vancomycin-resistant *S. aureus* was described in Michigan, USA, in a renal dialysis patient [63]. The utilization of growth promoters such as tylosin, macrolide and avoparcin has been related to the occurrence of erythromycin and vancomycin resistance in *S. aureus* [64] in the environment, which might have then leached to marine waters. Based on the phenotypic identification of MRSA, 50% and 73.3% of the isolates were potentially MRSA by both methods. The poor specificity of the phenotypic methods in this study was not surprising, as higher specificity and sensitivity of these phenotypic methods have mostly been recorded in clinical isolates [38].

In this study, all *S. aureus* isolates were multidrug resistant. This point is worth noting, as it potentially could lead to failure in treatment therapy, prolonged illnesses, increased expenses for health care, and in serious cases, risk of death if humans are infected with such strains [65]. The transmission of resistance (R-factor), a plasmid-mediated genetic determinant, may be credited with the development of MAR among these isolates [62]. Studies have shown an upward pattern in the incidences of *S. aureus* isolates with multiple antibiotic resistance [66–69]. It has also been reported that *S. aureus* isolates with multiple antibiotic resistance attributes have a negative impact on the treatment of staphylococcal infections, especially in elderly, children, and immune-compromised individuals [70].

Generally, a total of five out of 10 ARGs tested were detected, with a higher frequency of detection in beach water compared to sand isolates. The higher frequency of detection in seawater could be because water is exposed to a greater variety of potential contaminants than sand. These may include runoffs from pharmaceutical, hospital, and industrial waste as well as farmlands [71,72]. Sources may also include antibiotic-resistant bacteria from poorly treated or untreated sewage, as final effluents of waste water treatment plants that may leach into seawater [52].

The *blaZ* gene is responsible for the production of β- lactamase enzyme, which confers resistance to β- lactam antibiotics such as penicillin and ampicillin [73]. This gene was only detected in small proportions compared to its phenotypic detection. Molecular confirmation identified the *mec*A gene only in five (22.7%) of the MRSA isolates detected by at least one of the phenotypic methods. The presence of this gene encodes a penicillin-binding protein 2a (PBP2a), responsible for methicillin resistance in staphylococci, with this protein, rendering a reduced affinity for β-lactam antibiotics [74]. Various studies have reported the occurrence of methicillin-resistant *S. aureus* from water sources, animal-derived food and humans [55,75,76]. MRSA has also been previously reported from marine waters [21,24] and waste water treatment plants [77]. Oxacillin has been proposed as a proxy antibiotic for testing susceptibility not only to methicillin and to all β-lactams [46], which could explain why all oxacillin-resistant isolates were not carrying the *mec*A gene. Phenotypic resistance observed to oxacillin in this study was probably achieved through other mechanisms [78], which may include alteration of the penicillin binding proteins, which brings about hyper-production of methicillinase or beta-lactamase [37,79,80].

The mode of resistance of rifampin is inhibition of the process of RNA polymerase [81]. Mutations on the gene encoding the β-subunit of RNA polymerase (*rpo*B gene) account for rifampin resistance (Rif^r) [82–84]. In our study, this gene was detected in 45.8% (11/24) of the rifampicin-resistant *S. aureus* isolates. Erythromycin resistance in staphylococci is mainly facilitated by the *erm* genes, coding for erythromycin resistant methylase [85], with *erm*A and *erm*C reported as the most frequently detected *erm* gene associated with staphylococci in human infections [86]. Results from this study however, detected *erm*B as the only gene coding for erythromycin resistance. The high incidence (72.7%) of *tet*M in our study is similar to that (74.2%) earlier reported [87]. Another study has also reported the presence of both *tet*M and *tet*K gene from *S. aureus* isolates from public beaches [21].

The *fem*A gene was detected in 53.3% (16/30) of the confirmed *S. aureus* isolates. This gene is a chromosomally encoded factor in *Staphylococcus aureus*, which is crucial for the expression of advanced methicillin resistance, encoding proteins which influence the level of methicillin resistance [88]. Finding *fem*A gene in all *mec*A positive isolates is evidence that these isolates had a functional methicillin resistance. The detection of *fem*A together with *mec*A by PCR has long been considered a reliable indicator in the identification of MRSA [89].

5. Conclusions

This study is the first to report the occurrence of antibiotic resistant *S. aureus* on recreational beaches in the Eastern Cape Province, South Africa. Our results show that public beaches in the study area may be potential reservoirs for transmission of antibiotic resistant *S. aureus* to beach goers, particularly those with skin lesions. Results from this study are unlikely to be unique to the Eastern Cape or South Africa and further studies are needed to determine the distribution and level of antibiotic-resistant *S. aureus* in other public beaches.

Acknowledgments: The authors are grateful to the South African Institute for Aquatic Biodiversity (SAIAB) and the National Research Foundation of South Africa for financially supporting the research. We also acknowledge Gunda Spingies for critical English editing.

Author Contributions: Olufemi Emmanuel Akanbi and Anthony Otigbu collected the samples and performed the experiments; Justine Fri and Anna Clarke conceived and designed the experiment; Henry Akum Njom and Anna A. Clarke supervised the whole study; Olufemi Emmanuel Akanbi and Henry Akum Njom analyzed the data and wrote the manuscript.

Conflicts of Interest: The authors declare no conflict of interest.

References

1. Ho, J.; O'donoghue, M.M.; Boost, M.V. Occupational exposure to raw meat: A newly-recognized risk factor for *Staphylococcus aureus* nasal colonization amongst food handlers. *Int. J. Hyg. Environ. Health* **2014**, *217*, 347–353. [CrossRef] [PubMed]
2. O'Riordan, K.; Lee, J.C. *Staphylococcus aureus* capsular polysaccharides. *Clin. Microbiol. Rev.* **2004**, *17*, 218–234. [CrossRef] [PubMed]
3. Witte, W.; Strommenger, B.; Klare, I.; Werner, G. Antibiotic-resistant nosocomial pathogens. Part I: Diagnostic and typing methods. *Bundesgesundheitsblatt Gesundheitsforschung Gesundheitsschutz* **2004**, *47*, 352–362. [PubMed]
4. Shuval, H. Estimating the global burden of thalassogenic diseases: Human infectious diseases caused by wastewater pollution of the marine environment. *J. Water Health* **2003**, *1*, 53–64. [PubMed]
5. Yau, V.; Wade, T.J.; de Wilde, C.K.; Colford, J.M., Jr. Skin-related symptoms following exposure to recreational water: A systematic review and meta-analysis. *Water Qual. Expos. Health* **2009**, *1*, 79–103. [CrossRef]
6. Moore, P.C.L.; Lindsay, J.A. Genetic variation among hospital isolates of methicillin-sensitive *Staphylococcus aureus*: Evidence for horizontal transfer of virulence genes. *J. Clin. Microbiol.* **2001**, *39*, 2760–2767. [CrossRef] [PubMed]
7. Boyd, E.F.; Brüssow, H. Common themes among bacteriophage-encoded virulence factors and diversity among the bacteriophages involved. *Trends Microbiol.* **2002**, *10*, 521–529. [CrossRef]
8. Robinton, E.D.; Mood, E.W. A quantitative and qualitative appraisal of microbial pollution of water by swimmers: A preliminary report. *J. Hyg.* **1966**, *64*, 489–499. [CrossRef] [PubMed]
9. Charoenca, N.; Fujioka, R. Association of staphylococcal skin infections and swimming. *Water Sci. Technol* **1995**, *31*, 11–17.
10. WHO. Coastal and Fresh Waters. In *Guidelines for Safe Recreational Water Environments*; World Health Organization: Geneva, Switzerland, 2003; Volume 1.
11. Papadakis, J.A.; Mavridou, A.; Richardson, S.C.; Lampiri, M.; Marcelou, U. Bather-related microbial and yeast populations in sand and seawater. *Water Res.* **1997**, *31*, 799–804. [CrossRef]

12. Henrickson, S.E.; Wong, T.; Allen, P.; Ford, T.; Epstein, P.R. Marine swimming-related illness: Implications for monitoring and environmental policy. *Environ. Health Perspect.* **2001**, *109*, 645–650. [CrossRef] [PubMed]

13. Gabutti, G.; De Donno, A.; Bagordo, F.; Montagna, M.T. Comparative survival of faecal and human contaminants and use of *Staphylococcus aureus* as an effective indicator of human pollution. *Mar. Pollut. Bull.* **2000**, *40*, 697–700. [CrossRef]

14. Calderon, R.L.; Mood, E.W.; Dufour, A.P. Health effects of swimmers and nonpoint sources of contaminated waters. *Int. J. Environ. Health Res.* **1991**, *1*, 21–31. [CrossRef] [PubMed]

15. Seyfried, P.L.; Tobin, R.S.; Brown, N.E.; Ness, P.F. A prospective study of swimming-related illness. II. Morbidity and the microbiological quality of water. *Am. J. Public Health* **1985**, *75*, 1071–1075. [CrossRef] [PubMed]

16. Stotts, S.N.; Nigro, O.D.; Fowler, T.L.; Fujioka, R.S.; Steward, G.F. Virulence and antibiotic resistance gene combinations among *Staphylococcus aureus* isolated from coastal waters of Oahu, Hawaii. *J. Young Investig.* **2005**, *12*, 1–8.

17. McDougal, L.K.; Steward, C.D.; Killgore, G.E.; Chaitram, J.M.; McAllister, S.K.; Tenover, F.C. Pulsed-field gel electrophoresis typing of oxacillin-resistant *Staphylococcus aureus* isolates from the United States: Establishing a national database. *J. Clin. Microbiol.* **2003**, *41*, 5113–5120. [CrossRef] [PubMed]

18. Plano, L.R.W.; Garza, A.C.; Shibata, T.; Elmir, S.M.; Kish, J.; Sinigalliano, C.D.; Gidley, M.L.; Miller, G.; Withum, K.; Fleming, L.E.; et al. Shedding of *Staphylococcus aureus* and methicillin-resistant *Staphylococcus aureus* from adult and pediatric bathers in marine waters. *BMC Microbiol.* **2011**, *1*, 5. [CrossRef] [PubMed]

19. Elmir, S.M.; Wright, M.E.; Abdelzaher, A.; Solo-Gabriele, H.N.; Fleming, L.E.; Miller, G.; Rybolowik, M.; Shih, M.T.; Pillai, P.; Copoer, J.E.; et al. Quantitative evaluation of bacteria released by bathers in marine water. *Water Res.* **2007**, *41*, 3–10. [CrossRef] [PubMed]

20. Goodwin, K.D.; Pobuda, M. Performance of CHROMagar Staph aureus and CHROMagar MRSA for detection of *Staphylococcus aureus* in beach water and sand-comparison of culture, agglutination, and molecular analyses. *Water Res.* **2009**, *43*, 4802–4811. [CrossRef] [PubMed]

21. Soge, O.O.; Meschke, J.S.; No, D.B.; Roberts, M.C. Characterization of methicillin-resistant and methicillin-resistant coagulase-negative *Staphylococcus* spp. isolated from US West Coast public marine beaches. *J. Antimicrob. Chemother.* **2009**, *64*, 1148–1155. [CrossRef] [PubMed]

22. Tice, A.D.; Susan, J.R. Meeting the challenges of methicillin-resistant *Staphylococcus aureus* with outpatient parenteral antimicrobial therapy. *Clin. Infect. Dis.* **2010**, *51*, S171–S175. [CrossRef] [PubMed]

23. Shah, A.H.; Abdelzaher, A.M.; Phillips, M.; Hernandez, R.; Solo-Gabriele, H.M.; Kish, J.; Scorzetti, G.; Fell, J.W.; Diaz, M.R.; Scott, T.M.; et al. Indicator microbes correlate with pathogenic bacteria, yeasts and helminthes in sand at a subtropical recreational beach site. *J. Appl. Microbiol.* **2011**, *110*, 1571–1583. [CrossRef] [PubMed]

24. Goodwin, K.D.; Melody, M.; Yiping, C.; Darcy, E.; Melissa, M.; John, F.G. A multi-beach study of *Staphylococcus aureus*, MRSA, and enterococci in seawater and beach sand. *Water Res.* **2012**, *46*, 4195–4207. [CrossRef] [PubMed]

25. Deurenberg, R.H.; Stobberingh, E.E. The evolution of *Staphylococcus aureus*. *Infect. Genet. Evol.* **2008**, *8*, 747–763. [CrossRef] [PubMed]

26. Hartman, B.J.; Tomasz, A. Low-affinity penicillin-binding protein associated with beta-lactam resistance in *Staphylococcus aureus*. *J. Bacteriol.* **1984**, *158*, 513–516. [PubMed]

27. Tavares, A.L. Community-Associated Methicillin-Resistant *Staphylococcus aureus* (CA-MRSA) in Portugal: Origin, Epidemiology and Virulence. Ph.D. Thesis, Universidade Nova de Lisboa, Lisbon, Portugal, December 2014.

28. Werckenthin, C.; Cardoso, M.; Martel, J.L.; Schwarz, S. Antimicrobial resistance in staphylococci from animals with particular reference to bovine *Staphylococcus aureus*, porcine *Staphylococcus* hyicus and canine *Staphylococcus* intermedius. *Vet. Res.* **2001**, *32*, 341–362. [CrossRef] [PubMed]

29. Alekshun, M.N.; Levy, S.B. Molecular mechanisms of antibacterial multidrug resistance. *Cell* **2007**, *128*, 1037–1050. [CrossRef] [PubMed]

30. Tenover, F.C.; McDougal, L.K.; Goering, R.V.; Killgore, G.; Projan, S.J.; Patel, J.B.; Dunman, P.M. Characterization of a strain of community-associated methicillin-resistant *Staphylococcus aureus* widely disseminated in the United States. *J. Clin. Microbiol.* **2006**, *44*, 108–118. [CrossRef] [PubMed]

31. Wright, G.D. Bacterial resistance to antibiotics: Enzymatic degradation and modification. *Adv. Drug Deliv. Rev.* **2005**, *57*, 1451–1470. [CrossRef] [PubMed]

32. GOOGLE MAPS. Map of Eastern Cape Beaches. [Online]. Google, 2016. Available online: https://www.google.co.za/maps/search/eastern+cape+beaches+map+from+east+london+to+port+elizabeth/@33.9988012,22.5881698,2619m/data=!3m2!1e3!4b1?hl=en&authuser=0 (accessed on 18 June 2016).

33. Baums, I.B.; Goodwin, K.D.; Kiesling, T.; Wanless, D.; Fell, J.W. Luminex detection of faecal indicators in river samples, marine recreational water, and beach sand. *Mar. Pollut. Bull.* **2007**, *54*, 521–536. [CrossRef] [PubMed]

34. Soumet, C.; Ermel, G.; Fach, P.; Colin, P. Evaluation of different DNA extraction procedures for the detection of Salmonella from chicken products by polymerase chain reaction. *Lett. Appl. Microbiol.* **1994**, *19*, 294–298. [CrossRef] [PubMed]

35. Maes, N.; Magdalena, J.; Rottiers, S.; De Gheldre, Y.; Struelens, M. Evaluation of a triplex PCR assay to discriminate *Staphylococcus aureus* from coagulase-negative staphylococci and determine methicillin resistance from blood cultures. *J. Clin. Microbiol.* **2002**, *40*, 1514–1517. [CrossRef] [PubMed]

36. Bauer, A.W.; Kirby, W.M.; Sherris, J.C.; Turck, M. Antibiotic susceptibility testing by a standardized single disk method. *Am. J. Clin. Pathol.* **1966**, *45*, 493–496. [PubMed]

37. Clinical Laboratory Standards Institute. *Performance Standards for Antimicrobial Susceptibility Testing: Eighteenth Informational Supplement*; CLSI Document M100-S18; Clinical Laboratory Standards Institute: Wayne, PA, USA, 2014.

38. Veenemans, J.; Verhulst, C.; Punselie, R.; Van Keulen, P.H.J.; Kluytmans, J.A.J.W. Evaluation of brilliance MRSA 2 agar for detection of methicillin-resistant *Staphylococcus aureus* in clinical samples. *J. Clin. Microbiol.* **2013**, *51*, 1026–1027. [CrossRef] [PubMed]

39. Mason, W.J.; Blevins, J.S.; Beenken, K.; Wibowo, N.; Ojha, N.; Smeltzer, M.S. Multiplex PCR protocol for the diagnosis of Staphylococcal infection. *J. Clin. Microbiol.* **2001**, *39*, 3332–3338. [CrossRef] [PubMed]

40. Mehrotra, M.; Wang, G.; Johnson, W.M. Multiplex PCR for Detection of Genes for *Staphylococcus aureus* Enterotoxins, Exfoliative Toxins, Toxic Shock Syndrome Toxin 1, and Methicillin Resistance. *J. Clin. Microbiol.* **2000**, *38*, 1032–1035. [PubMed]

41. Schmitz, F.J.; Jones, M.E. Antibiotics for treatment of infections caused by MRSA and elimination of MRSA carriage. What are the choices? *Int. J. Antimicrob. Agents* **1997**, *9*, 1–19. [CrossRef]

42. Martineau, F.; Picard, F.J.; Lansac, N.; Menard, C.; Roy, P.H.; Ouellette, M.; Bergeron, M.G. Correlation between the resistance genotype determined by multiplex PCR assays and the antibiotic susceptibility patterns of *Staphylococcus aureus* and *Staphylococcus epidermidis*. *Antimicrob. Agents Chemother.* **2000**, *44*, 231–238. [CrossRef] [PubMed]

43. Aboshkiwa, M.; Rowland, G.; Coleman, G. Nucleotide sequence of the *Staphylococcus aureus* RNA polymerase *rpoB* gene and comparison of its predicted amino acid sequence with those of other bacteria. *Biochim. Biophys. Acta BBA Gene Struct. Express.* **1995**, *1262*, 73–78. [CrossRef]

44. Strommenger, B.; Kettlitz, C.; Werner, G.; Witte, W. Multiplex PCR assay for simultaneous detection of nine clinically relevant antibiotic resistance genes in *Staphylococcus aureus*. *J. Clin. Microbiol.* **2003**, *41*, 4089–4094. [CrossRef] [PubMed]

45. Nam, S.; Kim, M.J.; Park, C.; Park, J.G.; Lee, G.C. Detection and genotyping of vancomycin-resistant *Enterococcus* spp. by multiplex polymerase chain reaction in Korean aquatic environmental samples. *Int. J. Hyg. Environ. Health* **2012**, *216*, 421–427. [CrossRef] [PubMed]

46. Kuehnert, M.J.; Hill, H.A.; Kupronis, B.A.; Tokars, J.I.; Solomon, S.L.; Jernigan, D.B. Methicillin-resistant-*Staphylococcus aureus* hospitalizations, United States. *Emerg. Infect. Dis.* **2005**, *11*, 468. [CrossRef] [PubMed]

47. Baquero, F.J.M.; Rafael, C. Antibiotics and antibiotic resistance in water environments. *Curr. Opin. Biotechnol.* **2008**, *19*, 260–265. [CrossRef] [PubMed]

48. Oliveira, A.J.C.; de Franco, P.T.R.; Pinto, A.B. Antimicrobial resistance of heterotrophic marine bacteria isolated from seawater and sands of recreational beaches with different organic pollution levels in southeastern Brazil: evidences of resistance dissemination. *Environ. Monit. Assess.* **2010**, *169*, 375–384. [CrossRef] [PubMed]

49. Manivasagan, P.; Rajaram, G.; Ramesh, S.; Ashokkumar, S.; Damotharan, P. Occurrence and seasonal distribution of antibiotic resistance heterotrophic bacteria and physico-chemical characteristics of Muthupettai mangrove environment, southeast coast of India. *J. Environ. Sci. Technol.* **2011**, *4*, 139–149.
50. Mudryk, Z.; Perliński, P.; Skórczewski, P. Detection of antibiotic resistant bacteria inhabiting the sand of non-recreational marine beach. *Mar. Pollut. Bull.* **2010**, *60*, 207–214. [CrossRef] [PubMed]
51. Mudryk, Z. Occurrence and distribution antibiotic resistance of heterotrophic bacteria isolated from a marine beach. *Mar. Pollut. Bull.* **2005**, *50*, 80–86. [CrossRef] [PubMed]
52. Schwartz, T.W.; Kohenen, W.; Jansen, B.; Obst, U. Detection of antibiotic resistant bacteria and their resistance genes in wastewater, surface water and drinking water biofilms. *FEMS Microbiol Ecol.* **2003**, *43*, 325–335. [CrossRef] [PubMed]
53. David, M.Z.; Daum, R.S. Community-Associated Methicillin-Resistant *Staphylococcus aureus*: Epidemiology and Clinical Consequences of an Emerging Epidemic. *Clin. Microbiol. Rev.* **2010**, *23*, 616–687. [CrossRef] [PubMed]
54. Zetola, N.; Francis, J.S.; Nuermberger, E.L.; Bishai, W.R. Community-acquired meticillin-resistant *Staphylococcus aureus*: An emerging threat. *Lancet Infect. Dis.* **2005**, *5*, 275–286. [CrossRef]
55. Klevens, R.M.; Morrison, M.A.; Nadle, J.; Petit, S.; Gershman, K.; Ray, S.; Harrison, L.H.; Lynfield, R.; Dumyati, G.; Townes, J.M.; et al. Invasive methicillin-resistant *Staphylococcus aureus* infections in the United States. *JAMA* **2007**, *298*, 1763–1771. [CrossRef] [PubMed]
56. Boyle-Vavra, S.; Robert, S.D. Community-acquired methicillin-resistant *Staphylococcus aureus*: The role of Panton-Valentine leukocidin. *Lab. Investig.* **2007**, *87*, 3. [CrossRef] [PubMed]
57. Abdelzaher, A.M.; Wright, M.E.; Ortega, C.; Solo-Gabriele, H.M.; Miller, G.; Elmir, S.; Newman, X.; Shih, P.; Bonilla, J.A.; Bonilla, T.D.; et al. Presence of pathogens and indicator microbes at a non-point source subtropical recreational marine beach. *Appl. Environ. Microbiol.* **2010**, *76*, 724–732. [CrossRef] [PubMed]
58. Levin-Edens, E.; Bonilla, N.; Meschke, J.S.; Roberts, M.C. Survival of environmental and clinical strains of methicillin-resistant *Staphylococcus aureus* (MRSA) in marine and fresh waters. *Water Res.* **2011**, *45*, 5681–5686. [CrossRef] [PubMed]
59. Yamahara, K.M.; Sassoubre, L.M.; Goodwin, K.D.; Boehm, A.B. Occurrence and persistence of human pathogens and indicator organisms in beach sands along the California coast. *Appl. Environ. Microbiol.* **2011**, *78*, 1733–1745. [CrossRef] [PubMed]
60. Gundogan, N.; Citak, S.; Yucel, N.; Devren, A. A note on the incidence and the antibiotic resistance of *Staphylococcus aureus* isolated from meat and chicken samples. *Meat Sci.* **2005**, *69*, 807–810. [CrossRef] [PubMed]
61. Chambers, H.F.; DeLeo, F.R. Waves of resistance: *Staphylococcus aureus* in the antibiotic era. *Nat. Rev. Microbiol.* **2009**, *7*, 629–641. [CrossRef] [PubMed]
62. Akindolire, M.A. Detection and Molecular Characterization of Virulence Genes in Antibiotic Resistant *Staphylococcus aureus* from Milk in the North West Province. Ph.D. Thesis, North West University, Mafikeng, North West Province, South Africa, 2013.
63. Bartley, J. First case of VRSA identified in Michigan. *Infect. Control Hosp. Epidemiol.* **2002**, *23*, 480. [PubMed]
64. Boerlin, P.; Burnens, A.P.; Frey, J.; Kuhnert, P.; Nicolet, J. Molecular epidemiology and genetic linkage of macrolide and aminoglycoside resistance in *Staphylococcus intermedius* of canine origin. *Vet. Microbiol.* **2001**, *79*, 155–169. [CrossRef]
65. Tanwar, J.; Das, S.; Fatima, Z.; Hameed, S. Multidrug resistance: An emerging crisis. *Interdiscip. Perspect. Infect. Dis.* **2014**. [CrossRef] [PubMed]
66. Skórczewski, P.; Jan Mudryk, Z.; Miranowicz, J.; Perlinski, P.; Zdanowicz, M. Antibiotic resistance of Staphylococcus-like organisms isolated from a recreational sea beach on the southern coast of the Baltic Sea as one of the consequences of anthropogenic pressure. *Int. J. Oceanogr. Hydrobiol.* **2014**, *43*, 41–48. [CrossRef]
67. Ateba, C.N.; Mbewe, M.; Moneoang, M.S.; Bezuidenhout, C.C. Antibiotic-resistant *Staphylococcus aureus* isolated from milk in the Mafikeng Area, North West province, South Africa. *S. Afr. J. Sci.* **2010**, *106*, 1–6. [CrossRef]
68. Pesavento, G.; Ducci, B.; Comodo, N. Antimicrobial resistance profile of *Staphylococcus aureus* isolated from raw meat: A research for methicillin resistant *Staphylococcus aureus* (MRSA). *Food Control* **2007**, *18*, 196–200. [CrossRef]

69. Normanno, G.; La Salandra, G.; Dambrosio, A.; Quaglia, N.; Corrente, M.; Parisi, A.; Santagada, G.; Firinu, A.; Crisetti, E.; Celano, G. Occurrence, characterization and antimicrobial resistance of entero-toxigenic *Staphylococcus aureus* isolated from meat and dairy products. *Int. J. Food Microbiol.* **2007**, *115*, 290–296. [CrossRef] [PubMed]

70. Ito, T.; Okurna, K.; Ma, X.X.; Yuzawa, H.; Hiramatsu, K. Insights on antibiotic resistance of *Staphylococcus aureus* from its whole genome: Genomic island SCC. *Drug Resist. Updates* **2003**, *6*, 41–52. [CrossRef]

71. Guardabassi, L.; Petersen, A.; Olsen, J.E.; Dalsgaard, A. Antibiotic resistance in *Acinetobacter* sp. isolated from sewers receiving waste effluent from a hospital and a pharmaceutical plant. *Appl. Environ. Microbiol.* **1998**, *64*, 3499–3502. [PubMed]

72. Kümmerer, K. Drugs in the environment: Emission of drugs, diagnostic aids and disinfectants into wastewater by hospitals in relation to other sources—A review. *Chemosphere* **2001**, *45*, 957–969. [CrossRef]

73. Zscheck, K.K.; Murray, B.E. Genes involved in the regulation of β-lactamase production in enterococci and staphylococci, *Antimicrob. Agents Chemoth.* **1993**, *37*, 1966–1970. [CrossRef]

74. Lee, J.H. Methicillin (oxacillin)-resistant *Staphylococcus aureus* strains isolated from major food animals and their potential transmission to humans. *Appl. Environ. Microbiol.* **2003**, *69*, 6489–6494. [CrossRef] [PubMed]

75. Kamal, R.M.; Bayourni, M.A.; Abd El Aal, S.F.A. MRSA detection in raw milk, some dairy products and hands of dairy workers in Egypt, a mini-survey. *Food Control* **2013**, *33*, 49–53. [CrossRef]

76. Mirzaei, H.; Tofighi, A.; Sarabi, H.K.; Farajli, M. Prevalence of Methicillin-Resistant *Staphylococcus aureus* in Raw Milk and Dairy Products in Sarab by Culture and PCR Techniques. *J. Anim. Vet. Adv.* **2011**, *10*, 3107–3111.

77. Börjesson, S.; Melin, S.; Matussek, A.; Lindgren, P.E. A seasonal study of the *mec*A gene and *Staphylococcus aureus* including methicillin-resistant *S. aureus* in a municipal wastewater treatment plant. *Water Res.* **2009**, *43*, 925–932. [CrossRef] [PubMed]

78. Swenson, J.M.; Lonsway, D.; McAllister, S.; Thompson, A.; Jevitt, L.; Zhu, W.; Patel, J.B. Detection of mecA-mediated resistance using reference and commercial testing methods in a collection of *Staphylococcus aureus* expressing borderline oxacillin MICs. *Diagn. Microbiol. Infect. Dis.* **2007**, *58*, 33–39. [CrossRef] [PubMed]

79. Oliveira, D.C.; de Lencastre, H. Multiplex PCR strategy for rapid identification of structural types and variants of the mec element in methicillin-resistant *Staphylococcus aureus*. *Antimicrob. Agents Chemother.* **2002**, *46*, 2155–2161. [CrossRef] [PubMed]

80. Chambers, H.F. Methicillin resistance in staphylococci: Molecular and biochemical basis and clinical implications. *Clin. Microbiol. Rev.* **1997**, *10*, 781–791. [PubMed]

81. Wehrli, W. Rifampin: Mechanisms of action and resistance. *Rev. Infect. Dis.* **1983**, *5*, 407–411. [CrossRef]

82. Aubry-Damon, H.; Soussy, C.J.; Courvalin, P. Characterization of mutants in the *rpoB* gene that confer rifampin resistance in *Staphylococcus aureus*. *Antimicrob. Agents Chemother.* **1998**, *42*, 2590–2594. [PubMed]

83. Severinov, K.; Mustaev, A.; Severinova, E.; Kozlov, M.; Darst, S.A.; Goldfarb, A. The b-subunit rif-cluster I is only angstroms away from the active center of *Escherichia coli* RNA polymerase. *J. Biol. Chem.* **1995**, *270*, 29428–29432. [CrossRef] [PubMed]

84. Jin, D.J.; Gross, C.A. Mapping and sequencing of mutations in the *Escherichia coli rpoB* gene that lead to rifampicin resistance. *J. Mol. Biol.* **1988**, *202*, 45–58. [CrossRef]

85. Weisblum, B. Erythromycin resistance by ribosomes modification. *Antimicrob. Agents Chemother.* **1995**, *39*, 577–585. [CrossRef] [PubMed]

86. Nicola, F.G.; McDougal, L.K.; Biddle, J.W.; Tenover, F.C. Characterization of erythromycin- resistant isolates of *Staphylococcus aureus* recovered in the United States from 1958 through 1969. *Antimicrob. Agents Chemother.* **1998**, *42*, 3024–3027. [PubMed]

87. Simeoni, D.; Rizzotti, L.; Cocconcelli, P.; Gazzola, S.; Dellaglio, F.; Torriani, S. Antibiotic resistance genes and identification of staphylococci collected from the production chain of swine meat commodities. *Food Microbiol.* **2008**, *25*, 196–201. [CrossRef] [PubMed]

88. Maidhof, H.; Reinicke, B.; Blumel, P.; Berger-Bachi, B.; Labischinski, H. *femA*, Which Encodes a Factor Essential for Expression of Methicillin Resistance, Affects Glycine Content of Peptidoglycan in Methicillin-Resistant and Methicillin-Susceptible *Staphylococcus aureus* Strains. *J. Bacterial.* **1991**, *173*, 3507–3513. [CrossRef]

89. Kobayashi, N.; Wu, H.; Kojima, K.; Taniguchi, K.; Urasawa, S.; Uehara, N.; Omizu, Y.; Kishi, Y.; Yagihashi, A.; Kurokawa, I. Detection of *mec*A, *fem*A, and *fem*B genes in clinical strains of staphylococci using polymerase chain reaction. *Epidemiol. Infect.* **1994**, *113*, 259–266. [CrossRef] [PubMed]

International Journal of
*Environmental Research
and Public Health*

Article

Resistance of *Escherichia coli* in Turkeys after Therapeutic or Environmental Exposition with Enrofloxacin Depending on Flooring

Bussarakam Chuppava [1], Birgit Keller [1], Amr Abd El-Wahab [2], Jessica Meißner [3], Manfred Kietzmann [3] and Christian Visscher [1,*]

[1] Institute for Animal Nutrition, University of Veterinary Medicine Hannover, Foundation, Bischofsholer Damm 15, D-30173 Hanover, Germany; bussarakam.chuppava@tiho-hannover.de (B.C.); birgit.keller@tiho-hannover.de (B.K.)

[2] Department of Nutrition and Nutritional Deficiency Diseases, Faculty of Veterinary Medicine, Mansoura University, Mansoura 35516, Egypt; amrabdelwahab37@yahoo.de

[3] Institute for Pharmacology, Toxicology, and Pharmacy, University of Veterinary Medicine Hannover, Foundation, Bünteweg 17, D-30559 Hannover, Germany; jessica.meissner@tiho-hannover.de (J.M.); manfred.kietzmann@tiho-hannover.de (M.K.)

* Correspondence: christian.visscher@tiho-hannover.de; Tel: +49-511-856-7415

Received: 27 July 2018; Accepted: 11 September 2018; Published: 13 September 2018

Abstract: Gaining knowledge about the spread of resistance against antibacterial agents is a primary challenge in livestock farming. The purpose of this study was to test the effect of double antibiotic treatment (at days 10–14 and days 26–30) with enrofloxacin or solely environmental exposition (identical times, directly into the litter) on resistance against antibacterial agents in commensal *Escherichia coli* in comparison with the control (without treatment), depending on different flooring. A total of 720 Big 6 turkeys participated in three trials. Four different flooring designs were examined: An entire floor pen covered with litter, a floor pen with heating, a partially slatted flooring including 50% littered area, and a fully slatted flooring with a sand bath. A total of 864 *Escherichia coli* isolates were obtained from cloacal swabs and poultry manure samples at days 2, 9, 15, 21, and 35. The broth microdilution method (MIC) was used to determine the resistance of isolates to enrofloxacin and ampicillin. A double antibiotic treatment with enrofloxacin reduced the proportion of susceptible *Escherichia coli* isolates significantly in all flooring designs. Simulation of water losses had no significant effect, nor did the flooring design. Ampicillin-resistant isolates were observed, despite not using ampicillin.

Keywords: flooring design; Turkey; antibacterial resistance; enrofloxacin; commensal *E. coli*

---

1. Introduction

Resistance to antibacterial agents is an increasing problem in public health and veterinary medicine worldwide [1–3]. The major public health concern which has been expressed for several decades is still the potential for transmission of antibiotic-resistant bacteria from animals to humans [4]. Most of the amounts of antibiotics used (30–80%) in livestock farming are excreted by the animals directly into the environment via urine and feces because of partial metabolization of antibacterial agents and residue in manure [5–7]. Resistance to antibacterial agents in Gram-negative bacteria is on the rise in pathogens as well as in commensal bacterial flora, particularly in *Escherichia coli*. *E. coli* constitutes the majority of invasive Gram-negative isolates for humans in European countries [8]. The natural habitat of *E. coli* is the gastrointestinal tract of mammals and birds [9]. It is considered as an indicator bacteria for resistance detection. *E. coli* also has the ability to survive in and adapt to various extra intestinal habitats and to spread resistances between humans, animals, and the environment [10].

Antibacterial agents in livestock production have been either used to prevent diseases and promote animal growth or for therapeutic purposes [11,12]. The total sales of veterinary antibacterial agents during 2015 in the European Union (EU) amounted to approximately 8361 tons [13]. The average antibacterial consumption by humans (124 mg/kg) was lower than in animals (152 mg/kg) [3]. The resistance level of avian isolates to *E. coli* in Germany, for example, exceeded the level determined by the Federal Office of Consumer Protection and Food Safety for other veterinary pathogens in other animal species [14]. In the past, in relation to their respective fattening periods, in poultry, antibacterial agents have been used more often and for a longer duration compared with cattle and pigs [15].

Fluoroquinolones (FQ) have been classified as being critically important for human health and animal farms by the World Health Organization [4]. An unfavorable situation has arisen in Europe: Resistance to these antibiotics is widespread and the incidence of resistance increased significantly between 2012 and 2015 [3]. The application of FQ agents in poultry husbandry has led to increasing problems with resistance to antibacterial agents [16,17]. The level of fluoroquinolone consumption showed a significant correlation with antibiotic resistance in *E. coli* in livestock husbandry [3,18,19]. In turkeys, commensal and pathogenic *E. coli* are often resistant to quinolones, including enrofloxacin, and to β-lactams [2,13]. Commensal *E. coli* isolates gained from turkey meat in Germany showed higher rates of resistance to FQ than *E. coli* from broilers [20].

In commercial poultry meat production in Europe, turkeys are reared on littered concrete floor. During the fattening period, the primary litter material becomes mixed with poultry excreta, feathers, feed, and spilt drinking water [21], the resulting mixture being referred to as poultry manure. Therefore, close contact with their litter or rather manure is common for turkeys during their productive life. More than 95% of the dry matter in manure consists of excreta [22]. This material can contain residues of antibacterial agents as well as resistant bacteria [17]. On almost every farm (62.3%), *E. coli* can be isolated from manure [23]. The poultry environment has long been acclaimed as a potential source of antibiotic-resistant bacteria [5,17], acting as a possible reservoir for the dissemination of these organisms to humans via the food chain (poultry meat), person-to-person contact (food handlers), and environment (poultry waste disposal, organic fertilizers). A significant proportion of these antibiotics is excreted unchanged in animal urine and feces. These antibiotics can remain potent for a longer time in manure during storage [5,24].

Information concerning the effects of separating animals from their excreta on the development of resistance to antibacterial agents in commensal *E. coli* in rearing turkeys has only been described in the study by Chuppava et al. [25]. The aim of the present study was to evaluate the effect of double antibiotic treatment (at days 10–14 and at days 26–30) with enrofloxacin or solely environmental exposition (at days 10–14 and days 26–30 directly into the litter) on resistance against antibacterial agents in commensal *Escherichia coli* in comparison with the control (without treatment), depending on different flooring. The different types of flooring design were distinguished by means of the contact intensity of birds to their excreta.

2. Materials and Methods

The animal experiments were conducted in accordance with the corresponding German regulations and approved by the Ethics Committee of Lower Saxony for Care and Use of Laboratory Animals (LAVES) (Niedersächsisches Landesamt für Verbraucherschutz und Lebensmittelsicherheit; reference: 33.12-42502-04-15/2044).

2.1. Design of Experiments

A total of 720 female one-day-old turkeys (B.U.T. Big 6) were obtained from a commercial hatchery (Heidemark GmbH, Ahlhorn, Germany). A total of three independent experiments (T1–T3) were carried out. For each of these experiments, 240 birds were used.

Before starting the experiments in the second week after hatch, the birds were housed in dry and clean floor pens in a quarantine stable. Flooring was covered with wood shavings (GOLDSPAN®,

Goldspan GmbH and Co. KG, Goldenstedt, Germany). A commercially prepared pelleted diet was offered ad libitum (Best 3 Geflügelernährung GmbH, Twistringen, Germany).

Each experiment was started after the above described one-week adaptation period. For these experiments, specially manufactured boxes were used, twelve experimental pens (1.20 × 0.80 m) in total. Different flooring designs were used to establish different degrees of contact intensity of the animals to the manure (Figure 1).

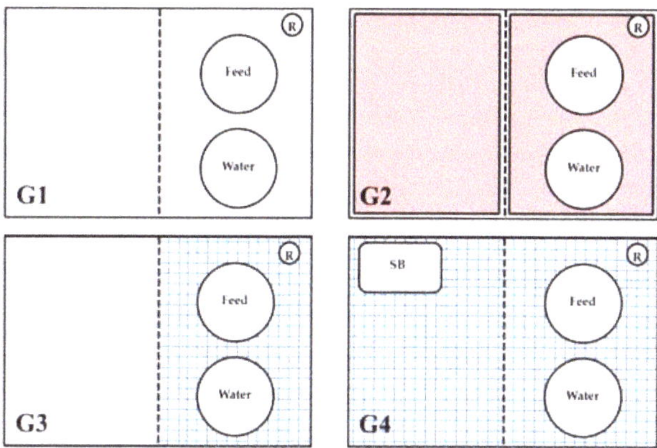

Figure 1. Flooring designs used in the study: **G1** = entire floor pen with litter; **G2** = identical to G1 and additionally having floor heating (in red); **G3** = plastic covered steel slats in 50% of the pen (in blue) as well as an area with litter; **G4** = fully-slatted flooring with plastic covered steel slats and a sand bath (900 cm^2). SB = sand bath, R = rope.

The first group served as a control. Animals were kept on dry wood shavings (G1—entire floor pen covered with litter). The second group was identically kept. The exception was an electrical floor heating system (Sauerland GmbH, Paderborn-Elsen, Germany) with an adjuster to control the temperature (G2—floor pen with litter with floor heating). Animals in these two groups continuously had full contact with manure. The pens in the third group (G3) were divided into two equal parts consisting of 50% solid flooring with wood shavings on the right-hand side and 50% plastic slatted flooring on the left-hand side. In the last group (G4), plastic slatted flooring with a sand bath (900 cm^2) was used, the bath being disinfected and the sand replaced on a daily basis. Animals in G4 had no contact with litter except possibly in the sand bath. Plastic covered steel slats consisted of holes (15 × 10 mm) and bridges (plastic covered steel; 3.5 mm wide; Big Dutchman International GmbH, Vechta, Germany). The excreta were stored under the slatted floor at a depth of approximately 30 cm without any material being removed during the trial, besides small amounts of material needed for the samples as described below.

The boxes were placed in a randomized sequence in blocks of four subgroups (G1–G4) in the same stable, as previously described [25]. Two boxes of each block were placed on the right-hand side and two on the left-hand side of a central corridor (~1.70 m width). Airing was provided by vacuum air ventilation. This system was installed in the ceiling in two rows above the pens. Wood shavings were used as bedding material (1 kg/m^2). Stocking densities reached a maximum of 25 kg/m^2. Hanging type feeders were used (Klaus Gritsteinwerk GmbH & Co. KG, Bünde, Germany) as well as bell drinkers (Ferdinand Stükerjürgen GmbH & Co. KG, Rietberg-Varensell, Germany).

Before commencing with the trials one week after hatch, stables and all materials had been disinfected. Also, tests had been performed to exclude the occurrence of Enterobacteriaceae. All birds were allocated to four groups, each with three identical subgroups (n = 20 birds). Rearing was done until day 36. The birds had unlimited access to fresh water and feed (commercial pelleted growing diet). The environmental temperature was gradually reduced from about 33 °C for the one-day-old birds to about 20 °C by day 36. Lights were continuously on between days 1 and 3 and the photoperiod from day 4 onwards amounted to 16 h of light and 8 h of darkness.

In T1 there was no antibiotic treatment. This experiment served as a nontreated control trial. In contrast, animals in T2 were medicated with Baytril® 10% in drinking water (10 mg enrofloxacin/kg body weight per day—corresponding to an addition of 0.5 mL Baytril® 10%/L of drinking water, in accordance with the recommended dosage; Bayer Vital GmbH, Leverkusen, Germany). In the last trial (T3), the birds were not treated with any antibiotic in drinking water. Spillage of drinking water containing enrofloxacin was simulated. The amount of water losses was calculated according to experience from former trials, comparing water intake in turkeys using drinking bowls and nipple drinkers (data not shown). Water containing enrofloxacin (dosage: 0.5 mL/L of Baytril® 10%, amount 240 mL per day) was sprayed into the litter or on the slatted flooring only in the feeding area. Both, in T2 and T3, five-day treatments were performed at days 10–14 and days 26–30.

At the end of day 21, eight out of 20 birds in each subgroup were dissected. Final dissection for all remaining turkeys (n = 12/box) was done at day 36. The stunning method (percussive blow to the head) was conducted in accordance with Annex I of the Council Regulation (EC) No. 1099/2009, Chapter I, Methods, Table 1—Mechanical Methods [26].

2.2. Collection of Cloacal Swabs and Manure Samples

Samples (864 in total) were taken before treatment, directly after antibiotic treatment and at the end of the trial. Cloacal swabs were collected at day 2 and manure samples at day 9 before treatment (BT: before treatment stage). After the enrofloxacin treatment (AT), at day 15, manure samples were taken and six days later, cloacal swabs were taken (day 21). Final sampling (ET) was done for both (manure and cloacal swabs) at day 35. Cloacal swabs were always collected from 24 animals per group, i.e., in total, 96 randomly selected animals. Six samples from each type of flooring design (two samples per pen), in total 24 samples of manure, were taken from two defined locations (feeding area and resting area) in every pen for all trial stages (BT, AT, and ET). Manure samples were taken with a plastic cup (6 cm in diameter) which removed the whole litter material at these locations right down to the floor. All samples were immediately transferred to the laboratory for following analyses.

2.3. Bacteriological Analyses and E. coli Isolation

The bacteriological investigations were carried out as previously described [25]. In brief: Cloacal swab samples were directly streaked on Gassner agar plates, following an incubation overnight at 37 °C. For manure samples 50 mL of peptone water (Oxoid, Wesel, Germany) as well as the manure sample itself (25 g each) were put into a sterile Whirl-Pak® Bag (Nasco, Fort Atkinson, WI, USA). Bags were mixed for three minutes with a Bag Mixer® 400 VW (Interscience, Saint Nom, France). Using a sterile loop, 10 µL of each mixed-sample was streaked on Gassner agar (Oxoid, Wesel, Germany) and incubated at 37 °C for 18–24 h.

One single blue color colony from each plate was selected and spread onto Columbia blood agar (Oxoid, Wesel, Germany) and Tryptone Bile X-glucuronide (TBX) agar (Oxoid, Wesel, Germany). Incubation was done overnight at 37 °C. Bluegreen colonies on TBX agar detected glucuronidase activity. The positive indole test with Kovac's indole reagent (Merck, Darmstadt, Germany) was used to confirm the diagnosis.

2.4. Antibacterial Susceptibility Testing

The guidelines of the Clinical and Laboratory Standards Institute (CLSI) and the manufacturer's recommendations were the basis for testing the resistance of *E. coli* isolates using the broth microdilution technique. Micronaut plates (Merlin, Bornheim-Hersel, Germany) with Mueller–Hinton Broth (Merlin, Bornheim-Hersel, Germany) were used to determine minimal inhibitory concentrations (MICs) of enrofloxacin (ENR) and ampicillin (AMP). Dried antibacterial agents in serial dilutions of enrofloxacin and ampicillin were placed in wells of these plates, as previously described by Chuppava et al. [25]. MIC values were determined by visually reading and interpreting the results. As the reference strain, *E. coli* ATCC 25922 was tested concurrently on each testing day.

2.5. Screening of Antibacterial Agents in Water Using High Performance Liquid Chromatography

An aliquot of collected water samples with enrofloxacin from days ten to 14 and days 26 to 30 was used for analyses. The concentration of enrofloxacin in the water was determined using high performance liquid chromatography (HPLC) via the method described by Scherz [27]. Exactly 100 µL of the sample was injected into the system with an autosampler (System Gold 508, Beckmann, Munich, Germany). A flow of 1 mL per minute was maintained by the System Gold 126 solvent module (Beckmann, Munich, Germany). A CC250/4 NUCLEODUR 100-5C 18ec (25 cm, Macherey-Nagel, Oensingen, Germany) column was used, this being connected to a precolumn (LiChroCART® 4-4, Li- Chrospher® 100 RP-18e, 5 µm, Merck, Darmstadt, Germany). A fluorescence detector (RF-551 Shimadzu, Nakagyo-ku, Japan) with 280 nm for excitation and 450 nm for emission was used for detection. The mobile phase consisted of 85% citrate buffer pH 3.0 (citric acid monohydrate: 1.80 g/L tri-sodium-citrate-dihydrate: 0.43 g/L). The concentration of enrofloxacin in the water samples was calculated with the external standard method.

2.6. Statistical Analyses

The data of resistance to antibacterial agents were performed using the SAS statistical software package version 7.1 (SAS Inst., Cary, NC, USA). MICs were summarized and reported as susceptible (S), intermediate (I), and resistant (R; the results were classified as 1 = S, 2 = I, or 3 = R), where CLSI veterinary breakpoints were available [28]. The analyses were made with these values for the categories. There are no intermediate values between classes one, two, and three. Therefore, a generally high standard deviation has to be tolerated. In the case of completely sensitive isolates at the beginning of the tests, the values are constant at one, i.e. the standard deviation is zero and can therefore not be seen graphically. Significant differences in the means of the resistance results between the four groups of flooring designs were tested using the repeated measures ANOVA (Fisher's Least Significant Difference (LSD)). This test was also used to determine the differences between the sampling stages and the frequency of resistance between the three trials.

3. Results

In total, 864 *E. coli* were isolated and analyzed. These isolates were obtained from 648 cloacal swabs and 216 manure samples at the BT, AT, and ET stages. In the water collected at days ten to 14 and days 26 to 30 in trial 2, the enrofloxacin concentration were 50.17 and 50.62 µg/mL, respectively; in trial 3, water contained 49.87 and 50.42 µg enrofloxacin/mL, respectively.

3.1. Differences in Resistance to Antibacterial Agents in E. coli between Sampling Points as Well as between Trials

Enrofloxacin-resistant *E. coli* isolated from cloacal swabs and manure samples were found at the beginning of trial 1 (T1) and showed significantly higher mean resistance rates than in the other trials (Table 1). In contrast, in trials 2 (T2) and 3 (T3), none of the *E. coli* isolates during the BT stage were resistant to enrofloxacin. There were no significant differences between trial 2 and trial 3 during this stage.

Significant differences could be found between the trials during the AT and ET stages (Table 1). Isolates from the cloacal swabs and manure samples from trial 2 showed the significantly highest resistance to enrofloxacin of the isolates after administering Baytril®, followed by mean values of trial 1 and trial 3 (cloacal swabs: 2.90, 1.98, and 1.00, respectively; manure samples: 2.63, 2.00, and 1.08, respectively; Table 1). Also, at the ET stage, the results of mean enrofloxacin resistance in trial 2 showed the same relationship to the other experiments (Table 1).

Table 1. Means of enrofloxacin-resistant *E. coli* isolates from cloacal swab and manure samples from turkeys.

Time of Sample Collection **	Enrofloxacin *					
	Cloacal Swab (N = 648) ***			Manure (N = 216) ***		
	BT	AT	ET	BT	AT	ET
T1	1.42 [A,b]	1.98 [B,a]	2.20 [B,a]	1.75 [A,a]	2.00 [B,a]	1.67 [B,a]
T2	1.00 [B,b]	2.90 [A,a]	2.99 [A,a]	1.00 [B,b]	2.63 [A,a]	2.92 [A,a]
T3	1.00 [B,a]	1.00 [C,a]	1.04 [C,a]	1.00 [B,a]	1.08 [C,a]	1.00 [C,a]

[A, B, C] means in the same column differ significantly between the experiments ($p < 0.05$); [a, b] means differ significantly between the stage of sampling within one experiment ($p < 0.05$); * MICs were summarized and reported as susceptible (S), intermediate (I), and resistant (R). Afterwards the results were classified as 1 = S, 2 = I, or 3 = R and means thereof were calculated; ** BT = before treatment; AT = after treatment; ET = end of trial. T1 = untreated antibiotic trial, T2 = treated antibiotic trial, T3 = trial with simulated water spillage containing antibiotic; *** Cloacal swabs: N = 648; per trial BT: n = 24, AT: n = 96, ET: n = 96; poultry manure: N = 216; per trial BT: n = 24, AT: n = 24, ET: n = 24. G1 = entire floor pen covered with litter; G2 = floor pen covered with litter and having floor heating; G3 = partially (50:50) slatted flooring including an area that was littered; G4 = fully slatted flooring with a sand bath (900 cm^2).

When comparing the sampling stages (Table 1), the means in enrofloxacin resistance were significantly different between trial 1 and trial 2 regarding the *E. coli* isolated from the cloacal swabs (Table 1). For the medicated group (T2), the number of samples with isolation of resistant *E. coli* in materials (cloacal swab and manure sample) significantly increased from the BT to AT stages upon exposure to enrofloxacin. Nevertheless, the *E. coli* from all samples showed no significant differences in the resistance between the AT and ET stages (Table 1).

The results of means in resistance of ampicillin resistant *E. coli* isolates in trials 1, 2, and 3 are presented in Table 2. *E. coli* isolates from cloacal swabs were 100% susceptible to ampicillin during the BT stage except in trial 1. In this trial, isolates showed a significantly higher resistance to ampicillin (G1 = 1.33, G2 = 1.00, G3 = 1.00, respectively; Table 2).

Table 2. Means of ampicillin-resistant *E. coli* isolates from cloacal swab and litter/excreta samples from turkeys.

Time of Sample Collection **	Ampicillin *					
	Cloacal Swab (N = 648) ***			Manure (N = 216) ***		
	BT	AT	ET	BT	AT	ET
T1	1.33 [A,a]	1.80 [B,a]	1.52 [B,a]	1.42 [A,a]	1.33 [B,a]	1.00 [B,a]
T2	1.00 [B,b]	1.31 [C,a]	2.00 [A,a]	1.00 [B,b]	2.13 [A,a]	1.67 [A,a]
T3	1.00 [B,b]	2.08 [A,a]	1.73 [B,a]	1.25 [AB,a]	1.17 [B,a]	1.83 [A,a]

[A, B, C] means in the same column differ significantly between the experiments ($p < 0.05$); [a, b] means differ significantly between the stage of sampling within one experiment ($p < 0.05$); * MICs were summarized and reported as susceptible (S), intermediate (I), and resistant (R). Afterwards the results were classified as 1 = S, 2 = I, or 3 = R and means thereof were calculated; ** BT = before treatment; AT = after treatment; ET = end of trial. T1 = untreated antibiotic trial, T2 = treated antibiotic trial, T3 = trial with simulated water spillage containing antibiotic; *** Cloacal swabs: N = 648; per trial BT: n = 24, AT: n = 96, ET: n = 96; poultry manure: N = 216; per trial BT: n = 24, AT: n = 24, ET: n = 24; G1 = entire floor pen covered with litter; G2 = floor pen covered with litter and having floor heating; G3 = partially (50:50) slatted flooring including an area that was littered; G4 = fully slatted flooring with a sand bath (900 cm^2).

During the AT stage, a significant difference between the three trials occurred in isolates from the cloacal samples. At this point in time, isolates in T3 showed the significantly highest means in enrofloxacin resistance in cloacal swabs, whereas in manure samples, T2-samples had the highest means. During the ET stage in trial 2, the means of ampicillin-resistant *E. coli* isolates from cloacal swabs were significantly higher than in the other trials (Table 2). In manure samples, no more ampicillin resistance was found in T1 during the ET stage.

The results of mean ampicillin resistance in T2 (animals treated twice with enrofloxacin) differed significantly between the sampling days regarding the *E. coli* strains isolated from the cloacal swabs and manure samples. There was a significant increase in means from the BT to the AT stage (Table 2). In trial 3, there was also a significant increase in means of resistance. The percentage of susceptible isolates changed from 100% susceptible isolates to 46% after simulation of water losses with water containing antibiotic. The significance between the AT and ET stages could not be found in all trials (Table 2).

3.2. Testing the Effect of Different Flooring Designs on the Resistance to Antibacterial Agents in E. coli

The mean values of resistance of *E. coli* isolates to enrofloxacin and ampicillin depending on sampling stage and flooring design are presented in Figures 2a–d and 3a–d.

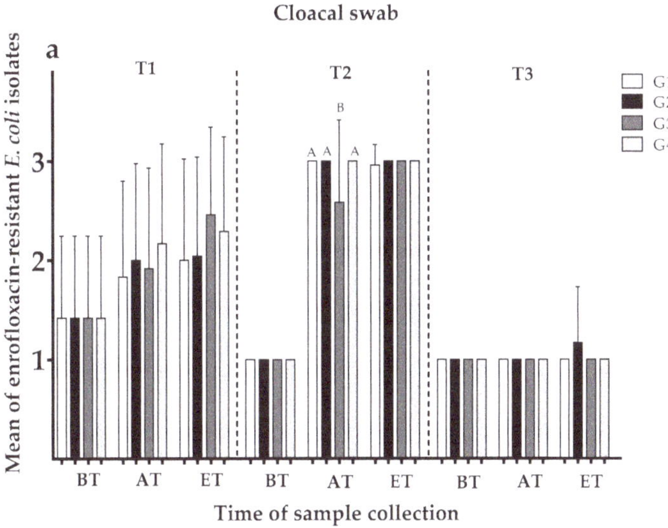

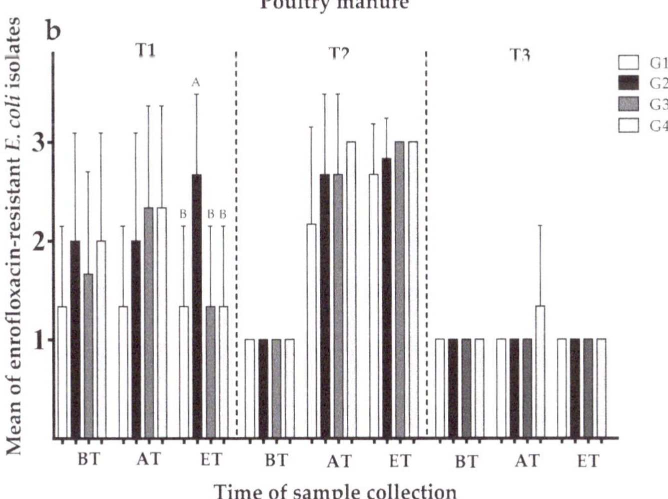

Figure 2. *Cont.*

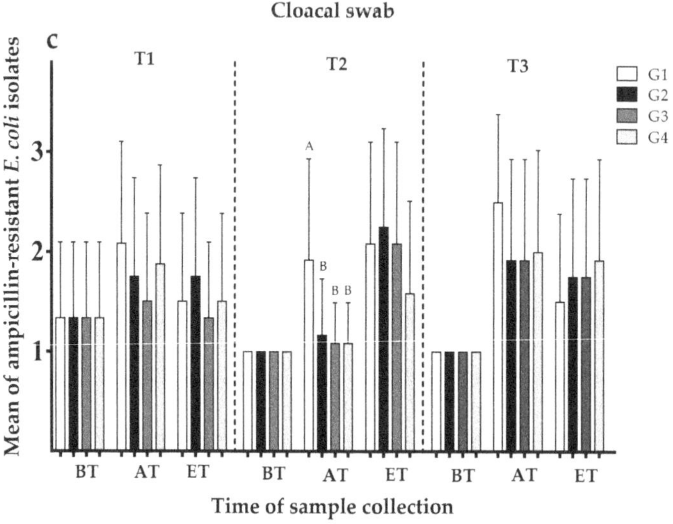

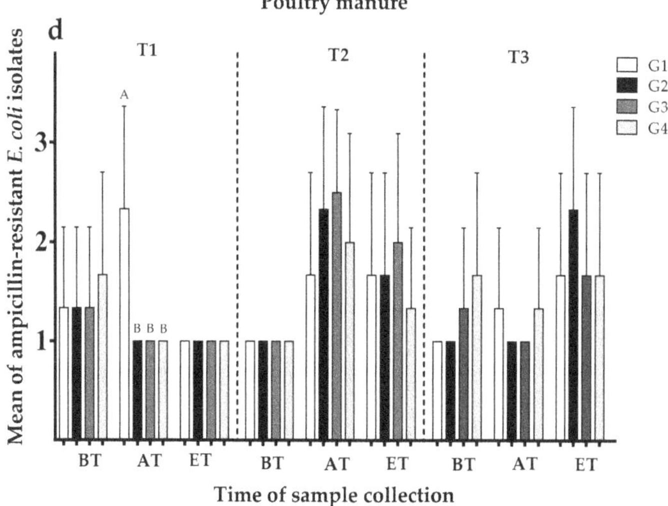

Figure 2. Means of susceptible (=1); intermediate (=2); and resistant (=3) *E. coli* isolates concerning enrofloxacin resistance in (**a**) cloacal swabs and (**b**) poultry manure samples as well as ampicillin resistance in (**c**) cloacal swabs and (**d**) poultry manure samples before treatment (BT), after treatment (AT) and at the end of trial (ET; cloacal swabs: N = 648; per trial BT: n = 24, AT: n = 96, ET: n = 96; poultry manure: N = 216; per trial BT: n = 24, AT: n = 24, ET: n = 24). T1 = no treatment with antibiotic; T2 = treatment of enrofloxacin via drinking water; and T3 = water (containing enrofloxacin) loss simulation trial. G1 = entire floor pen covered with litter; G2 = floor pen covered with litter and having floor heating; G3 = partially (50:50) slatted flooring including an area that was littered; and G4 = fully slatted flooring with a sand bath (900 cm^2). [A, B] means differ significantly between the groups at one sampling ($p < 0.05$).

cloacal swab

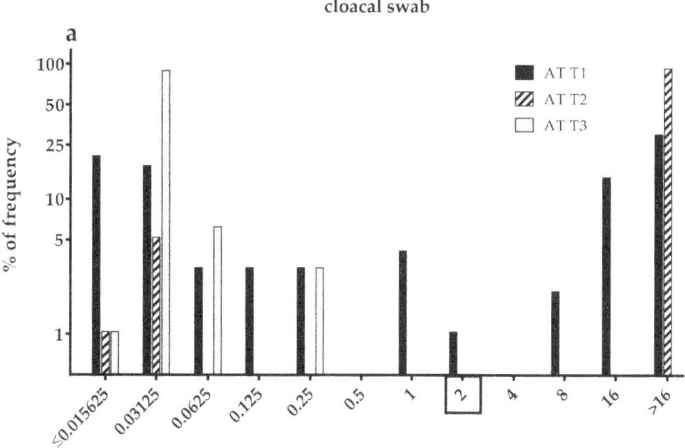

cloacal swab

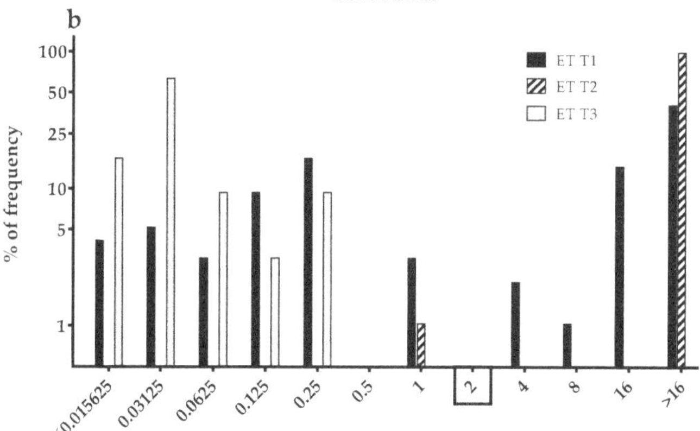

poultry manure

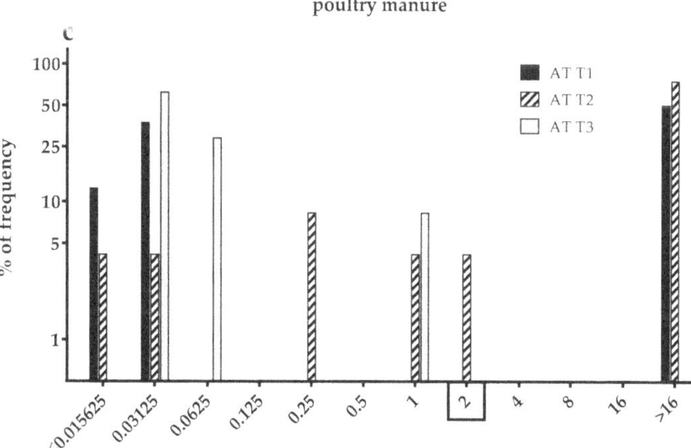

Figure 3. *Cont.*

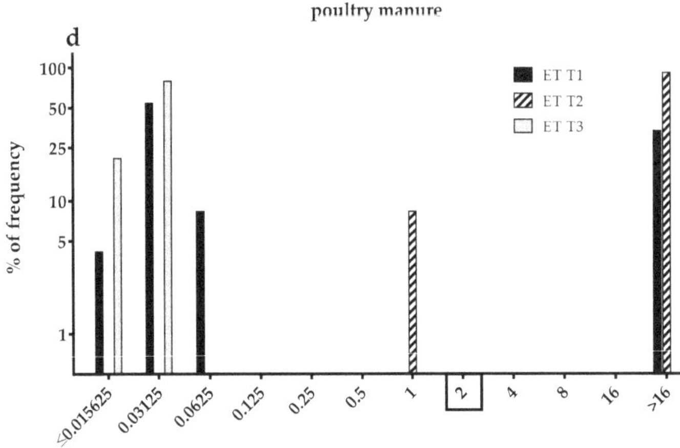

Figure 3. Percentage of frequency of enrofloxacin minimum inhibitory concentration (MIC) distribution in commensal *E. coli* isolates from (**a**) cloacal swabs after treatment (AT) and (**b**) end of trial (ET) as well as in (**c**) poultry manure samples during AT and (**d**) ET of untreated antibiotic (T1), treated twice with enrofloxacin via drinking water (T2) and simulated water spillage with water containing enrofloxacin (T3) in turkeys (cloacal swabs: N = 576; per trial AT: n = 96, ET: n = 96; poultry manure samples: N = 144; per trial AT: n = 24, ET: n = 24). Rectangle on the x-axis: Clinical Laboratory Standard Institute (CLSI) has determined a veterinary specific breakpoint of ≥2 µg/mL enrofloxacin for *E. coli* from chickens and turkeys.

3.2.1. Development of Enrofloxacin Resistance Depending on Group

Enrofloxacin resistance in *E. coli* isolates from all samples did not show any differences between the groups during the BT stage. During the AT stage in trial 2 (Figure 2a), *E. coli* isolated from cloacal swabs in G3 showed significantly lower means in resistance rates to enrofloxacin than the isolates collected from animals in other groups showing highest possible means (G3: 2.58; mean values of enrofloxacin resistance in *E. coli* for each group in detail in Supplementary Table S1a). During the ET stage in trial 1, G2 showed significantly higher means in resistance values of enrofloxacin in manure samples (Figure 2b). The *E. coli* isolates in T3 acquired from cloacal swabs and manure samples (Figure 2a,b) were susceptible to enrofloxacin and showed no significant differences between groups; 98% and 96%, respectively.

3.2.2. Development of Ampicillin Resistance Depending on Group

Regarding ampicillin resistance in all trials (Figure 2c,d), there were no significant differences between the groups concerning resistance in isolates from cloacal swabs and manure samples during the BT stages. In contrast, *E. coli* isolates from cloacal swabs during the AT stage in trial 2 (Figure 2c) from G1 demonstrated higher resistance means than in the other groups (G1: 1.92; mean values of ampicillin resistance in *E. coli* for each group in detail in Supplementary Table S1b). In trial 2, the results of means of ampicillin resistance from the manure samples during the AT stage also showed higher values for G1 than observed in either G2, G3, or G4 (2.33, 1.00, 1.00, and 1.00, respectively; Figure 2d; mean values of ampicillin resistance in *E. coli* for each group in detail in Supplementary Table S2b). There was no difference in means of ampicillin-resistant *E. coli* isolated from cloacal swabs and manure samples during the ET stage in all trials (Figure 2c,d).

3.3. Enrofloxacin MICs Distributions of the Commensal E. coli Isolates

The percentage of frequency of MICs distribution of the 576 commensal *E. coli* isolates from cloacal swabs and manure samples to enrofloxacin during the AT and ET stages are shown in Figure 3a–d. For *E. coli* from chickens and turkeys, the Clinical Laboratory Standard Institute (CLSI [28]) determined a veterinary specific breakpoint of ≥ 2 µg/mL for enrofloxacin.

When comparing the three trials (Figure 3a–d), a large number of resistant *E. coli* isolates during the AT and ET stages were found in trial 2, 93% and 99% in cloacal swabs, respectively (Figure 3a,b) and in manure samples, 79% and 92% during the AT and ET stages, respectively (Figure 3c,d). On the other hand, all *E. coli* isolates from trial 3 during the AT stage were susceptible to enrofloxacin (≤ 0.25 µg/mL).

Regarding the MICs distribution for enrofloxacin, susceptible isolates from cloacal swabs in trial 1 decreased gradually from 52% during the AT stage to 42% at the end of the study (Figure 3a,b). In contrast, the percentage of resistant *E. coli* isolated in trial 2 slightly increased from 93% to 99% during the AT and ET stages, respectively (Figure 3a,b). On the other hand, nearly all of the *E. coli* isolates from the cloacal swabs and manure samples in trial 3 had enrofloxacin MIC-values below the clinical breakpoint (MIC ≤ 2 µg/mL).

4. Discussion

The antibacterial agents used in the poultry for treatment or prophylaxis are implicated for the development of bacterial resistance [29]. Treatment in large groups of chickens is often done by oral administration [30]. Some studies reported that antibacterial agents or their metabolites are excreted in manure and residue can therefore be found in the environment [16,31,32].

4.1. Consequences of the Oral Administration of Antibacterial Agents

In the present study, one week after the first administration period (d21; AT stage), the amount of resistance was higher compared to the *E. coli* isolates before treatment. A higher rate of enrofloxacin-resistant *E. coli* after oral administration of antibacterial agents was observed in several studies [25,33,34]. Chuppava et al. [25] stated that a single treatment for five days with enrofloxacin led to markedly reduced ratios of susceptible *E. coli* isolates in cloacal swabs and manure samples. The highest proportion of cloacal swabs with resistant *E. coli* was found directly after treatment. Afterwards, a decrease in resistance to enroflaxacin was seen. Therefore, our results are in agreement with those of Chuppava et al. [25] who described a very rapid occurrence of FQ resistance among the commensal *E. coli* after enrofloxacin treatment in poultry. Nevertheless, no difference in resistance in *E. coli* isolates was found between the AT and ET stages after two consecutive treatments with enrofloxacin in this study because MIC values were already very high after one-time treatment.

As expected, after treatment with enrofloxacin, increased MIC values above 2 µg/mL occurred in *E. coli* isolates (T2) with high detection rates up to the end of the trial. Scherz et al. [27] showed that a long-term exposure (21 days) of the commensal flora of poultry to enrofloxacin leads to an amplification and selection of resistant—*E. coli* isolates. These isolates persist in the commensal microbiota. The transmission of *E. coli* isolates of animal origin between the animals in the same pen as well as into the environment may contribute directly to the spread of resistant bacteria in general and may also be a problem for public health [35].

Medication is the main reason for occurrence of resistance to antibacterial agents in *E. coli* [31]. Oral group treatments led to an environmental contamination with antibacterial agents. The application procedure itself or excreted feces from treated animals can be the source [36]. Due to the fact that the metabolic rate of antibiotics is low, 90% of the administered dose is excreted via feces [5]. Avian intestines can act as potential reservoirs of *E. coli* [37]. Thus, there is a higher risk for resistance to antibacterial agents spreading from birds to other birds or from birds to the environment. In other European countries, the higher occurrence of FQ resistance in broilers compared to turkeys has been suggested to depend on an overall use-dependent higher exposure to FQ [3]. It should be noted, however, that the fattening period in turkeys takes much longer under field conditions. Therefore, the resistance situation in the present investigations at the end of the experiment is not comparable with the resistance situation occurring within the normal fattening duration.

In agreement with our data, Jurado et al. [34] and Chuppava et al. [25] found a significant increase in the frequency of resistance to ampicillin in *E. coli* isolates from poultry after orally administering enrofloxacin. These findings may be due to the coselection of β-lactam resistance genes. As the transmissible genetic elements were not analyzed in our study, further studies are recommended in order to confirm the role of such elements in the spread of resistance genes in poultry for *E. coli*.

4.2. Effects of the Development of Resistance to Antibacterial Agents in E. coli by Water Loss Simulation (Indirect Administration)

To the best of our knowledge, using water loss simulations by spraying the water containing enrofloxacin exclusively into the litter or onto the slatted flooring in the drinking area in order to study the development of resistance to antibacterial agents has not been previously reported. In this study, it was hypothesized that excreted or metabolized enrofloxacin might alone influence the occurrence of resistance to antibacterial agents. However, in the present study, we could not verify the occurrence of enrofloxacin resistance due to spraying water with enrofloxacin directly into the animals' environment.

Earlier reports suggested that the carry-over effect of antibacterial agents like FQ as well as their active metabolites in the stable could foster the development of antibacterial resistance via oral ingestion by animals [27]. However, in the present investigation, we sprayed enrofloxacin containing water directly into the environment. In contrast, in the aforementioned study, subtherapeutic dosages (3% and 10% of the recommended dosage of 10 mg/kg body weight) were directly applied to drinking water for 21 days, which could explain the difference. The active dose may therefore have been significantly lower in our own experiments.

Chuppava et al. [25] stated from their experimental model that removing the animals from contaminated pens after antibiotic treatment might be the reason for the lower percentage of resistant *E. coli* isolates in the observed animals. Changing the environment was assumed to lead to a lower percentage of resistant *E. coli* isolates in manure. A lower exposure to resistant bacteria in manure as well as antibacterial agent residues was discussed as the cause for this observation. Additionally, in poultry, dirty or contaminated litter and other animal management parameters affect the microbial composition of the chicken gastrointestinal tract. This influence can be either directly, by providing a continuous source of bacteria, or indirectly, by influencing the physical condition and defence of the birds [37].

4.3. Effect of Different Types of Flooring Design on the Development of Resistant E. coli

Up to now, little is known about reducing the development of resistance to antibacterial agents by using different flooring designs simulating varying contact intensity between animals and manure. The development of enrofloxacin and ampicillin resistance in *E. coli* was almost independent of flooring design in the present study. Differences in antibacterial susceptibility of commensal *E. coli* isolates from turkeys depending on flooring design have been previously reported [25]. Chuppava et al. [25] mentioned that flooring design had hardly any effect on the development of resistance against antibacterial agents. Nevertheless, in fully slatted flooring systems, with animals having no contact to their litter, resistance to antibacterial agents still develops in the animals.

In T1, overall, the group with floor heating (G2; average floor temperature in all trials: G1 = 27.0/G2 = 30.5/G3 = 26.5/G4 = 26.0 °C) showed a significantly higher number of resistant *E. coli* isolates than the other groups. Previous studies showed that the resistance to antibacterial agents in animals can change when they are kept in a heat stress environment [25,38]. A high amount of enrofloxacin resistant isolates from cloacal swabs in fattening turkeys was already reported by Chuppava et al. [25] in a group with floor heating. Also, in swine, Moro et al. [38] found a significant increase in resistant *E. coli* isolates in the intestinal flora after the animals had been exposed to heat stress (environmental temperature: 34 °C).

A significantly higher prevalence of ampicillin resistance in *E. coli* isolates from excreta material from cloacal swabs and manure samples was found in the entire floor pen with litter (G1) even when no ampicillin had been administered to the animals and the pens and the stable had been tested and found to be free of *Enterobacteriaceae* at the start of the trial. Further genetic analyses were not conducted. Therefore, the reason for this difference remains unknown.

4.4. Natural Resistance to Antibacterial Agents Found in Day-Old Chickens

Turkey poults in this study had not been previously exposed to antibacterial agents. However, *E. coli* was isolated from day-old chicks' meconium in trial 1. Isolates showed resistance to enrofloxacin (48%) and ampicillin (42%). Similar results were reported in previous studies that found one-day-old chicks to be *E. coli* resistant to enrofloxacin [39] and 100% resistant to ampicillin [25]. It has to be mentioned that also other research groups observed high rates of resistance to antibacterial agents already before treatment as well as in the absence of treatment [40]. A vertical transmission of resistant isolates along the production pyramid can occur [3,41]. Also, contamination in the hatchery environment is possible [42]. Persoons et al. [43] stated that besides management, also hatchery-related factors can influence the occurrence of resistance to antibacterial agents. In newly hatched chicks, the common bacteria in the environment, whether antibacterial susceptible or resistant, colonize the intestines and become part of the intestinal normal microflora. Thus, contamination of chickens via vertical transmission could be a possible explanation for the resistance rates found in our study.

The natural enrofloxacin resistance observed in this present study increased strongly. This increase was higher than after one time treatment, as previously reported [25], despite the absence of antibacterial agent usage (T1). Chuppava et al. [25] suggested, according to their findings, that resistance could be reduced or increased, but not eliminated from the animals even with strict disinfection procedures during the experiment. From literature, it is known that a large number of animals carry resistant *E. coli*. These animals can shed huge numbers of resistant organisms. This could result in a rapid contamination of the other individuals in the same pen and in the stable environment [41]. Resistant bacteria can be ingested by birds from the environment. After entering their gut, these may cause the development of resistant *E. coli*. However, there are several possible mechanisms responsible for the development of quinolone resistance [44].

Therefore, further research is strongly recommended to analyze the genetic basis of resistance in the isolates in order to understand the resistance mechanism's origin, development and transfer.

5. Conclusions

In this study, resistance to enrofloxacin was detected at a very high frequency after treatments with enrofloxacin via drinking water. Therefore, the oral administration of enrofloxacin seems to be associated with a significant increase in the frequency of resistance to enrofloxacin in commensal *E. coli* isolates from turkeys. In addition, prevalence of isolates resistant to ampicillin rose significantly. Resistance to enrofloxacin was not detected when the antibacterial agent substance was indirectly sprayed with water into the environment of fattening turkeys. Flooring structure designs did not directly affect the development of resistance to antibacterial agents, or in groups where the animals had no contact to litter. The existence of resistant *E. coli* isolates in one-day-old birds strongly suggests vertical transmission from parent flocks as one possible explanation.

Furthermore, our results can provide useful information, prompting further studies on quinolone resistance mechanisms in commensal *E. coli* depending on different housing systems. However, we cannot consider all interactions when only one isolate is taken from a sample and then, by way of example, we try to deduce the complexity of the development of resistance. Therefore, research is needed to further investigate possible explanations regarding the mechanism behind the dissemination of enrofloxacin-resistant *E. coli* in fattening turkeys.

Supplementary Materials: The following are available online at http://www.mdpi.com/1660-4601/15/9/1993/s1, Table S1a: Means of enrofloxacin- resistant E. coli isolates from cloacal swab and manure samples from turkeys, Table S1b: Means of ampicillin-resistant E. coli isolates from cloacal swab and manure samples from turkeys.

Author Contributions: Conceptualization, C.V. and M.K.; Methodology, B.K., J.M., M.K. and C.V.; Validation, B.C., C.V., B.K. and J.M.; Formal Analysis, B.C., B.K., and C.V.; Investigation, B.C., C.V., A.A.E.-W. and J.M.; Resources, C.V. and M.K.; Data Curation, B.C. and C.V.; Writing-Original Draft Preparation, B.C. and C.V.; Writing-Review & Editing, B.C., C.V., B.K., A.A.E.-W., J.M. and M.K.; Visualization, B.C. and C.V.; Supervision, C.V.; Project Administration, C.V. and B.K.; Funding Acquisition, C.V. and M.K.

Funding: This project was supported by funds of the Federal Ministry of Food and Agriculture (BMEL, Germany) based on a decision of the Parliament of the Federal Republic of Germany via the Federal Office for Agriculture and Food (BLE, Germany) under the innovation support program. This publication was supported by the Deutsche Forschungsgemeinschaft and University of Veterinary Medicine Hannover, Foundation, Germany within the funding program Open Access Publishing.

Acknowledgments: We would like to thank Frances Sherwood-Brock for proof-reading the manuscript to ensure correct English.

Conflicts of Interest: The authors declare no conflicts of interest.

References

1. Kaesbohrer, A.; Schroeter, A.; Tenhagen, B.A.; Alt, K.; Guerra, B.; Appel, B. Emerging antimicrobial resistance in commensal *Escherichia coli* with public health relevance. *Zoonoses Public Health* **2012**, *59*, 158–165. [CrossRef] [PubMed]

2. Schroeter, A.; Kaesbohrer, A. *German Antimicrobial Resistance Situation in the Food Chain–DARlink*; Bundesinstitut fur Risikobewertung (BfR): Berlin, Germany, 2010; pp. 127–133.

3. ECDC; EFSA; EMA. Second joint report on the integrated analysis of the consumption of antimicrobial agents and occurrence of antimicrobial resistance in bacteria from humans and food-producing animals—joint interagency antimicrobial consumption and resistance analysis (JIACRA) report. *EFSA J.* **2017**, *15*, 4872.

4. WHO. Antimicrobial Resistance: Global Report on Surveillance. World Health Organization, 2014. Available online: http://apps.who.int/iris/bitstream/10665/112642/1/9789241564748_eng.pdf (accessed on 29 March 2018).

5. Kumar, K.; Gupta, S.C.; Chander, Y.; Singh, A.K. Antibiotic use in agriculture and its impact on the terrestrial environment. *Adv. Agron.* **2005**, *87*, 1–54.

6. Janusch, F.; Scherz, G.; Mohring, S.A.; Hamscher, G. Determination of fluoroquinolones in chicken feces—A new liquid–liquid extraction method combined with LC–MS/MS. *Environ. Toxicol. Pharmacol.* **2014**, *38*, 792–799. [CrossRef] [PubMed]

7. Riaz, L.; Mahmood, T.; Khalid, A.; Rashid, A.; Siddique, M.B.A.; Kamal, A.; Coyne, M.S. Fluoroquinolones (FQs) in the environment: A review on their abundance, sorption and toxicity in soil. *Chemosphere* **2018**, *191*, 704–720. [CrossRef] [PubMed]

8. Tacconelli, E.; Sifakis, F.; Harbarth, S.; Schrijver, R.; van Mourik, M.; Voss, A.; Sharland, M.; Rajendran, N.B.; Rodríguez-Baño, J.; Bielicki, J. Surveillance for control of antimicrobial resistance. *Lancet Infect. Dis.* **2017**, *18*, E99–E106. [CrossRef]

9. Projahn, M.; Daehre, K.; Semmler, T.; Guenther, S.; Roesler, U.; Friese, A. Environmental adaptation and vertical dissemination of esbl-/pampc-producing *Escherichia coli* in an integrated broiler production chain in the absence of an antibiotic treatment. *Microb. Biotechnol.* **2018**, *11*, 1–10. [CrossRef] [PubMed]

10. Varga, C.; Rajić, A.; McFall, M.E.; Reid-Smith, R.J.; Deckert, A.E.; Pearl, D.L.; Avery, B.P.; Checkley, S.L.; McEwen, S.A. Comparison of antimicrobial resistance in generic *Escherichia coli* and *Salmonella* spp. Cultured from identical fecal samples in finishing swine. *Can. J. Vet. Res.* **2008**, *72*, 181–187. [PubMed]

11. EFSA. The European Union summary report on antimicrobial resistance in zoonotic and indicator bacteria from humans, animals and food in 2013. *EFSA J.* **2015**, *13*, 4036.

12. ECDC; EFSA; EMA. First joint report on the integrated analysis of the consumption of antimicrobial agents and occurrence of antimicrobial resistance in bacteria from humans and food-producing animals. *EFSA J.* **2015**, *13*, 4006.

13. European Surveillance of Veterinary Antimicrobial Consumption (ESVAC). Sales of Veterinary Antimicrobial Agents in 30 European Countries in 2015. 2017. Available online: https://publications.europa.eu/en/publication-detail/-/publication/7656bd79-d970-11e7-a506-01aa75ed71a1/language-en (accessed on 29 March 2018).

14. GERMAP. *Report on the Consumption of Antimicrobials and the Spread of Antimicrobial Resistance in Human and Veterinary Medicine in Germany*; Antiinfectives Intelligence: Rheinbach, Germany, 2016.

15. Kreienbrock, L. *Ergebnisse aus VetCAb, Antibiotikaresistenz in der Lebensmittelkette*; Bundesinstitut für Risikobewertung (BfR): Berlin, Germany, 2013; pp. 52–54.

16. Diarra, M.S.; Malouin, F. Antibiotics in Canadian poultry productions and anticipated alternatives. *Front. Microbiol.* **2014**, *5*, 282. [CrossRef] [PubMed]

17. Furtula, V.; Farrell, E.; Diarrassouba, F.; Rempel, H.; Pritchard, J.; Diarra, M. Veterinary pharmaceuticals and antibiotic resistance of *Escherichia coli* isolates in poultry litter from commercial farms and controlled feeding trials. *Poult. Sci.* **2010**, *89*, 180–188. [CrossRef] [PubMed]

18. Dandachi, I.; Sokhn, E.S.; Dahdouh, E.A.; Azar, E.; El-Bazzal, B.; Rolain, J.-M.; Daoud, Z. Prevalence and characterization of multi-drug-resistant gram-negative *bacilli* isolated from Lebanese poultry: A nationwide study. *Front. Microbiol.* **2018**, *9*, 550. [CrossRef] [PubMed]

19. Chantziaras, I.; Boyen, F.; Callens, B.; Dewulf, J. Correlation between veterinary antimicrobial use and antimicrobial resistance in food-producing animals: A report on seven countries. *J. Antimicrob. Chemother.* **2013**, *69*, 827–834. [CrossRef] [PubMed]

20. Werckenthin, C. Escherichia coli Strains from Northwestern Lower Saxony Obtained as Part of the Zoonosis Monitoring (General Administrative Regulation on Zoonoses in the Food Chain). In *GERMAP 2015—Report on the Consumption of Antimicrobials and the Spread of Antimicrobial Resistance in Human and Veterinary Medicine in Germany*; Antiinfectives Intelligence: Rheinbach, Germany, 2016; pp. 141–142.

21. Font-Palma, C. Characterisation, kinetics and modelling of gasification of poultry manure and litter: An overview. *Energy Convers. Manag.* **2012**, *53*, 92–98. [CrossRef]

22. Kamphues, J.; Youssef, I.; El-Wahab, A.A.; Üffing, B.; Witte, M.; Tost, M. Influences of feeding and housing on foot pad health in hens and turkeys. *Übers. Tierernährg* **2011**, *39*, 147–193.

23. Bhushan, C.; Khurana, A.; Sinha, R.; Nagaraju, M. *Antibiotic Resistance in Poultry Environment: Spread of Resistance from Poultry Farm to Agricultural Field*; CSE: New Delhi, India, 2017.

24. Wu, Y.; Wang, Z.; Liao, X. Study on the excretion of enrofloxacin in chicken and its degradation in chicken feces. *Acta Vet. Zootech. Sin.* **2005**, *36*, 1069.

25. Chuppava, B.; Keller, B.; Meißner, J.; Kietzmann, M.; Visscher, C. Effects of different types of flooring design on the development of antimicrobial resistance in commensal *Escherichia coli* in fattening turkeys. *Vet. Microbiol.* **2018**, *217*, 18–24. [CrossRef] [PubMed]

26. Annex I of Council Regulation. European Commission (EC) No. 1099/2009 on the Protection of Animals at the Time of Killing. 2009; pp. 1–84. Available online: https://eur-lex.europa.eu/eli/reg/2009/1099/oj (accessed on 29 March 2018).

27. Scherz, G. Carryover of Subtherapeutic Antimicrobial Dosages of Enrofloxacin and the Influence on the Development of Antibiotic Resistance of Commensal *Escherichia Coli* in the Intestine of Poultry. Ph.D. Thesis, University of Veterinary Medicine, Hannover, Germany, 2013.

28. CLSI. *Performance Standards for Antimicrobial Disk and Dilution Susceptibility Tests for Bacteria Isolated from Animals*; Approved Standard-Fourth Edition; CLSI document Vet01-A4; Clinical and Laboratory Standards Institute: Wayne, PA, USA, 2013.

29. Witte, W. Medical consequences of antibiotic use in agriculture. *Science* **1998**, *279*, 996–997. [CrossRef] [PubMed]

30. Kietzmann, M.; Baeumer, W. Oral medication via feed and water pharmacological aspects. *Dtsch. Tierarztl. Wochenschr.* **2009**, *116*, 204–208. [PubMed]

31. Da Costa, P.M.; Belo, A.; Gonçalves, J.; Bernardo, F. Field trial evaluating changes in prevalence and patterns of antimicrobial resistance among *Escherichia coli* and *Enterococcus* spp. isolated from growing broilers medicated with enrofloxacin, apramycin and amoxicillin. *Vet. Microbiol.* **2009**, *139*, 284–292. [CrossRef] [PubMed]

32. Furtula, V.; Jackson, C.R.; Farrell, E.G.; Barrett, J.B.; Hiott, L.M.; Chambers, P.A. Antimicrobial resistance in *Enterococcus* spp. isolated from environmental samples in an area of intensive poultry production. *Int. J. Environ. Res. Public Health* **2013**, *10*, 1020–1036. [CrossRef] [PubMed]

33. Miranda, J.; Vázquez, B.; Fente, C.; Barros-Velázquez, J.; Cepeda, A.; Franco, C. Evolution of resistance in poultry intestinal *Escherichia coli* during three commonly used antimicrobial therapeutic treatments in poultry. *Poult. Sci.* **2008**, *87*, 1643–1648. [CrossRef] [PubMed]

34. Jurado, S.; Medina, A.; Ruiz-Santa-Quiteria, J.A.; Orden, J.A. Resistance to non-quinolone antimicrobials in commensal *Escherichia coli* isolates from chickens treated orally with enrofloxacin. *Jpn. J. Vet. Res.* **2015**, *63*, 195–200. [PubMed]

35. Schwarz, S.; Chaslus-Dancla, E. Use of antimicrobials in veterinary medicine and mechanisms of resistance. *Vet. Res.* **2001**, *32*, 201–225. [CrossRef] [PubMed]

36. Richter, A.; Hafez, H.M.; Böttner, A.; Gangl, A.; Hartmann, K.; Kaske, M.; Kehrenberg, C.; Kietzmann, M.; Klarmann, D.; Klein, G. Application of antibiotics in poultry. *Tierärztl. Praxis Großtiere* **2009**, *37*, 321–329.

37. Apajalahti, J.; Kettunen, A.; Graham, H. Characteristics of the gastrointestinal microbial communities, with special reference to the chicken. *Worlds Poult. Sci. J.* **2004**, *60*, 223–232. [CrossRef]

38. Moro, M.; Beran, G.; Griffith, R.; Hoffman, L. Effects of heat stress on the antimicrobial drug resistance of *Escherichia coli* of the intestinal flora of swine. *J. Appl. Microbiol.* **2000**, *88*, 836–844. [CrossRef] [PubMed]

39. Oloso, N.O.; Fagbo, S.; Garbati, M.; Olonitola, S.O.; Awosanya, E.J.; Aworh, M.K.; Adamu, H.; Odetokun, I.A.; Fasina, F.O. Antimicrobial Resistance in Food Animals and the Environment in Nigeria: A Review. *Int. J. Environ. Res. Public Health* **2018**, *15*, 1284. [CrossRef] [PubMed]

40. Ozaki, H.; Esaki, H.; Takemoto, K.; Ikeda, A.; Nakatani, Y.; Someya, A.; Hirayama, N.; Murase, T. Antimicrobial resistance in fecal *Escherichia coli* isolated from growing chickens on commercial broiler farms. *Vet. Microbiol.* **2011**, *150*, 132–139. [CrossRef] [PubMed]

41. Jiménez-Belenguer, A.; Doménech, E.; Villagrá, A.; Fenollar, A.; Ferrús, M.A. Antimicrobial resistance of *Escherichia coli* isolated in newly-hatched chickens and effect of amoxicillin treatment during their growth. *Avian Pathol.* **2016**, *45*, 501–507. [CrossRef] [PubMed]

42. Dierikx, C.M.; van der Goot, J.A.; Smith, H.E.; Kant, A.; Mevius, D.J. Presence of ESBL/AmpC-producing *Escherichia coli* in the broiler production pyramid: A descriptive study. *PLoS ONE* **2013**, *8*, e79005. [CrossRef] [PubMed]

43. Persoons, D.; Haesebrouck, F.; Smet, A.; Herman, L.; Heyndrickx, M.; Martel, A.; Catry, B.; Berge, A.; Butaye, P.; Dewulf, J. Risk factors for ceftiofur resistance in *Escherichia coli* from belgian broilers. *Epidemiol. Infect.* **2011**, *139*, 765–771. [CrossRef] [PubMed]

44. Hopkins, K.L.; Davies, R.H.; Threlfall, E.J. Mechanisms of quinolone resistance in *Escherichia coli* and Salmonella: Recent developments. *Int. J. Antimicrob. Agents* **2005**, *25*, 358–373. [CrossRef] [PubMed]

International Journal of
*Environmental Research
and Public Health*

Article

Plant Growth, Antibiotic Uptake, and Prevalence of Antibiotic Resistance in an Endophytic System of Pakchoi under Antibiotic Exposure

Hao Zhang [1,2,3], Xunan Li [2], Qingxiang Yang [1,2,3,*], Linlin Sun [2], Xinxin Yang [2], Mingming Zhou [2], Rongzhen Deng [2] and Linqian Bi [2]

[1] School of Environment, Henan Normal University, Xinxiang 453007, China; kele1564@126.com
[2] College of Life Sciences, Henan Normal University, Xinxiang 453007, China; 15516551517@163.com (X.L.); nlycds@163.com (L.S.); 15736966832@163.com (X.Y.); zm921011@163.com (M.Z.); dengrongzhen127@163.com (R.D.); 15736927022@163.com (L.B.)
[3] Key Laboratory for Microorganisms and Functional Molecules, University of Henan Province, Xinxiang 453007, China
[*] Correspondence: yangqx66@163.com; Tel./Fax: +86-373-3325-528

Received: 27 September 2017; Accepted: 28 October 2017; Published: 3 November 2017

Abstract: Antibiotic contamination in agroecosystems may cause serious problems, such as the proliferation of various antibiotic resistant bacteria and the spreading of antibiotic resistance genes (ARGs) in the environment or even to human beings. However, it is unclear whether environmental antibiotics, antibiotic resistant bacteria, and ARGs can directly enter into, or occur in, the endophytic systems of plants exposed to pollutants. In this study, a hydroponic experiment exposing pakchoi (*Brassica chinensis* L.) to tetracycline, cephalexin, and sulfamethoxazole at 50% minimum inhibitory concentration (MIC) levels and MIC levels, respectively, was conducted to explore plant growth, antibiotic uptake, and the development of antibiotic resistance in endophytic systems. The three antibiotics promoted pakchoi growth at 50% MIC values. Target antibiotics at concentrations ranging from 6.9 to 48.1 $\mu g \cdot kg^{-1}$ were detected in the treated vegetables. Additionally, the rates of antibiotic-resistant endophytic bacteria to total cultivable endophytic bacteria significantly increased as the antibiotics accumulated in the plants. The detection and quantification of ARGs indicated that four types, *tet*X, *bla*$_{CTX-M}$, and *sul*1 and *sul*2, which correspond to tetracycline, cephalexin, and sulfamethoxazole resistance, respectively, were present in the pakchoi endophytic system and increased with the antibiotic concentrations. The results highlight a potential risk of the development and spread of antibiotic resistance in vegetable endophytic systems.

Keywords: antibiotics; pakchoi; endophytic bacteria; antibiotic-resistant genes; hydroponic cultivation

1. Introduction

Antibiotic pollutants and their environmental impacts have become a mounting concern owing to their broad usage and persistence in the environment. A large range of veterinary and human antibiotics have been detected in soil, animal manure, sediment, municipal or industry wastewater, surface water, groundwater, and drinking water samples [1–6]. In agroecosystems, the contamination of various antibiotics, such as tetracyclines, sulfonamides, and fluoroquinolones, is a substantial problem globally, and especially in China [7–9]. A dominant source of agricultural antibiotic contamination is due to ~75% of the antibiotics ingested by animals passing unaltered through their digestive tracts, with the result that antibiotics are released in the field directly in feces or urine, or indirectly through the application of manure as fertilizer [10–12]. Another source is the irrigation of crops using wastewater containing antibiotics [13–15].

Once antibiotics are released into agricultural lands, crops are exposed to them due to their persistence, and the level of exposure depends on the physicochemical properties of the compounds, sorption potential, and environmental conditions [16,17]. Even if some antibiotics are degraded to a certain degree, most of them are replaced by ongoing use and release [18]. Under antibiotic contamination conditions, certain pharmaceutical compounds (such as tetracycline, oxytetracycline, sulfamethazine, sulfamethoxazole, tylosin, trimethoprim, ofloxacin, ciprofloxacin, and amoxicillin) can be absorbed by plants (such as wheat, corn, rice, lettuce, cabbage, spinach, carrot, cucumber, tomato, and potato) from the growth media through their roots and accumulate [13–15,19–22]. Although the human health implications of antibiotic pollutants in plant crops are largely unknown, several potential adverse impacts, including allergic reactions, chronic toxic effects as a result of prolonged exposure, and even the disruption of digestive system functions, have been speculated [16,23,24]. Thus, there is a growing concern that antibiotic pollution in food crops makes its way into food supply systems.

To date, the majority of research on the impact of antibiotic contamination in plants has focused on evaluating the toxicity of antibiotics to plants or detecting the ability of antibiotics to accumulate in plants. Limited knowledge is available regarding the potential effects of antibiotic stress on the development and spread of antibiotic resistance, including antibiotic-resistant bacteria and antibiotic-resistant genes (ARGs), in plant endophytic systems. There is a diverse range of endophytic bacteria, which includes pathogens, mutualists, and commensals that grow within the roots, vasculature, and aerial tissues of plants [25]. Recently, antibiotic resistance in endophytic bacteria isolated from medicinal plants has been reported [26,27]. Our previous research also reported a high prevalence of antibiotic-resistant endophytic bacteria (AREB), including some resistant to more than three different types of antibiotics, in various manure-fertilized vegetables, such as celery, pakchoi, and cucumber [28]. However, it is unclear whether the antibiotic resistance of endophytic bacteria can be impacted directly by antibiotic pollution in the environment, especially in the edible parts of vegetables.

To assess possible consequences, pakchoi (*Brassica chinensis* L.), a frequently consumed vegetable in China, was selected and planted in a hydroponic system and exposed to different antibiotics. Then, the antibiotic uptake and its effects on plant growth and the presence of AREB and ARGs in the endophytic system were investigated and evaluated. The findings will facilitate a more accurate assessment of the potential risks of antibiotic contamination to food quality and environmental health.

2. Materials and Methods

2.1. Chemicals and Reagents

Tetracycline (TC, >98.0%), cephalexin (CPL, >99.0%), and sulfamethoxazole (SMX, >99.5%) were purchased from Dr. Ehrenstorfer GmbH (Augsburg, Germany) and selected to represent the different classes of antibiotics (tetracyclines, β-lactams, and sulfonamides, respectively) based on their frequent usage in the local livestock farms in Xinxiang City, China [28]. Tetracyclines are broad-spectrum antibiotics that inhibit bacterial protein synthesis by preventing the attachment of aminoacyl-tRNA to the ribosomal acceptor (A) site. Resistance to tetracyclines has now emerged in many pathogenic bacteria due to genetic acquisition of *tet* genes, which include efflux genes, ribosomal protection genes, and enzymatic modification genes [29]. β-lactam antibiotics are the most widespread class of human antibacterials that inhibit bacteria by interfering with cell wall synthesis. The most major mechanism of bacterial resistance to β-lactam is the expression of β-lactamases that hydrolyze the antibiotic [30]. Sulfonamides, which are synthetic antibacterial drugs, inhibit bacterial folate biosynthesis by competing with the natural substrate *p*-amino-benzoic acid for binding to dihydropteroate synthase (DHPS), an enzyme in the folic acid synthesis pathway. Two genes, *sul*1 and *sul*2, mediated by transposons and plasmids, and expressing DHPS highly resistant to sulfonamide, have been found [31].

Methanol, acetonitrile, formic acid, and acetone of HPLC grade were purchased from Fisher Scientific (Fair Lawn, NJ, USA). Other chemicals were of analytical grade and obtained from Yaohua Chemical Reagent Factory (Tianjin, China). Ultrapure water was supplied using a Millipore Milli-Q system (Billerica, MA, USA). Oasis HLB cartridges (hydrophilic-lipophilic balance, 6 mL, 500 mg) purchased from Waters (Milford, MA, USA) were used for the extraction and purification of the target antibiotics. Individual stock standards were prepared by dissolving antibiotics separately in methanol and were stored at $-20\ ^\circ$C in brown vials.

2.2. Hydroponic Experimental Procedure

A hydroponic experiment with antibiotic treatments was performed in an experimental greenhouse in the College of Life Sciences, Henan Normal University, China, during the autumn of 2016. Seeds of pakchoi (*B. chinensis* L.) were used for this study. TC, CPL, and SMX were separately added into the hydroponic solution of the test system at two concentrations: 50% of minimum inhibitory concentration (MIC) of each antibiotic, and the MIC of each antibiotic [32]. Each of the MIC values in the hydroponic solution was set as the induced dose of resistant bacteria.

Prior to testing, the seeds were surface-sterilized in 0.1% sodium hypochlorite solution for 10 min and then rinsed with sterile deionized water [33]. Seeds on a piece of sterile filter paper were placed into 10 cm sterile Petri dishes, and 10 mL of sterile water was added. Then, the Petri dishes were covered with their lids and maintained in a dark incubator at $25 \pm 2\ ^\circ$C. After germination, the seeds were transferred to a plastic cuboid hydroponic tank ($45 \times 20 \times 17$ cm). All tanks were wiped with 75% ethanol and thoroughly rinsed with deionized water before first use. After that, each tank was filled with 12 L of Hoagland nutrient solution [34] or Hoagland nutrient solution supplemented with an antibiotic. The treatments were as follows: (1) control with no antibiotics added; (2) TC-treated (at concentrations of 8 and 16 mg$\cdot$L^{-1}, respectively); (3) CPL-treated (at concentrations of 32 and 64 mg$\cdot$L^{-1}, respectively); and (4) SMX-treated (at concentrations of 38 and 76 mg$\cdot$L^{-1}, respectively). Each treatment was designed with three replicates, and a total of 21 hydroponic tanks were used in the present study. Additionally, each tank had nine cylindrical holes (4 cm depth and 3 cm diameter) in its cover containing sponges ($3 \times 3 \times 2.5$ cm) in individual cylinders as a rooting medium. Five uniform seeds were planted per hole and irrigated with half-strength Hoagland solution every 2 days until the roots of the seedlings were immersed in the solution. Finally, only one or two strong seedlings were selected to leave in each hole, and they were grown directly in nutrient solution.

During the experiment the room conditions were maintained at $25 \pm 2\ ^\circ$C in daylight and $18 \pm 2\ ^\circ$C at night, with a relative humidity between 65% and 70%. Each planter was equipped with an electric aeration pump and was aerated for 2 h every day. Because of evapotranspiration from the vessels, lost water was supplemented with fresh nutrient solution without the addition of extra antibiotics. Pakchoi was harvested after 55 days of cultivation. Then, the sponges attached to the vegetables were trimmed off. The plants were rinsed first with tap water and then with deionized water and dried on adsorbent paper. The growth parameters and abundance of endophytic bacteria were measured immediately. Antibiotic analyses and DNA extraction from endophytic bacteria were completed within two weeks of sampling.

2.3. Measurements of Growth Parameters

All plants were harvested at the end of the test, and then, plant heights, root lengths, and fresh biomasses of 10 plants from each treatment were measured and recorded. The growth inhibition rate was calculated using following formula:

$$\% \text{ inhibition} = (M_0 - M_t)/M_0 \times 100$$

where M_0 indicates the measurement of the control treatment and M_t indicates the measurement of the antibiotic treatment.

2.4. Antibiotic Analyses in Plants

Samples of the edible pakchoi portions (stem and leaf) were selected and freeze-dried for 24 h until moisture was no longer present [35]. Then, the freeze-dried samples were ground thoroughly, and the amount of antibiotics in the plant tissues were determined using ultrasonic extraction, solid-phase extraction, and liquid chromatography-mass spectrometry. The extraction method and clean-up procedure used was already described for the analysis of Chinese white cabbage, water spinach, and other crops [15]. Thereafter, the target compounds from treated samples were analyzed using an ultra-performance liquid chromatography-tandem mass spectrometer (Waters, Milford, MA, USA) equipped with an electrospray ionization source in multiple-reaction monitoring mode. Details of the quantitative analysis were described by Gros et al. [36].

2.5. Enumeration of Total Cultivable Endophytic Bacteria (TCEB) and AREB

To isolate endophytic bacteria, the edible portions of fresh pakchoi were immersed in 3% hydrogen peroxide for 30 min, followed by rinses with sterile deionized water (3 min $\times$ 3 times). Then, they were immersed in 70% ethanol for 1 min and rinsed as before [37]. Finally, surface-sterilized samples were dried using sterilized filter papers. To ensure the complete surface disinfection, 100 μL of the last wash water was spread on meat-peptone agar and cultivated at 30 °C for 3 days to check for colony growth [38]. Samples with no bacterial growth were considered successfully sterilized. For each experimental treatment, the disinfected vegetable was cut with a sterile scalpel into pieces and ground together with quartz sand in a sterile mortar. Then, 3 g of ground tissue was mixed with 10 mL of sterile water and the mixture was diluted to 10^{-3}. Each 100 μL of diluted suspension was spread on meat-peptone agar and on corresponding antibiotic-containing agars (TC, CPL, and SMX at concentrations of 16, 64, and 76 mg·L^{-1}, respectively) for cultivation at 28 °C for 3 days. Each sample was replicated three times. The colony-forming units (CFUs) of TCEB and AREB (endophytic bacteria resistant to TC, CPL, and SMX, respectively) were enumerated.

2.6. DNA Extraction, PCR Detection, and ARGs Quantification

The surface-sterilized edible pakchoi portions were cut into pieces and ground with liquid nitrogen before extraction under sterile conditions. Total DNA was extracted using PowerPlant DNA Isolation Kit (MoBio Laboratories, San Diego, CA, USA) following the manufacturer's instructions [37]. The concentrations and qualities of the extracted DNA samples were determined using a NanoDrop ND-2000 spectrophotometer (Thermo Scientific, Waltham, MA, USA) and agarose gel electrophoresis, respectively.

PCR detection assays were used to screen for the presence or absence of 23 types of ARGs in the antibiotic-treated samples, including 12 tetracyclines-resistant genes (*tet*A, *tet*C, *tet*G, *tet*K, *tet*L, *tet*M, *tet*O, *tet*Q, *tet*T, *tet*W, *tet*B/P, and *tet*X), 5 sulfonamides-resistant genes (*sul*1, *sul*2, *sul*3, *dfr*A1, and *dfr*A7), and 6 β-lactams-resistant genes (*bla*$_{ampC}$, *bla*$_{VIM}$, *bla*$_{CTX-M}$, *bla*$_{TEM}$, *bla*$_{SHV}$, and *bla*$_Z$). PCR detection assays were performed as previously described [39]. Primers and annealing temperatures are described in Table S1.

The positive ARGs and eubacterial 16S rRNA gene were quantified by fluorescence quantitative PCR (qPCR) using a LightCycler real-time PCR system (Roche, Basel, Switzerland) with SYBR Green I. Details of the primers are listed in Table S2. Plasmids carrying target genes in the pMD19-T vector (TaKaRa, Ostu Shiga, Japan) were constructed to produce the standard curves [40], which consisted of at least five orders of magnitude ($R^2 > 0.99$) (Table S3). The 20 μL reactions contained 10 μL of SYBR Premix Ex Taq (TaKaRa), 0.2 μM of each primer, 2 μL of template DNA, and 7.2 μL of ddH$_2$O. The reaction program was set as follows: initial denaturation at 95 °C for 30 s, 40 cycles at 95 °C for 5 s, annealing temperature for 30 s and 72 °C for 30 s, then a melt curve stage with temperature ramping from 60 °C to 95 °C.

2.7. Statistical Analysis

Ten plant samples from each treatment were used to test the effect of antibiotics on plant growth. For other analyses, three repetitions, each of which was the mixture of different plant parts from six plants, were performed. The mean values and standard deviations (SDs) of all data were calculated using Microsoft Office Excel 2016 (Microsoft, Redmond, WA, USA). Statistical analyses were performed using the software SPSS 21.0 (IBM, Armonk, NY, USA). Duncan's multiple comparisons were used to determine the significant differences ($p < 0.05$) between treatments.

3. Results

3.1. Effects of Antibiotic Exposure on Pakchoi Growth

The effects of antibiotics on pakchoi growth were assessed by analyzing the growth parameters. The detected values of plant height, root length, and fresh biomass are shown in Table S4. Figure 1 shows the changes of pakchoi growth at different antibiotic types and doses. Compared with the control, the growth indices increased when exposed to 50% MIC levels of antibiotics ($p < 0.0001$ for the plant heights and $p < 0.001$ for the fresh biomass values). When the exposure dose was increased to MIC levels, this growth-promoting effect was maintained in CPL-treated plants. However, the TC- and SMX-treated plants were significantly inhibited ($p < 0.01$), as indicated by the growth parameters being less than those of the controls. According to the detection of growth inhibition rates (data shown in Table S4), fresh plant biomass was the most affected parameter (fresh biomass > root length > plant height) under antibiotic exposure. SMX showed the greatest impact on pakchoi growth in the three antibiotic types and CPL had the least inhibition on the plants.

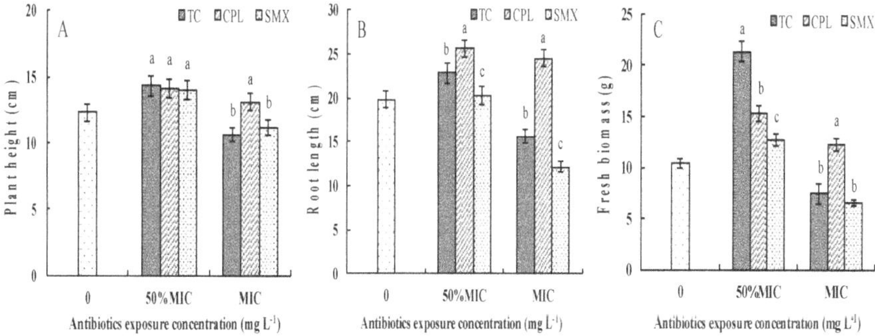

Figure 1. Effects of different dosages and types of antibiotic exposure on the growth of pakchoi under hydroponic condition. (**A**) Plant height; (**B**) Root length; (**C**) Fresh biomass. Values are mean ± SD (n = 10). Different letters on the top of the error bars indicate statistically difference among the treatments ($p < 0.05$). TC, tetracycline; CPL, cephalexin; SMX, sulfamethoxazole; MIC, minimum inhibitory concentration; SD, standard deviation.

3.2. Antibiotic Uptake by Pakchoi

The edible portions of pakchoi samples, both controls and antibiotic-exposed plants, were separated and the concentrations of TC, CPL, and SMX within plants were determined to evaluate the uptake of antibiotics by plants, as shown in Figure 2. The results indicated a concentration range from 6.9 to 11.8 µg·kg⁻¹ for TC (Figure 2A), 26.4 to 48.1 µg·kg⁻¹ for CPL (Figure 2B), and 18.1 to 35.3 µg·kg⁻¹ for SMX (Figure 2C) in plants, respectively. However, no antibiotic accumulation was detected in the controls. Obviously, the antibiotic concentrations in the vegetables increased as the antibiotic dose increased in the culture solution. The mean CPL concentration in the CPL-treated samples was higher than those of the other antibiotic residuals in their corresponding treated samples when the exposure concentrations were at MIC levels.

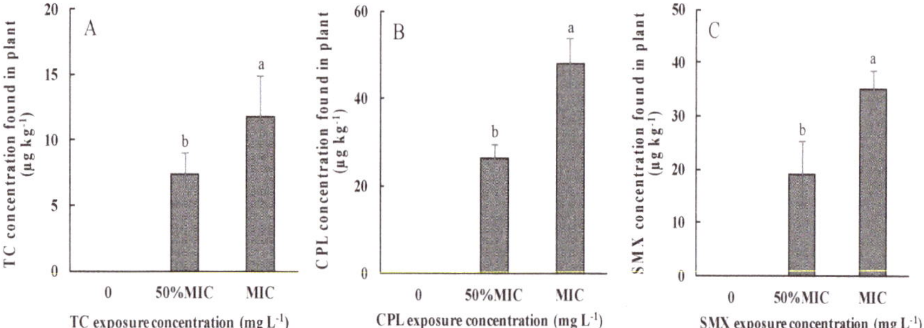

Figure 2. Accumulation of antibiotics in hydroponic pakchoi under different dosages of antibiotic exposure. (**A**) TC exposure treatment; (**B**) CPL exposure treatment; (**C**) SMX exposure treatment. Values are mean ± SD (*n* = 3). Different letters on the top of the error bars indicate statistical difference among the treatments (*p* < 0.05).

3.3. Effects of Antibiotic Exposure on AREB in Pakchoi

The CFUs of TCEB and AREB in plant tissues under different antibiotic treatment conditions were determined, and the rates of AREB to TCEB were calculated, as shown in Figure 3. Compared with the controls, the cultivable AREB levels in TC-, CPL-, and SMX-treated samples reached 0.61–0.85×10^3, 4.63–5.36×10^3, and 4.89–5.18×10^3 CFU·g^{-1}, respectively, which were higher than those in the control samples (0.23×10^3, 3.77×10^3, and 1.26×10^3 CFU·g^{-1}). These changes in AREB abundance resulted in dramatic increases in the ratios of AREB to TCEB from 0.23%, 3.77%, and 1.26% of TC, CPL, and SMX resistance, respectively, in the controls, to 0.79–1.23% (Figure 3A), 6.41–8.29% (Figure 3B), and 6.00–7.43% (Figure 3C), respectively, in the corresponding antibiotic-treated plants.

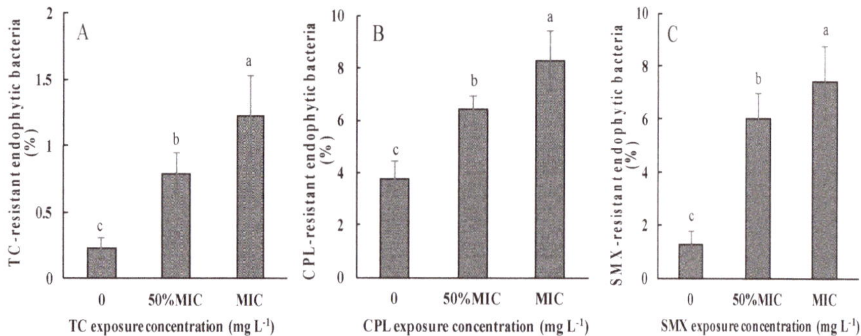

Figure 3. Rates of AREB to TCEB in pakchoi under different dosages of antibiotic exposure. (**A**) TC exposure treatment; (**B**) CPL exposure treatment; (**C**) SMX exposure treatment. Values are mean ± SD (*n* = 3). Different letters on the top of the error bars indicate statistical difference among the treatments (*p* < 0.05).

3.4. Abundance of ARGs in the Pakchoi Endophytic System

In total, 23 ARGs corresponding to three antibiotics were detected in antibiotic-treated vegetables using the PCR technique. Among them, only one *tet* gene (*tetX*), one β-lactamase gene (*bla*CTX-M), and two sul genes (*sul1* and *sul2*) responsible for TC, CPL, and SMX resistance, respectively, were present in the corresponding antibiotic-treated samples. Thus, further quantification using qPCR was conducted to monitor their responses to different treatment doses. To minimize the differences in background bacterial abundances and DNA extraction efficiency, 16S rRNA gene was also quantified and the absolute numbers of the above four quantified ARGs were normalized to that of the ambient 16S rDNA (Figure 4). In the control plants without antibiotic treatment, the values of the *tet*X gene were under the detection limit but the other three ARGs were detected at ~10^{-6} copies/16S rRNA gene copies. For the *tet*X gene, the relative abundance continuously increased as the TC dose increased. Sul and bla genes, on the whole, showed similar changes during the planting period. The four ARGs all reached their highest relative abundances of 10^{-5} to 10^{-4} copies/16S rRNA gene copies at the MIC exposure levels, which were one to two orders of magnitude greater than those in the control samples. Thus, the variation trends of the *tet*X, *sul*1, *sul*2, and *bla*CTX-M genes during different antibiotic treatments demonstrated great approximations.

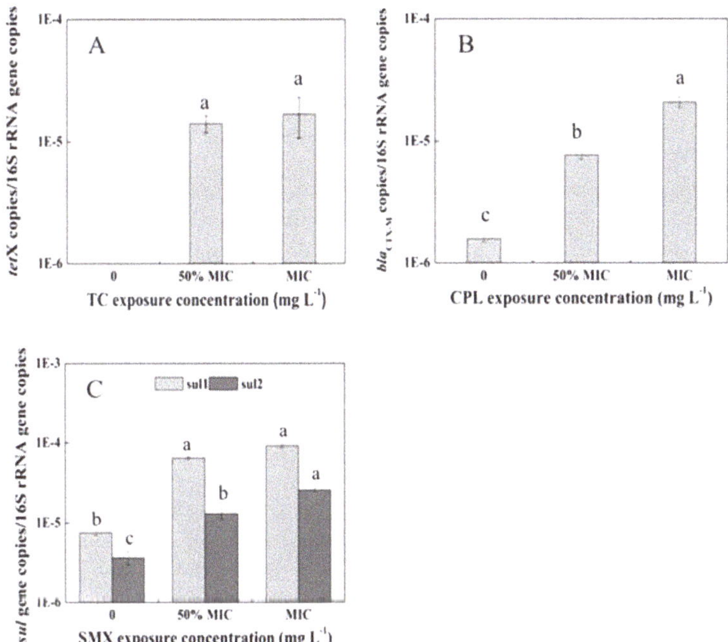

Figure 4. Abundance of antibiotic resistance genes (ARGs) in the endophytic system of pakchoi under different dosages of antibiotic exposure. (**A**) *tet*X gene in TC-treated plants; (**B**) *bla*CTX-M gene in CPL-treated plants; (**C**) *sul*1 and *sul*2 genes in SMX-treated plants. Values are mean ± SD (*n* = 3). Different letters on the top of the error bars indicate statistical difference among the treatments (*p* < 0.05).

4. Discussion

Plants are an important component of terrestrial ecosystems and are a potential pathway for antibiotic transport because of their absorption capacity [33]. Our previous studies indicated that cephalosporin, tetracyclines, and sulfonamides were the most frequently used antimicrobial agents in livestock farms in China, and high ratios of AREB occurred in the livestock manure fertilized field vegetables [28]. The transformation of these antibiotics and their induction of antibiotic resistance in soil or water environments have been frequently reported [6,41,42]. Further study through pot planting experiments confirmed different compositions of AREB presence in vegetable endophytic systems [43]. Therefore, TC, CPL, and SMX were selected to explore their accumulation and induction in pakchoi endophytic systems in this study. To simplify the conditions, a series of hydroponic cultures of pakchoi were used. Based on the Clinical and Laboratory Standards Institute (CLSI) standards [32] and the bacterial community composition in the pakchoi endophytic system [43], the highest MIC values of TC, CPL, and SMX for different genera of endophytic bacteria were selected and set as the levels of antibiotic exposure in the present study. From our results, we can see that although the exposure doses of different drugs were greater than their practical occurrence in agroecosystems, the pakchoi still showed natural growth throughout the planting process. Moreover, according to our previous investigation [28], sometimes animal manure containing high concentrations of antibiotic residue also will be used for field plants. Therefore, this study provided direct evidence for the effects of antibiotics on plant growth and the development of antibiotic resistance, especially under different treatment doses. The changes in growth parameters indicated the phytotoxic levels of each antibiotic at different doses. Generally, at 50% MIC levels, antibiotics stimulated growth, increasing plant fresh biomass production. However, at MIC levels, the antibiotics acted as inhibitors, reducing yields and inducing metabolic disturbances. Previous research had indicated that low antibiotic concentrations are beneficial for plant growth, whereas high antibiotic concentrations can induce toxicity [44]. In a comparison of the three antibiotics, CPL has the lowest toxicity to pakchoi. This may be due to β-lactam's specific actions on bacterial cell wall components, which are targets that do not exist in plant cells [45]. Therefore, the growth inhibition rate was lowest under CPL exposure when compared with controls.

The detection of the three antibiotics in pakchoi tissues indicates the uptake and transfer of antibiotics from the water environment to the vegetable, which is similar to previous results [14,18]. The bioaccumulation of antibiotics in plants can vary depending on plant species and antibiotic class [15,20]. Usually, ionization, as well as the properties of sorption and water solubility, can directly affect how plants uptake pharmaceuticals [21]. CPL was noted to have accumulated to the highest concentration among the study compounds at the MIC exposure level, which may reflect the greater absorbency of CPL compared with the other compounds. Furthermore, the concentrations of antibiotics in pakchoi did not increase unlimitedly as antibiotic dose increased. The probable reasons include: (1) the saturation level of antibiotic accumulation was reached; (2) the incorporated antibiotics were stored in the plant cells, in which they can be degraded; and (3) the degradation of antibiotics was accelerated by the release of plant enzymes during sample grinding [20]. Nevertheless, vegetables that have accumulated antibiotics from contaminated environments will be consumed by humans, and then might be absorbed by the human body, resulting in increased antibiotic resistance, including the emergence of multidrug-resistant bacteria, which leads to antibiotic treatment failures [13,46].

Few studies have explored the influence of antibiotic pollution on endophytic systems in vegetables. Yet vegetables contaminated by antibiotics may contribute to the development of AREB. The present study showed that the rates of AREB occurrence and the relative abundances of ARGs increased in pakchoi endophytic systems after exposure to the three antibiotics. The trends of these changes were comparable with the results from our previous study, in which a distinct increase in some AREBs was shown in manure- or organic fertilizer-amended pakchoi samples [43]. Two possible factors may contribute to such increases. Firstly, during the pakchoi cultivation, there would be a natural rhizosphere microbial consortium forming in the nutrient solution. However, spiked antibiotics as a selective pressure could influence the microbial community compositions and induce the occurrence of high ratios of resistance. This has been proved in many other environments even at much lower concentrations than their MIC values [47–50]. Thus, a special microbial consortium would be established corresponding to different treatments of antimicrobial agents. As we know, environmental bacteria, especially rhizosphere bacteria, are an important source for plant endophytic bacteria [51], which could enter through the tissues to the plant endophytic systems, thus resulting in high occurrence of AREB in the plant. Secondly, the AREB can be persistent in the plant endophytic systems. In the present study, the accumulated antibiotics in plants, although below their corresponding minimum MIC values for various species of bacteria, might also provide a selection pressure to the endophytic bacteria, thereby providing the AREB survival advantages in the polluted environments. The qPCR also demonstrated that the abundance of ARGs in the endophytic bacteria corresponding to the three antibiotics continuously increased as the antibiotic uptake in the pakchoi increased.

Among the four detected ARGs, the presence of *tet*X, a special enzymatic modification gene for the degradation of TCs, may be related to the low detected TC concentration in pakchoi. In addition, the absence of the *tet*X gene in control samples and the persistence of it in the treated samples may indicate the potential transfer of bacteria carrying the *tet*X gene from the environment to plant. Notably, the third-generation drug tigecycline has been used in clinical treatments due to its broad spectrum of antibacterial activity (especially inhibiting multiple antibiotic-resistant bacteria and super bacteria) [52–54]. However, the bacterial strains containing the *tet*X gene isolated from patients are still resistant to tigecycline [55]. Therefore, the prevalence of the *tet*X gene in edible pakchoi should be highly concerning. The other ARGs, *sul*1, *sul*2, and *bla*$_{\text{CTX-M}}$, are widely present in various environmental media [56–58], and act as the most prevalent mechanisms of sulfonamide and β-lactam resistance, respectively. In particular, the *sul*1 gene is normally found linked to other resistance genes in the Tn21 type integron, while *sul*2 is usually located on small plasmids of the IncQ family [59]. qPCR showed that the abundance levels of these ARGs increased in antibiotic contaminated environments, indicating their enrichment and transmission under antibiotic selection pressure.

Previous studies have demonstrated the antibiotic uptake [22] and the presence of resistant human pathogens or opportunistic pathogens in vegetables planted in manure-amended soil [43]. Thus, accumulated antibiotics in vegetables and the prevalence of antibiotic resistance in endophytic systems might be disseminated to humans when these vegetables are consumed. Consequently, evaluating the biological responses of terrestrial crops to antibiotics, especially frequently consumed vegetables, is important. However, compared with the soil environment, this hydroponic cultivation system is just a simple model to evaluate the influences of antibiotics on plant growth and plant endophytic bacteria. Further research is required to study the community compositions of AREB corresponding to different types of antibiotic exposure under soil cultivation systems. The results will provide basic information for an integrative risk assessment of antibiotic application and food security.

5. Conclusions

The present study investigated the growth of pakchoi and the antibiotic resistance in its endophytic system under TC, CPL, and SMX exposure. Pakchoi was shown to absorb antibiotics from the hydroponic culture environment. The absorption was selective toward different antibiotics, and the absorption amount was related to the antibiotic concentration. The accumulated antibiotics in the plant influenced the growth of the plant and increased the levels of AREB and ARGs, even at sub-inhibitory doses, which should be noted due to considerations surrounding the possible transfer of ARGs through the food chain.

Supplementary Materials: The following are available online at www.mdpi.com/1660-4601/14/11/1336/s1, Table S1: PCR primers, annealing temperatures, and resistance mechanisms, Table S2: qPCR primers and annealing temperatures used in the present study, Table S3: qPCR standard curves for 16S rRNA gene and antibiotic resistance genes, Table S4: Root length, plant height, and fresh biomass values of hydroponic pakchoi under different dosages of antibiotic treatment.

Acknowledgments: The authors would like to acknowledge the financial support from the National Natural Science Foundation of China (NSFC 21477035 and U1504219), the Key Science and Technology Project of Henan Province (142102210447), and the Specialized Research Fund for the Doctoral Program of Higher Education (20134104110006).

Author Contributions: Hao Zhang conducted the experiments, performed the data analysis, and drafted and edited the manuscript. Xunan Li performed the data analysis. Qingxiang Yang designed the study, provided the academic guidance for the work, and contributed to the critical revision for the paper. Linlin Sun, Xinxin Yang, Mingming Zhou, Rongzhen Deng, and Linqian Bi conducted the experiments.

Conflicts of Interest: The authors declare no conflict of interest.

References

1. Awad, Y.M.; Kim, S.C.; El-Azeem, S.A.M.A.; Kim, K.H.; Kim, K.R.; Kim, K.; Jeon, C.; Lee, S.S.; Ok, Y.S. Veterinary antibiotics contamination in water, sediment, and soil near a swine manure composting facility. *Environ. Earth Sci.* **2014**, *71*, 1433–1440. [CrossRef]
2. Hou, J.; Wan, W.N.; Mao, D.Q.; Wang, C.; Mu, Q.H.; Qin, S.Y.; Luo, Y. Occurrence and distribution of sulfonamides, tetracyclines, quinolones, macrolides, and nitrofurans in livestock manure and amended soils of Northern China. *Environ. Sci. Pollut. Res.* **2015**, *22*, 4545–4554. [CrossRef] [PubMed]
3. Hu, X.G.; Zhou, Q.X.; Luo, Y. Occurrence and source analysis of typical veterinary antibiotics in manure, soil, vegetables and groundwater from organic vegetable bases, northern China. *Environ. Pollut.* **2010**, *158*, 2992–2998. [CrossRef] [PubMed]
4. Yang, J.F.; Ying, G.G.; Zhao, J.L.; Tao, R.; Su, H.C.; Chen, F. Simultaneous determination of four classes of antibiotics in sediments of the Pearl Rivers using RRLC–MS/MS. *Sci. Total Environ.* **2010**, *408*, 3424–3432. [CrossRef] [PubMed]
5. Zhou, L.J.; Ying, G.G.; Liu, S.; Zhao, J.L.; Chen, F.; Zhang, R.Q.; Peng, F.Q.; Zhang, Q.Q. Simultaneous determination of human and veterinary antibiotics in various environmental matrices by rapid resolution liquid chromatography–electrospray ionization tandem mass spectrometry. *J. Chromatogr. A* **2012**, *1244*, 123–138. [CrossRef] [PubMed]
6. Zhou, L.J.; Ying, G.G.; Liu, S.; Zhao, J.L.; Yang, B.; Chen, Z.F.; Lai, H.J. Occurrence and fate of eleven classes of antibiotics in two typical wastewater treatment plants in South China. *Sci. Total Environ.* **2013**, *452–453*, 365–376. [CrossRef] [PubMed]
7. Li, Y.W.; Wu, X.L.; Mo, C.H.; Tai, Y.P.; Huang, X.P.; Xiang, L. Investigation of sulfonamide, tetracycline, and quinolone antibiotics in vegetable farmland soil in the Pearl River Delta area, southern China. *J. Agric. Food Chem.* **2011**, *59*, 7268–7276. [CrossRef] [PubMed]
8. Li, C.; Chen, J.Y.; Wang, J.H.; Ma, Z.H.; Han, P.; Luan, Y.X.; Lu, A.X. Occurrence of antibiotics in soils and manures from greenhouse vegetable production bases of Beijing, China and an associated risk assessment. *Sci. Total Environ.* **2015**, *521–522*, 101–107. [CrossRef] [PubMed]
9. Wei, R.C.; Ge, F.; Zhang, L.L.; Hou, X.; Cao, Y.N.; Gong, L.; Chen, M.; Wang, R.; Bao, E.D. Occurrence of 13 veterinary drugs in animal manure-amended soils in Eastern China. *Chemosphere* **2016**, *144*, 2377–2383. [CrossRef] [PubMed]
10. Badea, M.N.; Diacu, E.; Radu, V.M. Influence of antibiotics on copper uptake by plants. *Rev. Chim.* **2013**, *64*, 684–687.
11. Chitescu, C.L.; Nicolau, A.I.; Stolker, A.A. Uptake of oxytetracycline, sulfamethoxazole and ketoconazole from fertilised soils by plants. *Food Addit. Contam. Part A* **2013**, *30*, 1138–1146. [CrossRef] [PubMed]
12. Zhang, H.B.; Zhou, Y.; Huang, Y.J.; Wu, L.H.; Liu, X.H.; Luo, Y.M. Residues and risks of veterinary antibiotics in protected vegetable soils following application of different manures. *Chemosphere* **2016**, *152*, 229–237. [CrossRef] [PubMed]
13. Franklin, A.M.; Williams, C.F.; Andrews, D.M.; Woodward, E.E.; Watson, J.E. Uptake of three antibiotics and an antiepileptic drug by wheat crops spray irrigated with wastewater treatment plant effluent. *J. Environ. Qual.* **2015**, *45*, 546. [CrossRef] [PubMed]
14. Hussain, S.; Naeem, M.; Chaudhry, M.N.; Iqbal, M.A. Accumulation of residual antibiotics in the vegetables irrigated by pharmaceutical wastewater. *Expo. Health* **2016**, *8*, 107–115. [CrossRef]
15. Pan, M.; Wong, C.K.C.; Chu, L.M. Distribution of antibiotics in wastewater-irrigated soils and their accumulation in vegetable crops in the Pearl River Delta, Southern China. *J. Agric. Food Chem.* **2014**, *62*, 11062–11069. [CrossRef] [PubMed]
16. Azanu, D.; Mortey, C.; Darko, G.; Weisser, J.J.; Styrishave, B.; Abaidoo, R.C. Uptake of antibiotics from irrigation water by plants. *Chemosphere* **2016**, *157*, 107–114. [CrossRef] [PubMed]
17. Dolliver, H.; Kumar, K.; Gupta, S. Sulfamethazine uptake by plants from manure-amended soil. *J. Environ. Qual.* **2007**, *36*, 1224–1230. [CrossRef] [PubMed]
18. Li, W.H.; Shi, Y.L.; Gao, L.H.; Liu, J.M.; Cai, Y.Q. Occurrence of antibiotics in water, sediments, aquatic plants, and animals from Baiyangdian Lake in North China. *Chemosphere* **2012**, *89*, 1307–1315. [CrossRef] [PubMed]

19. Ahmed, M.B.M.; Rajapaksha, A.U.; Lim, J.E.; Vu, N.T.; Kim, I.S.; Kang, H.M.; Lee, S.S.; Ok, Y.S. Distribution and accumulative pattern of tetracyclines and sulfonamides in edible vegetables of cucumber, tomato, and lettuce. *J. Agric. Food Chem.* **2015**, *63*, 398–405. [CrossRef] [PubMed]

20. Chowdhury, F.; Langenkämper, G.; Grote, M. Studies on uptake and distribution of antibiotics in red cabbage. *J. Verbr. Lebensm.* **2016**, *11*, 61–69. [CrossRef]

21. Herklotz, P.A.; Gurung, P.; Vanden, H.B.; Kinney, C.A. Uptake of human pharmaceuticals by plants grown under hydroponic conditions. *Chemosphere* **2010**, *78*, 1416–1421. [CrossRef] [PubMed]

22. Kang, D.H.; Gupta, S.; Rosen, C.; Fritz, V.; Singh, A.; Chander, Y.; Murray, H.; Rohwer, C. Antibiotic uptake by vegetable crops from manure-applied soils. *J. Agric. Food Chem.* **2013**, *61*, 9992–10001. [CrossRef] [PubMed]

23. Gullberg, E.; Cao, S.; Berg, O.G.; Ilbäck, C.; Sandegren, L.; Hughes, D.; Andersson, D.I. Selection of resistant bacteria at very low antibiotic concentrations. *PLoS Pathog.* **2011**, *7*, 1002158. [CrossRef] [PubMed]

24. Schuijt, T.J.; Poll, T.; Vos, W.M.; Wiersinga, W.J. The intestinal microbiota and host immune interactions in the critically ill. *Trends Microbiol.* **2013**, *21*, 221–229. [CrossRef] [PubMed]

25. Danhorn, T.; Fuqua, C. Biofilm formation by plant-associated bacteria. *Annu. Rev. Microbiol.* **2007**, *61*, 401–422. [CrossRef] [PubMed]

26. Arunachalam, C.; Gayathri, P. Studies on bioprospecting of endophytic bacteria from the medicinal plant of *Andrographis paniculata* for their antimicrobial activity and antibiotic susceptibility pattern. *Int. J. Curr. Pharm. Res.* **2010**, *2*, 63–68.

27. Pal, A.; Paul, A.K. Bacterial endophytes of the medicinal herb *Hygrophila spinosa* T. Anders and their antimicrobial activity. *Br. J. Pharm. Res.* **2013**, *3*, 795–806. [CrossRef]

28. Yang, Q.X.; Ren, S.W.; Niu, T.Q.; Guo, Y.H.; Qi, S.Y.; Han, X.K.; Liu, D.; Pan, F. Distribution of antibiotic-resistant bacteria in chicken manure and manure-fertilized vegetables. *Environ. Sci. Pollut. Res.* **2014**, *21*, 1231–1241. [CrossRef] [PubMed]

29. Chopra, I.; Roberts, M. Tetracycline antibiotics: Mode of action, applications, molecular biology, and epidemiology of bacterial resistance. *Microbiol. Mol. Biol. Rev.* **2001**, *65*, 232–260. [CrossRef] [PubMed]

30. Fernandes, R.; Amador, P.; Prudencio, C. β-Lactams: Chemical structure, mode of action and mechanisms of resistance. *Rev. Med. Microbiol.* **2013**, *24*, 7–17. [CrossRef]

31. Skold, O. Resistance to trimethoprim and sulfonamides. *Vet. Res.* **2001**, *32*, 261–273. [CrossRef] [PubMed]

32. Clinical and Laboratory Standards Institute (CLSI). *Performance Standards for Antimicrobial Susceptibility Testing*; Twenty-Fourth Informational Supplement, M100-S24; CLSI Press: Wayne, PA, USA, 2014; pp. 44–128.

33. Pan, M.; Chu, L.M. Phytotoxicity of veterinary antibiotics to seed germination and root elongation of crops. *Ecotoxicol. Environ. Saf.* **2016**, *126*, 228–237. [CrossRef] [PubMed]

34. Yin, A.G.; Yang, Z.Y.; Ebbs, S.; Yuan, J.G.; Wang, J.B.; Yang, J.Z. Effects of phosphorus on chemical forms of Cd in plants of four spinach (*Spinacia oleracea* L.) cultivars differing in Cd accumulation. *Environ. Sci. Pollut. Res.* **2016**, *23*, 5753–5762. [CrossRef] [PubMed]

35. Jones-Lepp, T.L.; Sanchez, C.A.; Moy, T.; Kazemi, R. Method development and application to determine potential plant uptake of antibiotics and other drugs in irrigated crop production systems. *J. Agric. Food Chem.* **2010**, *58*, 11568–11573. [CrossRef] [PubMed]

36. Gros, M.; Rodríguez-Mozaz, S.; Barceló, D. Rapid analysis of multiclass antibiotic residues and some of their metabolites in hospital, urban wastewater and river water by ultra-high-performance liquid chromatography coupled to quadrupole-linear ion trap tandem mass spectrometry. *J. Chromatogr. A* **2013**, *1292*, 173–188. [CrossRef] [PubMed]

37. Miller, K.I.; Qing, C.; Sze, D.M.Y.; Neilan, B.A. Investigation of the biosynthetic potential of endophytes in traditional Chinese anticancer herbs. *PLoS ONE* **2012**, *7*, e35953. [CrossRef] [PubMed]

38. Wang, F.H.; Qiao, M.; Chen, Z.; Su, J.Q.; Zhu, Y.G. Antibiotic resistance genes in manure-amened soil and vegetables at harvest. *J. Hazard. Mater.* **2015**, *299*, 215–221. [CrossRef] [PubMed]

39. Yang, Q.X.; Tian, T.T.; Niu, T.Q.; Wang, P.L. Molecular characterization of antibiotic resistance in cultivable multidrug-resistant bacteria from livestock manure. *Environ. Pollut.* **2017**, *229*, 188–198. [CrossRef] [PubMed]

40. Liu, M.M.; Ding, R.; Zhang, Y.; Gao, Y.X.; Tian, Z.; Zhang, T.; Yang, M. Abundance and distribution of Macrolide-Lincosamide-Streptogramin resistance genes in an anaerobic-aerobic system treating spiramycin production wastewater. *Water Res.* **2014**, *63*, 33–41. [CrossRef] [PubMed]

41. An, J.; Chen, H.W.; Wei, S.H.; Gu, J. Antibiotic contamination in animal manure, soil, and sewage sludge in Shenyang, northeast China. *Environ. Earth. Sci.* **2015**, *74*, 5077–5086. [CrossRef]

42. Park, S.B.; Kim, S.J.; Kim, S.C. Evaluating plant uptake of veterinary antibiotics with hydroponic method. *Korean J. Soil Sci. Fertil.* **2016**, *49*, 242–250. [CrossRef]

43. Yang, Q.X.; Zhang, H.; Guo, Y.H.; Tian, T.T. Influence of chicken manure fertilization on antibiotic-resistant bacteria in soil and the endophytic bacteria of pakchoi. *Int. J. Environ. Res. Public Health* **2016**, *13*, 662. [CrossRef] [PubMed]

44. Migliore, L.; Cozzolino, S.; Fiori, M. Phytotoxicity to and uptake of flumequine used in intensive aquaculture on the aquatic weed, *Lythrum salicaria* L. *Chemosphere* **2000**, *40*, 741–750. [CrossRef]

45. Pollock, K.; Barfield, D.G.; Shields, R. The toxicity of antibiotics to plant cell cultures. *Plant Cell Rep.* **1983**, *2*, 36–39. [PubMed]

46. World Health Organization (WHO). *Presented at the 3rd Meeting of the WHO Advisory Group on Integrated Surveillance of Antimicrobial Resistance, Oslo, Norway, 14–17 June 2011*; World Health Organization: Geneva, Switzerland, 2012.

47. Zhang, H.B.; Luo, Y.M.; Wu, L.H.; Huang, Y.J.; Christie, P. Residues and potential ecological risks of veterinary antibiotics in manures and composts associated with protected vegetable farming. *Environ. Sci. Pollut. Res.* **2015**, *22*, 5908–5918. [CrossRef] [PubMed]

48. Heuer, H.; Solehati, Q.; Zimmerling, U.; Kleineidam, K.; Schloter, M.; Müller, T.; Focks, A.; Thiele-Bruhn, S.; Smalla, K. Accumulation of sulfonamide resistance genes in arable soils due to repeated application of manure containing sulfadiazine. *Appl. Environ. Microbiol.* **2011**, *77*, 2527–2530. [CrossRef] [PubMed]

49. Liu, A.; Fong, A.; Becket, E.; Yuan, J.; Tamae, C.; Medrano, L.; Maiz, M.; Wahba, C.; Lee, C.; Lee, K.; et al. Selective advantage of resistant strains at trace levels of antibiotics: A simple and ultrasensitive color test for detection of antibiotics and genotoxic agents. *Antimicrob. Agents Chemother.* **2011**, *55*, 1204–1210. [CrossRef] [PubMed]

50. Sandegren, L. Selection of antibiotic resistance at very low antibiotic concentrations. *Ups. J. Med. Sci.* **2014**, *119*, 103–107. [CrossRef] [PubMed]

51. Compant, S.; Clément, C.; Sessitsch, A. Plant growth-promoting bacteria in the rhizo- and endosphere of plants: Their role, colonization, mechanisms involved and prospects for utilization. *Soil Biol. Biochem.* **2010**, *42*, 669–678. [CrossRef]

52. Aminov, R.I. Evolution in action: Dissemination of tet(X) into pathogenic microbiota. *Front. Microbiol.* **2013**, *4*, 192. [CrossRef] [PubMed]

53. Bradford, P.A. Tigecycline: A first in class glycylcycline. *Clin. Microbiol. Newslett.* **2004**, *26*, 163–168. [CrossRef]

54. Kumarasamy, K.K.; Toleman, M.A.; Walsh, T.R.; Bagaria, J.; Butt, F.; Balakrishnan, R.; Chaudhary, U.; Doumith, M.; Giske, C.G.; Irfan, S.; et al. Emergence of a new antibiotic resistance mechanism in India, Pakistan, and the UK: A molecular, biological, and epidemiological study. *Lancet Infect. Dis.* **2010**, *10*, 597–602. [CrossRef]

55. Moore, I.F.; Hughes, D.W.; Wright, G.D. Tigecycline is modified by the flavin-dependent monooxygenase TetX. *Biochemistry* **2005**, *44*, 11829–11835. [CrossRef] [PubMed]

56. Koczura, R.; Mokracka, J.; Taraszewska, A.; Łopacinska, N. Abundance of Class 1 integron-integrase and sulfonamide resistance genes in river water and sediment is affected by anthropogenic pressure and environmental factors. *Microb. Ecol.* **2016**, *72*, 909–916. [CrossRef] [PubMed]

57. Lee, J.; Shin, S.G.; Jang, H.M.; Kim, Y.B.; Lee, J.; Kim, Y.M. Characterization of antibiotic resistance genes in representative organic solid wastes: Food waste-recycling wastewater, manure, and sewage sludge. *Sci. Total Environ.* **2017**, *579*, 1692–1698. [CrossRef] [PubMed]

58. Makowska, N.; Koczura, R.; Mokracka, J. Class 1 integrase, sulfonamide and tetracycline resistance genes in wastewater treatment plant and surface water. *Chemosphere* **2016**, *144*, 1665–1673. [CrossRef] [PubMed]

59. Sköld, O. Sulfonamide resistance: Mechanisms and trends. *Drug Resist. Updat.* **2000**, *3*, 155–160. [CrossRef] [PubMed]

International Journal of
*Environmental Research
and Public Health*

Article

Antibiotic Resistance in an Indian Rural Community: A 'One-Health' Observational Study on Commensal Coliform from Humans, Animals, and Water

Manju Raj Purohit [1,2,*], Salesh Chandran [1,3], Harshada Shah [3], Vishal Diwan [1,4,5], Ashok J. Tamhankar [1,6] and Cecilia Stålsby Lundborg [1]

[1] Department of Public Health Sciences, Global Health—Health Systems and Policy (HSP): Medicines Focusing Antibiotics, Karolinska Institutet, 17177 Stockholm, Sweden; saleshp@gmail.com (S.C.); vishal.diwan@ki.se (V.D.); ejetee@gmail.com (A.J.T.); cecilia.stalsby.lundborg@ki.se (C.S.L.)
[2] Department of Pathology, R.D. Gardi Medical College, Ujjain 456006, India
[3] Department of Microbiology, R.D. Gardi Medical College, Ujjain 456006, India; sharshada1955@yahoo.com
[4] International Centre for Health Research, Ujjain Charitable Trust Hospital and Research Centre, Ujjain 456006, India
[5] Department of Public Health and Environment, R.D. Gardi Medical College, Ujjain 456006, India
[6] Indian Initiative for Management of Antibiotic Resistance, Department of Environmental Medicine, R.D. Gardi Medical College, Ujjain 456006, India
* Correspondence: manjuraj.purohit64@gmail.com; Tel.: +91-942-509-3572

Academic Editor: Paul B. Tchounwou
Received: 9 January 2017; Accepted: 30 March 2017; Published: 6 April 2017

Abstract: Antibiotic-resistant bacteria are an escalating grim menace to global public health. Our aim is to phenotype and genotype antibiotic-resistant commensal *Escherichia coli (E. coli)* from humans, animals, and water from the same community with a 'one-health' approach. The samples were collected from a village belonging to demographic surveillance site of Ruxmaniben Deepchand (R.D.) Gardi Medical College Ujjain, Central India. Commensal coliforms from stool samples from children aged 1–3 years and their environment (animals, drinking water from children's households, common source- and waste-water) were studied for antibiotic susceptibility and plasmid-encoded resistance genes. *E. coli* isolates from human ($n = 127$), animal ($n = 21$), waste- ($n = 12$), source- ($n = 10$), and household drinking water ($n = 122$) carried 70%, 29%, 41%, 30%, and 30% multi-drug resistance, respectively. Extended spectrum beta-lactamase (ESBL) producers were 57% in human and 23% in environmental isolates. Co-resistance was frequent in penicillin, cephalosporin, and quinolone. Antibiotic-resistance genes $bla_{CTX-M-9}$ and $qnrS$ were most frequent. Group D-type isolates with resistance genes were mainly from humans and wastewater. Colistin resistance, or the *mcr-1* gene, was not detected. The frequency of resistance, co-resistance, and resistant genes are high and similar in coliforms from humans and their environment. This emphasizes the need to mitigate antibiotic resistance with a 'one-health' approach.

Keywords: antibiotic resistance; community; environment; India; coliforms; commensal

1. Introduction

Antibiotic resistance represents a significant and complex global health problem. Global consumption of antibiotics has increased by nearly 40% in the last decade [1]. Apart from fundamental applications in clinical settings, very large amounts of antibiotics are used in agriculture, the food industry, and aquaculture [2]. Due to incomplete metabolism and the environmental spread of unused antibiotics, they enter the ecosystem, serving as a potent stimulus to elicit a bacterial adaptation response to develop antibiotic resistance and genes [3,4]. The accumulation of antibiotics in the environment facilitates the spread of antibiotic resistance genes. Various resistance mechanisms are continuously emerging and spreading globally, which threatens our ability to treat common infectious diseases, resulting in increased death, disability, and costs. The World Health Assembly, in 2015, thus adopted a global action plan on antimicrobial resistance focussing on bacterial resistance [5].

There is a worldwide concern about the emergence of antibiotic resistance in bacteria carried by healthy individuals, so-called commensal bacteria. Commensal bacteria from the gut microbes, e.g., coliforms, may play a crucial role in the spread of resistance within a community. Surveillance data shows that resistance in *Escherichia coli* is generally consistently highest for antimicrobial agents that have been in use the longest time in human and veterinary medicine [6]. *E. coli* is also considered an indicator bacteria of antibiotic resistance. Animal and human fecal flora and the environment, including water sources, serve as natural habitats and reservoirs of antibiotic-resistant bacteria and resistance genes. Antibiotic resistance in wastewater, surface water, and drinking water is well documented [7,8].

Thus, within the community, resistant bacteria circulated from person to person or from animals and environment to person, or vice versa. The epidemiology of antibiotic-resistant microorganisms at the human-animal-environmental interface involves complex and largely unpredictable systems that include transmission routes of resistant bacteria, as well as resistance genes and the impact of antibiotic-selective pressures in various reservoirs (animals, humans, and the environment). Though the presence and patterns of antibiotic resistant commensal indicator bacteria *E. coli* isolates from humans, animals, and water have been studied in isolation, it is now recognized that they need to be studied together, i.e., using the 'one-health' approach [5]. Thus, our aim is to determine and compare the antibiotic resistance pattern among commensal coliforms and *E. coli* from humans, animals, and water from the same community.

2. Materials and Methods

2.1. Study Setting and Sample Collection

The present study is a part of an ongoing project that has been described in detail previously [9]. In brief, the study was conducted in Ujjain district of Madhya Pradesh, India. We selected the village from the demographic surveillance site of Ruxmaniben Deepchand (R.D.) Gardi Medical College having poor literacy and living standards (Table 1) as described in [9]. The children aged between 1 and 3 years in the village at the commencement of the study, i.e., September 2014 were identified. Trained research assistants visited selected children's homes and informed the children's parents/guardians about the study. All children whose parents consented for their children to participate were included in the study. Stool samples from selected children and drinking water samples from their households were collected. Stool samples from five different animals (cattle, hen, dog, goat, and horse), which commonly share their environment with children, two common drinking-water sources, and two waste-water samples from the village were also collected, as depicted in Figure 1. All of the collected samples were transported within five hours to the Central Research Laboratory at R.D. Gardi Medical College. All samples could be collected within three days from the village. The village health worker in a predesigned format noted basic socio-demographic details by interviewing the head of the family.

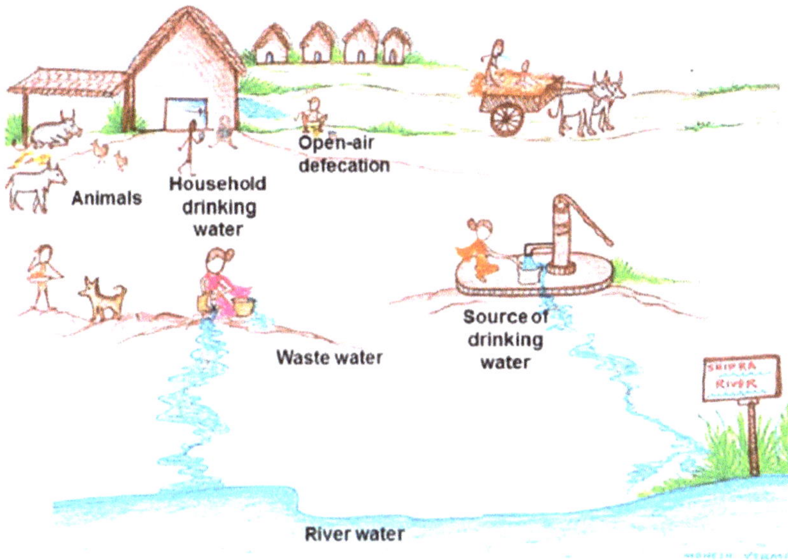

Figure 1. "One-health" approach.

2.2. Identification of Coliforms and Confirmation of E. coli

Microbiological processing of the samples was started as soon as the samples were received in the laboratory. The samples were processed on selective and differential HiCrome coliform chromogenic agar (HiMedia Laboratories Pvt. Ltd., Mumbai, India) to identify *E. coli* (blue-violet colony) and non-*E. coli* coliforms (*Citrobacter freundii* and *Enterobacter cloacae*—salmon to red, *Klebsiella pneumoniae*—light pink, *Salmonella enteritidis* and *Shigella flexneri*—colorless) as described in detail [9]. The presumptive *E. coli* were confirmed by PCR (mentioned in detail below). Briefly, stool samples were inoculated at 37 °C for 24 h directly on the chromogenic agar while the water samples were first filtered through membranes [9], followed by inoculation of the membranes on agar plate. In water samples, colony-forming units (CFUs) per unit volume of sample were estimated for total coliforms and *E. coli* to provide a snapshot of the abundance of coliforms and the *E. coli* load in the samples tested. Six *E. coli* and two colonies from every type of non-*E. coli* were isolated, purified, stored, and processed for antibiotic susceptibility testing and DNA extraction.

2.3. Antibiotic Susceptibility Testing

All the pure and confirmed six *E. coli* and HiCrome coliform agar categorized non-*E. coli* isolates from each sample were analyzed for the susceptibility to colistin, ampicillin, ceftriaxone, cefepime, ciprofloxacin, tetracycline, tigecycline, meropenem, imipenem, gentamicin, amikacin, sulphamethoxazole, cotrimoxazole, nalidixic acid, and nitrofurantoin (all purchased from HiMedia Laboratories Pvt. Ltd., Mumbai, India) by the Kirby-Bauer disc diffusion method as described in [9]. The results of inhibitory zones of the antibiotic susceptibility testing procedure were interpreted as detailed previously [9] using the Clinical and Laboratory Standards Institute (CLSI) criteria. The isolates were categorized as the number of resistant isolates per antibiotic type per sample (out of six isolates), phenotypically-confirmed beta-lactamase producers (only where beta-lactamase production is indicated as a possible mechanism explaining observed resistance) by the combined disc diffusion method(isolates resistant to either ceftazidime (HiMedia Laboratories Pvt. Ltd., Mumbai, India) or ceftriaxone (third generation cephalosporin), the presence of co-resistance (phenotypic resistance to two or more antibiotics of same or different group per isolate), and multidrug resistance (MDR) (MDR co-resistance involving three or more antibiotics of three different groups) in each sample. *E. coli* reference strain ATCC 25922 was used for quality control. Intermediate resistant isolates were categorized as resistant.

2.4. Amplification of Genes

The total bacterial DNA from *E. coli* isolates was extracted using the alkaline lysis method [10]. The genetic confirmation of *E. coli* was done through PCR with genus-specific oligonucleotide primers [11]. β-lactamase-encoding (bla_{CTXM}, bla_{TEM}, and bla_{SHV}); plasmid-mediated quinolone resistance (*qnrA, qnrS, qnrS, aac(6′)-Ib-cr*, and *qepA*), carbapenem resistant (*VIM, NDM, IMP*) and colistin resistant (*mcr-1*) genes were amplified and identified with previously-described primers [12,13] for all *E. coli* isolates. The phylogenetic grouping of all *E. coli* was performed based on *chuA, vjaA*, and *TspE4C2* genes which were amplified by multiplex PCR as described in detail elsewhere [14]. All of the amplified PCR products were visualized using a gel documentation system for all *E. coli* isolates.

2.5. Data Analysis

Drug susceptibility and gene detection data were generated, and entered into IBM SPSS Statistics 23.0 (SPSS Inc., Chicago, IL, USA). Descriptive statistics, frequencies, and bivariate analyses (cross-tabulations) for the susceptibility pattern of *E. coli* isolates from different samples were calculated. Multiple linear regression analysis was performed to assess the effect of demographic features on antibiotic resistance. Resistance to different antibiotics was included as dependent variables, and age, sex, and other demographic parameters were included as independent variables in the model to adjust for confounding variables. Differences were considered statistically significant at $p < 0.05$. Results were also noted for the variation in coliforms load (in terms of CFU per 100 mL) in drinking water and wastewater. The results of the susceptibility pattern of *E. coli* and non-*E. coli* were correlated with the corresponding pattern and between human and environmental samples.

2.6. Ethical Consideration

Ethical issues: Ethics permission for the study was obtained from the Institutional Ethics Committee of R.D. Gardi Medical College, Ujjain (India) (No. 2013/07/17-311). Parents/guardians were explained about the purpose of the study, about voluntary participation, and were assured by researchers to maintain confidentiality. Oral and written informed consent was taken, thereafter. Children identified as having need of medical care were referred and treated at the Department of Pediatrics at C.R. Gardi Hospital.

3. Results

3.1. Study Samples

A total of 24 children were identified according to the inclusion criterion from the selected village. Stool samples from 22 children and drinking water from their respective home were collected. Samples from two children (one not at home, one did not passed motion by the time of collection) could not be collected even after two follow-up visits. Table 1 shows the demographic details of the families of the children from whom isolates were obtained. All of the isolates identified as blue-violet colonies on HiCrome agar were confirmed by PCR as *E. coli* while other bacterial isolates identified on HiCrome were considered together as the non-*E. coli* group and processed for antibiotic susceptibility testing.

The number and source of the samples and the number and types of coliforms isolated and studied from each sample is shown in Table 2.

Table 1. Socio-demographic characteristics of the families of included children (n = 22) in a village in Central India.

Variable	Number (%)
Family type	
Nuclear family	6
Joint family	16
Total number of family members	162
Male	84/162 (52)
Female	78/162 (48)
Number of children-	
Up to five years of age	46
Male	25/46 (54)
Female	21/46 (46)
Between one and three years of age	24
Male	15
Female	9
Highest education of family member	
Primary education (up to 5th grade)	5
Middle	11
Secondary	2
Illiterate	144
Occupation of head of family	
Job	1
Farmer	12
Labor/self employed	6
Unemployed	3
Type of house	
Kuchcha	11
Pucca/semi-pucca	1/10
Total number of livestock in all households	75
Source of drinking water	
Piped water into dwelling	1
In-house tube wells/bore hole	1
Hand pump	10
Unprotected dug well	1

Types of house: walls, roof, and floors are made of bamboo, mud, grass, reeds, thatch, plastic/polythene, loosely-packed stone, etc., in Kachcha houses, stones, bricks packed with lime or cement mortar or concrete, in pucca houses, while in Semi-Pucca houses walls and roof are of concrete or un-burnt bricks, but the floor is made of mud or non-concrete items.

Table 2. Samples and commensal coliforms isolated from human and animal stool and water samples collected from a village in Central India.

Source of Samples	Number of Samples	Number of *E. coli*	Number of Non-*E. coli*
Children stool	22	127	67
Dog stool	1	6	2
Hen stool	1	6	6
Goat stool	1	3	0
Horse stool	1	6	4
Source-water	2	10	14
Waste-water	2	12	7
Household drinking water	22	122	143
Total	52	292	243

3.2. Antibiotic Resistance Pattern of E. coli in Various Sources

All (six) isolates from one child and two household drinking-water samples were susceptible to all drugs and no isolates from any of the samples showed resistance to all antibiotics. The overall percentage of resistant isolate is significantly higher in samples from humans compared to those from the environment ($p = 0.04$). The percentage of resistance for individual antibiotics is also high in humans, except for gentamycin, amikacin, and tigecycline (Figure 2A). Nearly 70% of human stool had co-resistance *E. coli*, of which 57% (73/127) were extended spectrum beta-lactamase (ESBL) producers, and 33% were MDR isolates. In animals, 19% isolates were fully susceptible, 29% co-resistant, 23% ESBL producers and 14% MDR. Co-resistance was more frequent (MDR 41% and ESBL producer 33%) in wastewater isolates than in source water and household drinking water (MDR in 30% and ESBL producer 24%) isolates (Figure 2B). The load of resistant isolates (described as <3 or ≥3 resistant isolates among six collected *E. coli* isolates per sample), in each sample is significantly higher ($p = 0.001$) in human stool than in household drinking-water samples (Table 3), but the resistant isolates from drinking-water were distributed in a higher number of samples. The samples from nuclear families significantly showed less resistance ($p = 0.05$) than in samples from joint families. The resistance pattern of a child and his/her respective household drinking water was not significantly ($p = 0.05$) dissimilar.

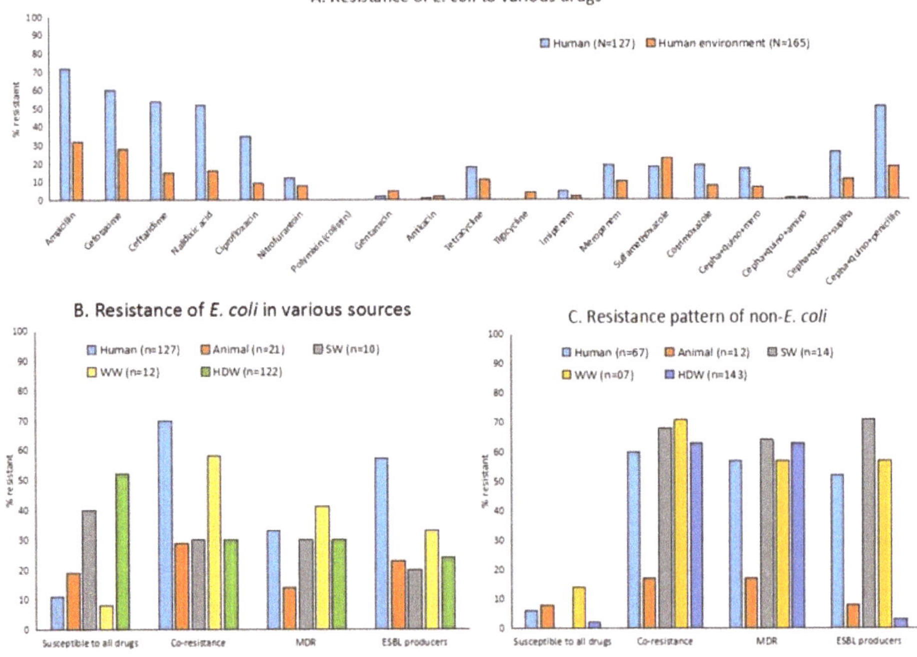

Figure 2. Antibiotic resistance pattern to tested antibiotics in *E. coli* and non-*E. coli* isolates from various sources in a rural setting of Central India. (**A**) Percentage of resistance to various drugs of *E. coli* from human and environmental samples; (**B**) Pattern of resistant *E. coli* isolated from various sources; (**C**) Pattern of resistant non-*E. coli* isolated from various sources. SW: source-water; WW: wastewater; HDW: household drinking water; MDR: multidrug resistance; ESBL: extended spectrum beta-lactamase producers.

3.3. Antibiotic Resistance Pattern of E. coli to Various Antibiotic Groups

There was no resistance to the polymyxin (colistin) group in any of the sample types. There was high resistance frequency to penicillins, quniolones, and cephalosporins in human (23%–77%) and environmental (12%–25%) isolates. The MDR combinations having penicillin + cephalosporins + quinolones and sulfonamides + cephalosporin + quinolones groups of drugs were more common than the cephalosporin + quinolone + aminoglycosides or carbapenem combinations (Figure 2A). Most of the isolates from all the sources showed resistance simultaneously to ceftazidime, cefotaxime, cefapime, ampicillin, tetracycline, and co-trimoxazole.

3.4. Antibiotic Resistant Genes in E. coli

In human stool, plasmid-mediated cephalosporin-coding genes of the $bla_{CTX-M-1}$ group was predominant, especially the gene $bla_{CTX-M-1}$. In environmental samples $bla_{CTX-M-1}$ and $bla_{CTX-M-9}$ genes were also common (Table 4). Plasmid-mediated quinolone resistance genes (i.e., *qnrA*, *qnrS*, *qnrS*) were detected in 34% human and in 9% environmental quinolone resistant isolates. The *qnrS* gene was most common in human (23/72), and only three household drinking waters were carrying quinolone-resistant genes from environmental isolates (n = 33) (Table 4). The coexistence of $bla_{CTX-M-1}$ and *qnrS* genes were also common (n = 12). Carbapenemases encoding genes *NDM-1*, *VIM*, and *IMP* were not detected in any of the carbapenem resistant isolates (n = 35) and colistin-resistant gene *mcr-1* was also not detected in any of the isolates (n = 292).

The majority (56%–100%) of cephalosporin, quinolone, and carbapenem resistant *E. coli* isolates belonged to phylogenetic group A and B1 (considered as commensal) and 0%–40% belonged to D (considered as extra-intestinal virulent). Human samples carried significantly higher numbers (30%–52%) of group D isolates than environmental samples (0%–40%) (Table 5). Isolates, which showed susceptibility to all drugs, belonged equally to groups A, B1, B2, and D in human samples, but in environmental samples these isolates mainly belonged to the A or B1 groups. The majority (82%) of isolates carrying resistant genes belonged to phylogenetic group A and B1 and the rest (18%) were categorized into group D. Human stool and wastewater were the source of most of the group D *E. coli* isolates.

Table 3. Distribution of various antibiotic resistant *E. coli* isolates in human and drinking water collected from households in a village in Central India.

Name of Antibiotic Tested	Human Stool (n = 22)		Household Drinking Water (n = 20)	
	Resistant *E. coli* Isolates in Samples (n) *			
	<3	≥3	<3	≥3
Ampicillin	6	12	15	3
Ceftazidime	7	11	13	1
Cefotaxime	7	11	11	-
Nalidixic acid	5	9	6	2
Ciprofloxacin	5	6	7	-
Nitrofurantoin	1	1	2	-
Gentamicin	2	1	2	-
Amikacin	1	-	2	-
Tetracycline	1	3	8	-
Tigicycline	3	1	2	-
Imipenem	-	1	-	-
Meropenem	5	2	1	-
Sulfamethoxazole	2	4	4	2
Cotrimoxazole	2	4	4	1

HDW: household drinking-water; *: p = 0.001

Table 4. Antibiotic resistant genes in commensal *E. coli* isolated from samples from humans and their shared environment from a village in Central India.

Cephalosporin Resistant Isolates	Cephalosporin Resistance Genes		
	CTX-M1	*CTX-M2*	*CTX-M9*
HS (*n* = 73)	62	0	0
HDW (*n* = 26)	11	0	0
AS (*n* = 6)	4	0	0
SW (*n* = 5)	0	0	1
WW (*n* = 5)	0	0	5
Quinolone Resistant Isolates	**Quinolone Resistance Genes**		
	qnrA	*qnrS*	*qnrS*
HS (*n* = 72)	0	2	23
HDW (*n* = 13)	0	0	3
AS (*n* = 2)	0	0	0
SW (*n* = 1)	0	0	0
WW (*n* = 8)	0	0	0

HS: human stool; AS: animal stool; SW: source-water; WW: wastewater.

Table 5. Phylogenetic grouping of resistant commensal *E. coli* isolates collected from various samples from a village in Central India.

Phylogenetic Group	A *n* = 135	B1 *n* = 55	B2 *n* = 13	D *n* = 92
Cephalosporin-Resistant Isolates				
HS (*n* = 73)	35	17	0	21
HDW (*n* = 26)	16	3	0	7
AS (*n* = 6)	4	0	1	1
SW (*n* = 5)	4	0	1	0
WW (*n* = 5)	0	0	0	5
Quinolone-Resistant Isolates				
HS (*n* = 72)	28	13	1	30
HDW (*n* = 13)	9	1	0	3
AS (*n* = 2)	2	0	0	0
SW (*n* = 1)	1	0	0	0
WW (*n* = 8)	1	0	0	7
Meropenem-Resistant Isolates				
HS (*n* = 19)	10	1	0	8
HDW (*n* = 8)	6	2	0	0
AS (*n* = 1)	1	0	0	0
SW (*n* = 1)	1	0	0	0
WW (*n* = 0)	0	0	0	0
Susceptible to All Drugs				
HS (*n* = 21)	2	8	6	5
HDW (*n* = 9)	1	6	1	1
AS (*n* = 2)	0	2	0	0
SW (*n* = 3)	1	0	2	0
WW (*n* = 1)	0	1	0	0

3.5. Non-E. coli Coliforms and AST Pattern

We have also detected many non-*E coli* coliforms (Table 2). We found higher numbers and types of non-*E. coli* coliforms from water samples than in stool samples (human and animal). The number of suggested total coliforms as grown on HiCrome media in terms of number of *E. coli* (identified as blue-violet colonies) and different non-*E. coli* (identified as different color colonies) CFU/unit volume in MDR-positive water samples is shown in Table 6. Only 6% of non-*E. coli* coliform isolates from human stool were susceptible to all tested drugs, while 57% and 52% isolates were MDR and ESBL producers (Figure 2C), respectively. Animal stool carried lower MDR and ESBL producers as compared to isolates from other sources.

Table 6. Load of commensal non-*E. coli* and *E. coli* isolates from various water samples carrying multi-drug resistant *E. coli*.

Sample *	Total Coliform Count/100 mL	Total *E. coli*/100 mL n (%)	Total-*E. coli* = Non-*E. coli* n (%) (Calculated)
1	1630	260 (16)	1370 (84)
2	1400	40 (3)	1360 (97)
3	520	100 (19)	400 (81)
4	498	64 (13)	434 (87)
5	430	40 (8)	390 (92)
6	414	14 (3)	400 (97)
7	152	3 (2)	149 (98)
8	150	1 (0.66)	149 (99.4)
9	134	34 (25)	100 (75)
10	48	20 (41)	28 (59)
11	365,000,000	15,000,000 (41)	215,000,000 (59)
12	204,000,000	32,000,000 (16)	1,720,000,000 (84)
13	3650	150 (4)	3500 (96)
14	3	0	3 (100)

*: The samples 1–10 are from household drinking water, 11–12 from village waste-water, and 13–14 are from source drinking water.

4. Discussion

We studied antibiotic resistance and selected antibiotic resistance genes in human stool together with their shared and neighboring environment in a rural community from Central India with a 'one-health' approach. We found that the antibiotic resistance pattern and its genetic make-up are essentially the same in commensal bacteria from humans and their environment. The percentage of resistant isolates, including MDR (Figure 1A,B), is higher in humans than in the environment (animal stool and water samples), but the load (number of resistant isolates/sample) is higher in the environment than in humans. The appearance of antibiotic-resistant bacteria in healthy individuals and their environment should be evaluated together to accomplish effective antibiotic resistance control.

The antibiotic resistance profile including certain patterns of co-resistance and MDR (i.e., cephalosporin-quinolone-penicillin, sulphonamide + tetracycline + cephalosporin, quinolones + carbapenem + sulfonamide or + tetracycline) in *E. coli* obtained from humans, animals, source- and household-drinking water are high (57%–69%) in our study area. The presence of co-resistance and MDR signifies that there might be high use of antibiotics inhuman and non-human use in the community. The non-human use of highly-important antibiotics contributes to the resistance against a range of antibiotics [1,2,15]. Van den Bogaard et al. and others have shown that the selective pressure on the commensal microflora due to antibiotic misuse determine the frequency and pattern of resistance in a population [16]. The relatively cheap and commonly prescribed drugs commonly favour high co-resistance [17,18].

We found similar patterns of co-resistance, MDR, and gene carriage in various sources. Nearly 90% of MDR *E. coli* isolates are carrying plasmid-encoded (bla_{CTX-M1}, bla_{CTX-M9}, *qnrS*, and *qnrS*) genes, which may indicate the possible spread of the resistance genes between diverse sources. This is similar to another study from India [19]. $_{CTX-M}$–producing *E. coli* is the dominant MDR *E. coli* in all parts of Asia and of major clinical significance [20]. The patterns of antibiotic use in the community favor the persistence of plasmids carrying antibiotic resistance genes. The intestine is considered as a 'hot spot' for the transfer of resistance genes between bacteria as the exposure of frequently-used antibiotics to a high density of bacteria favours evolution and dissemination of antibiotic resistance by cell-to-cell contact [21,22]. Additionally, the existing various species of MDR bacteria, as we noticed in MDR non-*E. coli* coliform species, (Table 6, Figure 1C), might also be contributing to the spread of antibiotic resistance genes in the intestine with *E. coli*.

The resistant isolates are distributed in a higher percentage of drinking-water samples compared to human samples. In rural communities, the high level of bacterial contamination is reported in source-water to the extent that it lacks the criteria of safe-water supply for domestic purposes [23]. Studies illustrate that surface water contamination occurs mainly from livestock operations and human sewage and that decreasing livestock access to surface water reduced the fecal coliforms levels by an average of 94% [24]. Treatment processes of water, however, might further result in a selective increase of antibiotic-resistant bacteria and might, therefore, increase the occurrence of multidrug-resistant organisms [11,25]. It has also been observed that the microbiological quality of water in vessels in households is lower than that at the source, suggesting that bacterial contamination is widespread during collection, transport, storage, and drawing of water [26].

In our study, phylogenetic group D (extra-intestinal virulent) *E. coli* isolates with resistant genes are more often found from human stool than from environmental samples (30%–52% vs. 0%–24%). It has been reported that co-location of genes in plasmids not only results in resistance to multiple antibiotics, but also in the increased presence of virulence determinants, which facilitates infections [27]. Indeed, the exposure of commensal bacteria to antibiotics increases the carriage level of resistant organisms that might result in the transmission of resistance to a virulent organism [28]. Johnson et al. [29] reported the horizontal transfer of antibiotic resistance not only between isolates from one source to another, but also from resistant to susceptible isolates in the same source. The number of virulent strains carrying resistant genes in human commensal samples is a matter of public health concern, as it may give rise to infection with an increased risk of treatment failure.

We have not identified any *E. coli* or non-*E. coli* isolates (including all forms of MDR strains) with colistin resistance or *mcr-1* gene carriage. With the emergence of MDR and extensive drug resistant (XDR) strains of Gram-negative bacteria, colistin is considered as one of the few last resort antibacterial agents. Recently, sporadic clinical cases infected with colistin-resistant *E. coli* carrying the *mcr-1* gene has been described in India [30,31]. The plasmid-mediated *mcr-1* gene to colistin resistance is a matter of global alarm as its spread within the human commensal flora could lead to epidemics of virtually untreatable infections. Measures with the 'one-health' approach, such as colistin susceptibility testing of MDR isolates from patients, testing of food, animal, environmental isolates, and the reduction of colistin use in food-producing animals would be crucial for effective minimization of *mcr-1*-positive commensal dissemination in the community and healthcare facilities.

Our study has some methodological limitations. The study, being from a village, cannot be generalized. There is no reason, however, to believe that the situation in this village is very different from many other villages with similar low socio-economic levels in India. Additionally, in our study, none of the carbapenem-resistant isolates (six imipenem resistant and 29 meropenem resistant isolates) from all sources are carrying any of the tested (*NDM-1*, *VIM*, and *IMP*) carbapenemases encoding genes. Studies showed the presence of *OXA*-48 and *NDM*-1 genes in clinical isolates from India [32,33]. However, in another study from our setting, we did not find any of these genes in either clinical or in hospital waste water [34]. We, however, cannot rule out some different resistance mechanisms in these isolates, which we have not tested. Although our study involves a limited number of animals and sewage water samples, the comparison of multiple types of environmental samples with apparently healthy human samples from community provides us better understanding about the current scenario of antibiotic resistance at the community level. This is required in scientific research for establishing effective measures to mitigate resistance in clinically relevant bacteria.

5. Conclusions

We found similar and widespread antibiotic resistance, co-resistance, MDR, and their genetic make-up in commensal bacteria from humans and their environment. The percentage of antibiotic resistance is higher in humans than in the environment, but the load (number of resistant isolates/sample) is higher in the environment than in humans. The study, thus, raises a number of important public health concerns. Firstly, community-based studies should be conducted to quantify attributes of antibiotic resistance to design an effective stewardship program; secondly, there should be a multi-sectorial national alliance with all key stakeholders to discourage non-therapeutic use of antibiotics; and, lastly, a strengthening of antimicrobial policy and antibiotic stewardship in India.

Acknowledgments: This study is a part of Swedish Research Council project number K2013-70X-20514-07-5. We extend our sincere thanks to the medical director, V.K. Mahadik, and the management of R.D. Gardi Medical College for facilitating this study. We are grateful for the excellent data collection and support in the field offered by Vivek Parashar and Shweta Khare and their team.

Author Contributions: All authors were involved in the conceptualization and design of the study. Manju Raj Purohit initiated the formulation of the study, responsible for the microbiological and molecular work and analyses, preparation of the manuscript, coordinated, and drafted the first version of the manuscript. Salesh Chandran was involved in the microbiological and molecular work, Harshada Shah was microbiology advisor, Vishal Diwan was responsible for data collection, and Ashok J Tamhankar was senior advisor for the environmental part. All authors have given intellectual input in the development of the manuscript, read, and approved the final version of the manuscript. Cecilia Stålsby Lundborg is the principle investigator of the project to which this study belongs.

Conflicts of Interest: The authors declare that they have no competing interests.

References

1. Van Boeckel, T.P.; Gandra, S.; Ashok, A.; Caudron, Q.; Grenfell, B.T.; Levin, S.A.; Laxminarayan, R. Global antibiotic consumption 2000 to 2010: An analysis of national pharmaceutical sales data. *Lancet Infect. Dis.* **2014**, *14*, 742–750. [CrossRef]

2. Van Boeckel, T.P.; Brower, C.; Gilbert, M.; Grenfell, B.T.; Levin, S.A.; Robinson, T.P.; Teillant, A.; Laxminarayan, R. Global trends in antimicrobial use in food animals. *Proc. Natl. Acad. Sci. USA* **2015**, *112*, 5649–5654. [CrossRef] [PubMed]

3. Zhang, X.X.; Zhang, T.; Zhang, M.; Fang, H.H.; Cheng, S.P. Characterization and quantification of class 1 integrons and associated gene cassettes in sewage treatment plants. *Appl. Microbiol. Biotechnol.* **2009**, *82*, 1169–1177. [CrossRef] [PubMed]

4. Zhang, X.X.; Zhang, T.; Fang, H.H. Antibiotic resistance genes in water environment. *Appl. Microbiol. Biotechnol.* **2009**, *82*, 397–414. [CrossRef] [PubMed]

5. World Health Organization. Antimicrobial Resistance Draft Global Action Plan on Antimicrobial Resistance. Available online: http://apps.who.int/gb/ebwha/pdf_files/WHA68/A68_20-en.pdf?ua=1 (accessed on 23 December 2015).

6. U.S. Food and Drug Administration. *National Antimicrobial Resistance Monitoring System—Enteric Bacteria (NARMS): Executive Report*; USFDA: Rockville, MD, USA, 2010.

7. Łuczkiewicz, A.; Jankowska, K.; Fudala-Ksiazek, S.; Olańczuk-Neyman, K. Antimicrobial resistance of fecal indicators in municipal wastewater treatment plant. *Water Res.* **2010**, *44*, 5089–5097. [CrossRef] [PubMed]

8. Schwartz, T.; Kohnen, W.; Jansen, B.; Obst, U. Detection of antibiotic-resistant bacteria and their resistance genes in wastewater, surface water, and drinking water biofilms. *FEMS Microbiol. Ecol.* **2003**, *43*, 325–335. [CrossRef] [PubMed]

9. Stålsby Lundborg, C.; Diwan, V.; Pathak, A.; Purohit, M.R.; Shah, H.; Sharma, M.; Mahadik, V.K.; Tamhankar, A.J. Protocol: A 'One health' two year follow-up, mixed methods study on antibiotic resistance, focusing children under 5 and their environment in rural India. *BMC Public Health* **2015**, *30*, 1321–1332. [CrossRef] [PubMed]

10. Sambrook, J.F.E.; Maniatis, T. *Molecular Cloning: A Laboratory Manual*; Cold Spring Harbor Laboratory Press: New York, NY, USA, 1989.

11. Iqbal, S.; Robinson, J.; Deere, D.; Saunders, J.R.; Edwards, C.; Porter, J. Efficiency of the polymerase chain reaction amplification of the *uid* gene for detection of *Escherichia coli* in contaminated water. *Lett. Appl. Microbiol.* **1997**, *24*, 498–502. [CrossRef] [PubMed]

12. Chandran, S.P.; Diwan, V.; Tamhankar, A.J.; Joseph, B.V.; Rosales-Klintz, S.; Mundayoor, S.; Lundborg, C.S.; Macaden, R. Detection of carbapenem resistance genes and cephalosporin, and quinolone resistance genes along with *oqxAB* gene in *Escherichia coli* in hospital wastewater: A matter of concern. *J. Appl. Microbiol.* **2014**, *117*, 984–995. [CrossRef] [PubMed]

13. Liu, Y.Y.; Wang, Y.; Walsh, T.R.; Yi, L.X.; Zhang, R.; Spencer, J.; Doi, Y.; Tian, G.; Dong, B.; Huang, X.; et al. Emergence of plasmid-mediated colistin resistance mechanism *MCR-1* in animals and human beings in China: A microbiological and molecular biological study. *Lancet Infect. Dis.* **2016**, *16*, 161–168. [CrossRef]

14. Clermont, O.; Bonacorsi, S.; Bingen, E. Rapid and simple determination of the *Escherichia coli* phylogenetic group. *Appl. Environ. Microbiol.* **2000**, *66*, 4555–4558. [CrossRef] [PubMed]

15. Hammerum, A.M.; Heuer, O.E. Human health hazards from antimicrobial-resistant *Escherichia coli* of animal origin. *Clin. Infect. Dis.* **2009**, *48*, 916–921. [CrossRef] [PubMed]

16. Van den Bogaard, A.E.; Bruinsma, N.; Stobberingh, E.E. The effect of banning avoparcin on VRE carriage in The Netherlands. *J. Antimicrob. Chemother.* **2000**, *46*, 146–147. [CrossRef] [PubMed]

17. Abo-State, M.A.; Mahdy, H.M.; Ezzat, S.M.; Abd El Shakour, E.H.; El-Bahnasawy, M.A. Antimicrobial resistance profiles of Enterobacteriaceae isolated from Rosetta branch of River Nile. *Egypt. Appl. Sci. J.* **2012**, *19*, 1234–1243.

18. Sivri, N.; Sandalli, C.; Ozgumus, O.B.; Colakoglu, F.; Dogan, D. Antibiotic Resistance Profiles of Enteric Bacteria Isolated from Kucukcekmece Lagoon (Istanbul–Turkey). *Turk. J. Fish. Aquat. Sci.* **2012**, *12*, 699–707. [CrossRef]

19. Hussain, A.; Ranjan, A.; Nandanwar, N.; Babbar, A.; Jadhav, S.; Ahmed, N. Genotypic and phenotypic profiles of *Escherichia coli* isolates belonging to clinical sequence type 131 (ST131), clinical non-ST131, and fecal non-ST131 lineages from India. *Antimicrob. Agents Chemother.* **2014**, *58*, 7240–7249. [CrossRef] [PubMed]

20. Sidjabat, H.E.; Paterson, D.L. Multidrug-resistant *Escherichia coli* in Asia: Epidemiology and management. *Expert Rev. Anti-Infect. Ther.* **2015**, *13*, 575–591. [CrossRef] [PubMed]

21. Schjørring, S.; Krogfelt, K.A. Assessment of bacterial antibiotic resistance transfer in the gut. *Int. J. Microbiol.* **2011**, *2011*. [CrossRef] [PubMed]

22. Sørensen, S.J.; Bailey, M.; Hansen, L.H.; Kroer, N.; Wuertz, S. Studying plasmid horizontal transfer in situ: A critical review. *Nat. Rev. Microbiol.* **2005**, *3*, 700–710. [CrossRef] [PubMed]

23. Ichor, T.; Umeh, E.U.; Duru, E.E. Microbial Contamination of Surface Water Sources in Rural Areas of Guma Local Government Area of Benue State, Nigeria. *Med. Sci. Public Health* **2014**, *2*, 43–51.

24. Hagedorn, C.; Robinson, S.L.; Filtz, J.R.; Grubbs, S.M.; Angier, T.A.; Reneau, R.B., Jr. Determining sources of fecal pollution in a rural Virginia watershed with antibiotic resistance patterns in fecal *Streptococci*. *Appl. Environ. Microbiol.* **1999**, *65*, 5522–5531. [PubMed]

25. Zhang, Y.; Marrs, C.F.; Simon, C.; Xi, C. Wastewater treatment contributes to selective increase of antibiotic resistance among *Acinetobacter* spp. *Sci. Total Environ.* **2009**, *407*, 3702–3706. [CrossRef] [PubMed]

26. Wright, J.; Gundry, S.; Conroy, R. Household drinking water in developing countries: A systematic review of microbiological contamination between source and point-of-use. *Trop. Med. Int. Health* **2004**, *9*, 106–117. [CrossRef] [PubMed]

27. Pitout, J.D. Extraintestinal Pathogenic *Escherichia coli*: A Combination of Virulence with Antibiotic Resistance. *Front. Microbiol.* **2012**, *19*, 9. [CrossRef] [PubMed]

28. Tenover, F.C.; McGowan, J.E., Jr. Reasons for the emergence of antibiotic resistance. *Am. J. Med. Sci.* **1996**, *311*, 9–16. [CrossRef]

29. Johnson, T.J.; Wannemuehler, Y.; Johnson, S.J.; Stell, A.L.; Doetkott, C.; Johnson, J.R.; Kim, K.S.; Spanjaard, L.; Nolan, L.K. Comparison of extraintestinal pathogenic *Escherichia coli* strains from human and avian sources reveals a mixed subset representing potential zoonotic pathogens. *Appl. Environ. Microbiol.* **2008**, *74*, 7043–7050. [CrossRef] [PubMed]

30. Kumar, M.; Saha, S.; Subudhi, E. More Furious Than Ever: *Escherichia coli*-Acquired Co-resistance toward Colistin and Carbapenems. *Clin. Infect. Dis.* **2016**, *63*, 1267–1268. [PubMed]

31. Gupta, M.; Lakhina, K.; Kamath, A.; Vandana, K.E.; Mukhopadhyay, C.; Vidyasagar, S.; Varma, M. Colistin-resistant *Acinetobacter baumannii* ventilator-associated pneumonia in a tertiary care hospital: An evolving threat. *J. Hosp. Infect.* **2016**, *94*, 72–73. [CrossRef] [PubMed]

32. Ckakraborty, A.; Adhikari, P.; Shenoy, S.; Baliga, S.; Bhat, G.; Rao, S.; Biranthabail, D.; Saralaya, V. Molecular characterization and clinical significance of New Delhi metallo-beta-lactamases-1 producing *Escherichia coli* recovered from a South Indian tertiary care hospital. *Indian J. Pathol. Microbiol.* **2015**, *58*, 323–327. [CrossRef] [PubMed]

33. Mittal, G.; Gaind, R.; Kumar, D.; Kaushik, G.; Gupta, K.B.; Verma, P.K.; Deb, M. Risk factors for fecal carriage of carbapenemase producing Enterobacteriaceae among intensive care unit patients from a tertiary care center in India. *BMC Microbiol.* **2016**, *16*, 138–148. [CrossRef] [PubMed]

34. Purohit, M.R.; Chandran, S.P.; Diwan, V.; Harshada, S.; Tamhankar, A.J.; Lundborg, C.S. Detection of carbapenem resistance genes and cephalosporin, and quinolone resistance genes in *Escherichia coli* in hospital and community wastewater. (Manuscript in preparation).

International Journal of
***Environmental Research
and Public Health***

Article

Multiresistant Bacteria Isolated from Activated Sludge in Austria

Herbert Galler [1], Gebhard Feierl [1], Christian Petternel [2], Franz F. Reinthaler [1], Doris Haas [1], Juliana Habib [1], Clemens Kittinger [1], Josefa Luxner [1] and Gernot Zarfel [1,*]

[1]   Institute of Hygiene, Microbiology and Environmental Medicine, Medical University of Graz, 8010 Graz, Austria; he.galler@medunigraz.at (H.G.); gebhard.feierl@medunigraz.at (G.F.); franz.reinthaler@medunigraz.at (F.F.R.); doris.haas@medunigraz.at (D.H.); juliana.habib@medunigraz.at (J.H.); clemens.kittinger@medunigraz.at (C.K.); josefa.luxner@medunigraz.at (J.L.)

[2]   Institute of Laboratory Diagnostics and Microbiology, Klinikum-Klagenfurt am Wörthersee, 9020 Klagenfurt, Austria; christian.petternel@kabeg.at

*   Correspondence: gernot.zarfel@medunigraz.at; Tel.: +43-316-385-73604

Received: 15 December 2017; Accepted: 5 March 2018; Published: 9 March 2018

Abstract: Wastewater contains different kinds of contaminants, including antibiotics and bacterial isolates with human-generated antibiotic resistances. In industrialized countries most of the wastewater is processed in wastewater treatment plants which do not only include commercial wastewater, but also wastewater from hospitals. Three multiresistant pathogens—extended spectrum β-lactamase (ESBL)-harbouring Enterobacteriaceae (Gram negative bacilli), methicillin resistant *Staphylococcus aureus* (MRSA) and vancomycin resistant Enterococci (VRE)—were chosen for screening in a state of the art wastewater treatment plant in Austria. Over an investigation period of six months all three multiresistant pathogens could be isolated from activated sludge. ESBL was the most common resistance mechanism, which was found in different species of Enterobacteriaceae, and in one *Aeromonas* spp. Sequencing of ESBL genes revealed the dominance of genes encoding members of CTX-M β-lactamases family and a gene encoding for PER-1 ESBL was detected for the first time in Austria. MRSA and VRE could be isolated sporadically, including one EMRSA-15 isolate. Whereas ESBL is well documented as a surface water contaminant, reports of MRSA and VRE are rare. The results of this study show that these three multiresistant phenotypes were present in activated sludge, as well as species and genes which were not reported before in the region. The ESBL-harbouring Gram negative bacilli were most common.

Keywords: ESBL; MRSA; VRE; sewage sludge; PER-1

---

1. Introduction

Antibiotics in the environment represent a growing concern as their presence can promote the selection of antibiotic resistant bacteria (ARB) that pose a serious public health threat. ARB can further spread resistance genes in the environment by the mechanism of horizontal gene transfer through which environmental bacteria can then mediate pathogens to acquire antibiotic resistance genes [1–4]. Among the various sources accounting for the spread of ARB, organic wastes, including wastes of municipal and agricultural origin, have been widely reported to be potent reservoirs of ARB- harbouring genes for multidrug resistance. Previous studies have pointed out that numerous ARB and resistant genes have been detected in sewage sludge from municipal wastewater treatment plants (WWTPs) [4,5]. One predominant antibiotic resistance mechanism is the presence of Extended Spectrum β-Lactamases (ESBLs). ESBLs are of great microbiological and clinical importance in Enterobacteriaceae, especially *Escherichia coli* and *Klebsiella* spp. and other non-fermenting bacteria such as *Acinetobacter* spp. and *Pseudomonas aeruginosa* [6,7]. The presence of ESBL in surface water has been frequently demonstrated all over the world, which leads to the conclusion that if the bacteria in the water are able to host ESBL genes, then there will be ESBL in the population [8–11]. The spread of ESBL is enhanced by the localization of most of the ESBL genes on mobile genetic elements which allow the transmission of resistance genes to strains and species which are better adapted to the surface water environment. As a consequence of this, environmental bacteria can acquire resistance genes from e.g., strains of clinical origin [8–11]. Methicillin resistant *Staphylococcus aureus* (MRSA) originates from the clinical setting, as hospital acquired (HA)-MRSA. Nevertheless, MRSA strains started to spread among the healthy human population (so called community acquired CA-MRSA) and livestock (LA-MRSA) within the last decades like ESBL [12–14]. MRSA detections from environmental reservoirs, including surface water, are very rare compared to multiresistant Gram negative bacteria isolation. Although the population of Staphylococci flushed into the wastewater is high, the survival of Staphylococci in water environment seems to be much lower than that of Gram negative bacilli. Therefore reports of MRSA from this reservoir are mainly restricted to areas of high human influence, e.g., hospital waste water effluent [15–17]. Vancomycin resistant Enterococci (VRE) are one of the first documented antibiotic resistant bacteria with primary origin in animal farming. The rise of VRE was caused by the use of the glycopeptide avoparcin as a growth promoter from 1975 on. Although glycopeptide use was banned in livestock production in the European Union (1996) VRE are still present in animals and can also be found in hospital settings [18–20]. Hence VRE are present in waste and surface water, it seems that they are detected mostly sporadically. Furthermore, the number of studies covering this topic is limited. The aim of the present study was to investigate the presence of multidrug resistant bacteria such as ESBL-producing Enterobacteriaceae, MRSA and VRE in activated sludge in the second largest commercial WWTP in Austria.

2. Materials and Methods

2.1. Sample Collection

Activated sludge samples were collected in the period between September 2011 and February 2012, twice a month (except January) from the basin of the incoming untreated waste water at a sewage treatment plant (>500,000 population equivalent, wastewater load 1200 L/min) at the area of Graz, Styria/Austria. Wastewater entry into this treatment plant contained mainly domestic waste water and wastewater from hospitals in the area. The sludge samples were collected using sterile wide-mouth bottles. They were transported to the laboratory in a cooling box, where they were immediately stored in a refrigerator at 4–8 °C until processing within 24 h. In total, eleven sludge samples were collected in six measuring series.

2.2. Strain Isolation and Identification

Sludge samples were homogenized by vortexing for two minutes. For qualitative analysis, an amount of 1 mL from the homogenized sludge sample was suspended in 9 mL sterile saline solution (0.9% NaCl). In order to reduce the bacterial concentration, a decimal dilution series with saline solution was prepared.

ESBL isolation: 0.1 mL of each homogenized sludge sample was plated on chromID™ ESBL Agar (bioMérieux, Marcy-l'Etoile, France) and incubated for 24 h at 37 °C. Following incubation, ESBL positive colonies were determined based on the colour reaction of the ESBL-media (according to the manufacturer's protocol). Additionally 0.1 mL of the sludge samples was incubated (24 h, 37 °C) in thioglycolate nutrient broth for enrichment, then 10 µL of the material was inoculated on ESBL-media and incubated for 24 h at 37 °C [21].

MRSA isolation: 0.1 mL of the homogenized solutions were plated on oxacillin agar (OXOID Ltd., Basingstoke, UK) and incubated for 48 h at 37 °C. Following incubation, MRSA positive colonies were determined based on the colour reaction of the OXA-media. Blue colonies were presumed to be MRSA.

VRE isolation: For selective enrichment of VRE, an amount of 1 mL from the homogenized sludge sample was inoculated in 9 mL BBL™ Enterococcosel™ broth (BD, Sparks, MD, USA) containing 6 mg/L Vancomycin. Enterococci growing in the media turn the colour of the media from light brown to dark brown or black. In order to reduce the bacterial concentration, a decimal dilution series with saline solution was prepared. Subsequently, 0.1 mL from each of the homogenized solutions were plated on chromID™ VRE Agar (bioMérieux, Marcy-l'Etoile, France) and incubated for 24 h at 37 °C. VRE positive colonies were determined based on the colour reaction of the VRE-media (according to the manufacturer's protocol).

To obtain pure cultures, colonies growing on selective-media were transferred to blood agar (24 h, 37 °C). Identification was done using the Vitek® MS (bioMérieux, Marcy-l'Etoile, France), an automated microbial identification system using matrix-assisted laser desorption/ionization time-of-flight mass spectrometry (MALDI-TOF-MS) and the biochemical-based VITEK®2 system (bioMérieux, Marcy-l'Etoile, France).

2.2.1. Characterisation of ESBL Harbouring Gram Negative Bacilli

Identified Enterobacteriaceae were characterized for their resistance pattern by susceptibility testing according to EUCAST (EUCAST V2.0, 2012) [22], with ampicillin (AM), amoxicillin/clavulanic acid (AMC), piperacillin/tazobactam (TZP), cefalexin (CN), cefuroxime (CXM), cefoxitin (FOX), cefotaxime (CTX), ceftazidime (CAZ), cefepime (FEP), imipenem (IPM), meropenem (MEM), gentamicin (GM), trimethoprim/sulfamethoxazole (SXT), nalidixic acid (NA), ciprofloxacin (CIP), moxifloxacin (MOX), tetracycline (TE) and chloramphenicol (C) BD BBL™ Sensi-Disc™ paper discs (BD, Sparks, MD, USA). The inhibition zone diameters were interpreted according to EUCAST guidelines, except Enterobacteriaceae tested for tetracycline, chloramphenicol and nalidixic acid, which were evaluated by the Clinical Laboratory Standards Institute (CLSI, 2011) guidelines [23]. There are no interpretation guidelines for zone diameters of these three antibiotics according to EUCAST.

E. coli 25299 was used as reference. The inhibition zone diameters were interpreted according to EUCAST guidelines. The antimicrobials tested and resistance breakpoints applied can be found in the Supplementary Materials (Table S1).

All isolates were screened for ESBL gene families, $bla_{\text{CTX-M-1group}}$, $bla_{\text{CTX-M-2group}}$, $bla_{\text{CTX-M-9group}}$, bla_{GES}, bla_{SHV}, bla_{TEM}, and bla_{VEB} by PCR and sequencing as described previously [24,25]. False-positive (not estimated Enterobacteriaceae) strains growing on ESBL-media with a green or brownish colour were identified as Pseudomonadales and Aeromonadales; we decided to include them in the study and therefore these strains were also screened for ESBL genes. Identified Pseudomonadales were characterized for their resistance pattern by susceptibility testing according to EUCAST (EUCAST V2.0, 2012); piperacilin/tazobactam (TZP), ceftazidime (CAZ), cefepime (FEP), meropenem (MEM), imipenem (IPM), amikacin (AN), gentamicin (GM), tobramycin (NN), ciprofloxacin (CIP) and levofloxacin (LEV).

2.2.2. Determination of VRE

After isolation and identification of suspected VRE colonies antibiotic susceptibility was determined for ampicillin (AM), vancomycin (VA), teicoplanin (TEC), linezolid (LZD), tigecycline (TGC) and trimethoprim/sulfamethoxazole (SXT) by disc diffusion test according to the EUCAST guidelines (EUCAST V2.0, 2012). *E. faecalis* DSM20478 was used as reference. The minimal inhibition concentration (MIC) for 22 antibiotics was assigned by VITEK®2 using the AST-P586 card (bioMérieux, Marcy-l'Etoile, France). Resistance to the glycopeptides vancomycin and teicoplanin was confirmed by Etest (bioMérieux, Marcy-l'Etoile, France) according to the manufacturer's instructions. The detection of the vancomycin resistance genes (*vanA/vanB*) was performed by real time PCR applying the Light cycler VRE Detection Kit (Roche, Branchburg, NJ, USA).

2.2.3. Determination of MRSA

MRSA isolates were characterized for their resistance pattern by susceptibility testing according to EUCAST (EUCAST V2.0, 2012), tested with penicillin (P), cefoxitin (FOX), tetracycline (TE), erythromycin (E), clindamycin (CL), norfloxacin (NOR), amicazin (AN), gentamicin (GM), trimethoprim/sulfamethoxazole (SXT), fusidic acid (FA), rifampicin (RIF), linezolid (LZD) and mupirocin (MUP) using BD BBL™ Sensi-Disc™ paper discs (BD, Sparks, MD, USA). *Staphylococcus aureus* DSM799 was used as reference. PCR amplification was used to determine SCC*mec* type and presence of the Panton-Valentine-Leukozidin (PVL)-gene [26,27]. *Spa* typing was performed as described previously [28].

3. Results

All eleven investigated sludge samples revealed at least one kind of the screened multiresistant bacteria. In detail, ten of the eleven samples were positive for ESBL-harbouring Enterobacteriacea (82%), three samples were positive for MRSA (27%) and four samples for VRE (36%).

3.1. ESBL Gram Negative Bacilli Isolates

In total, 117 Enterobacteriaceae were screened for multidrug resistance phenotypically. Genetic analysis revealed 32 different positive isolates consisting of 21 *E. coli*, seven *Klebsiella pneumoniae*, three *Enterobacter* sp. and one *Raoultella ornithinolytica* (Table 1). Members of the CTX-M gene family were the most predominant ESBL genes.

The most detected ESBL gene was $bla_{CTX-M-15}$, which was present in twelve (28.6%) of the 32 isolates followed by $bla_{CTX-M-1}$, which was found in six (14.3%) isolates. In addition, five (11.9%) of the isolates harboured the $bla_{CTX-M-14}$, three (7.1%) $bla_{CTX-M-3}$, and one (2.4%) the $bla_{CTX-M-38}$ gene. The non-CTX-M ESBL genes bla_{SHV-2} and bla_{SHV-12} were detected in four isolates from activated sludge.

Bacteria with ESBL phenotypes frequently carry additional antibiotic resistances. For the purpose of phenotypic differentiation, all ESBL *E. coli* isolates were tested for their susceptibility to 19 antibiotics. The antibiotic resistances of each of the investigated isolates are listed in Table 1.

No ESBL-producing Enterobacteriaceae showed resistance to tigecycline, amikacin and the carbapenems imipenem and meropenem. Penicillin-inhibitor combinations such as amoxicillin/clavulanic acid (53.1%, 17 of 32) and piperacillin/tazobactam (9.4%, 3 of 32) showed reduced efficacy against the ESBL producing Enterobacteriaceae. The cephamycin cefoxitin revealed resistance to eight (25%) isolates.

The most common co-resistance rates among the ESBL producing Enterobacteriaceae isolates to non-beta lactam antibiotics were detected for the quinolones, nalidixic acid 75% (24 of 32), ciprofloxacin 56.3% (18 of 32) and moxifloxacin 53.1% (17 of 32). The co-resistance for tetracycline was as high as 53.1% (17 of 32) and for the drug combination trimethoprim/sulfamethoxazole 50% (16 of 32). Co-resistance rates to aminoglycoside compounds were low with 34.4% (11 of 32) for gentamicin and 0% for amikacin.

Two ESBL-producing isolates were resistant to three antibiotics and 26 of the isolates were resistant to more than three antibiotic classes, which lead to a number of 28 ESBL isolates that could be assigned as multidrug resistant (Table 1).

Additional 25 Pseudomonadales were isolated from the ESBL screening plates but genetic analysing showed no positive confirmation for ESBL genes. Only one *Aeromonas* spp. isolate was tested positive for the ESBL gene bla_{PER-1}. This isolate revealed resistance to ceftazidime and meropenem.

Table 1. Detected resistance genes and resistance pattern of all isolates. Non *E. coli* Enterobacteriaceae are automatically set resistant to AM according EUCAST.

Isolate ID	Sample	Date	Species	Resistance Genes	Resistance Pattern [a]
ESBL-01	KS1	2011-09	*E. coli*	$bla_{CTX-M-15}$, bla_{TEM-1} [b]	AM, AMC, CN, CXM, FOX, CTX, GM, SXT, CIP, MXF, CAZ, FEP, TE, NA
ESBL-02	KS5	2011-10	*E. coli*	$bla_{CTX-M-15}$	AM, AMC, CN, CXM, CTX, CIP, MXF, CAZ, FEP, TE, NA
ESBL-03	KS5	2011-10	*E. coli*	$bla_{CTX-M-1}$, bla_{TEM-1}	AM, CN, CXM, CTX, GM, SXT, TE, C
ESBL-04	KS5	2011-10	*E. coli*	bla_{TEM-1}	AM, CN, CXM, CTX, SXT, CIP, CAZ, FEP, TE, NA, C
ESBL-05	KS5	2011-10	*E. coli*	$bla_{CTX-M-1}$	AM, AMC, CN, CXM, CTX, SXT, CIP, MXF, NA, C
ESBL-06	KS5	2011-10	*E. coli*	$bla_{CTX-M-1}$	AM, CN, CXM, CTX, FEP
ESBL-07	KS6	2011-11	*E. coli*	$bla_{CTX-M-15}$	AM, CN, CXM, CTX, SXT, CIP, MXF, CAZ, FEP, TE, NA, C
ESBL-08	KS6	2011-11	*E. coli*	$bla_{CTX-M-3}$	AM, AMC, CN, CXM, CTX, SXT, CIP, MXF, FEP, TE, NA, C
ESBL-09	KS6	2011-11	*E. coli*	$bla_{CTX-M-3}$, bla_{TEM-1}	AM, AMC, CN, CXM, FOX, CTX
ESBL-10	KS7	2011-12	*E. coli*	$bla_{CTX-M-14}$, bla_{TEM-1}	AM, CN, CXM, CTX, CIP, MXF, NA
ESBL-11	KS7	2011-12	*E. coli*	$bla_{CTX-M-3}$, bla_{TEM-1}	AM, CN, CXM, CTX, FEP, NA
ESBL-12	KS7	2011-12	*E. coli*	$bla_{CTX-M-1}$	AM, CN, CXM, CTX, GM, SXT, CIP, MXF, CAZ, FEP, TE, NA
ESBL-13	KS7	2011-12	*E. coli*	$bla_{CTX-M-15}$, bla_{TEM-1}	AM, CN, CXM, CTX, FEP
ESBL-14	KS7	2011-12	*E. coli*	$bla_{CTX-M-1}$, bla_{TEM-1}	AM, CN, CXM, CTX, SXT, FEP, TE
ESBL-15	KS7	2011-12	*E. coli*	$bla_{CTX-M-14}$	AM, AMC, CN, CXM, CTX, TE, NA
ESBL-16	KS7	2011-12	*E. coli*	$bla_{CTX-M-14}$, bla_{TEM-1}	AM, AMC, CN, CXM, CTX, TE, NA
ESBL-17	KS8	2011-12	*E. coli*	$bla_{CTX-M-14}$, bla_{TEM-1}	AM, AMC, CN, CXM, CTX, CIP, MXF, CAZ, TE, NA
ESBL-18	KS9	2012-01	*E. coli*	$bla_{CTX-M-38}$	AM, AMC, CN, CXM, FOX, CTX, GM, CIP, MXF, TZP, CAZ, FEP, TE, NA
ESBL-19	KS10	2012-02	*E. coli*	$bla_{CTX-M-1}$	AM, CN, CXM, CTX, SXT, FEP, TE
ESBL-20	KS11	2012-04	*E. coli*	$bla_{CTX-M-15}$, bla_{SHV-11} [b], bla_{TEM-1}	AM, CN, CXM, CTX, CAZ, FEP
ESBL-21	KS11	2012-04	*E. coli*	$bla_{CTX-M-14}$	AM, CN, CXM, CTX, SXT, CIP, MXF, NA
ESBL-22	KS1	2011-09	*E. kobei*	bla_{SHV-2}	AM, AMC, CN, CXM, FOX, CTX, GM, CAZ, FEP, NA, C
ESBL-23	KS4	2011-09	*E. kobei*	bla_{SHV-2}	AM, AMC, CN, CXM, FOX, CTX, GM, CAZ, NA, C
ESBL-24	KS9	2012-01	*E. cloacae*	$bla_{CTX-M-15}$, bla_{TEM-1}	AM, AMC, CN, CXM, FOX, CTX, GM, SXT, CIP, MXF, TZP, CAZ, TE, NA
ESBL-25	KS2	2011-09	*K. pneumoniae*	$bla_{CTX-M-14}$, bla_{SHV-2}	AM, CN, CXM, CTX, GM, SXT, FEP
ESBL-26	KS6	2011-11	*K. pneumoniae*	bla_{SHV-12}	AM, CN, CXM, CTX, CIP, MXF, CAZ, NA, C
ESBL-27	KS7	2011-12	*K. pneumoniae*	$bla_{CTX-M-15}$, bla_{SHV-1} [b], bla_{TEM-1}	AM, AMC, CN, CXM, CTX, GM, SXT, CIP, MXF, CAZ, TE, NA
ESBL-28	KS7	2011-12	*K. pneumoniae*	$bla_{CTX-M-15}$, bla_{SHV-1}, bla_{TEM-1}	AM, AMC, CN, CXM, CTX, GM, SXT, CIP, MXF, CAZ, FEP, TE, NA
ESBL-29	KS8	2011-12	*K. pneumoniae*	$bla_{CTX-M-15}$, bla_{SHV-11}	AM, AMC, CN, CXM, FOX, CTX, SXT, CIP, MXF, CAZ, FEP, NA, C
ESBL-30	KS9	2012-01	*K. pneumoniae*	$bla_{CTX-M-15}$, bla_{SHV-1}, bla_{TEM-1}	AM, CN, CXM, CTX, CIP, MXF, NA, TE
ESBL-31	KS11	2012-04	*K. pneumoniae*	$bla_{CTX-M-15}$, bla_{SHV-11}	AM, AMC, CN, CXM, CTX, GM, SXT, CIP, MXF, CAZ, TE, NA
ESBL-32	KS11	2012-04	*R. ornithinolytika*	bla_{SHV-2}	AM, AMC, CN, CXM, FOX, CTX, TZP, CAZ, NA
ESBL-33	KS7	2011-12	*Aeromonas.* sp.	bla_{PER-1}	CAZ, MEM

Table 1. *Cont.*

Isolate ID	Sample	Date	Species	Resistance Genes	Resistance Pattern [a]
VRE-01	KS4	2001-09	*E. faecium*	*van*A	AM, TEC, VA, SXT
VRE-02	KS7	2011-12	*E. faecium*	*van*A	AM, TEC, VA, SXT
VRE-03	KS8	2011-12	*E. faecium*	*van*A	AM, TEC, VA, SXT
VRE-04	KS10	2012-02	*E. faecium*	*van*A	AM, TEC, VA
MRSA-01	KS3	2011-09	*S. aureus*	*mec*A	P, FOX, E, NOR, GM
MRSA-02	KS5	2011-10	*S. aureus*	*mec*A	P, FOX
MRSA-03	KS6	2011-11	*S. aureus*	*mec*A	P, FOX, E, CC, NOR, GM

[a] AM, ampicillin; AMC, amoxicillin/clavulanic acid; TZP, piperacillin/tazobactam; CN, cephalexin; CXM, cefuroxime; FOX, cefoxitin; CTX, cefotaxime; CAZ, ceftazidime; FEP, cefepime; MEM, meropenem; CIP, ciprofloxacin; MXF, moxifloxacin; GM, gentamicin; SXT, trimethoprim/sulfamethoxazole; TE, tetracycline; NA, nalidixic acid; C, chloramphenicol; TEC, teicoplanin; VA, vancomycin; P, penicillin; E, erythromycin; NOR, norfloxacin. [b] Resistance genes bla_{TEM-1}, bla_{SHV-1} and bla_{SHV-11} encoding non-extended-spectrum-β-lactamases.

3.2. MRSA

Three MRSA isolates from three different activated sludge samples were detected. All three isolates harboured the mecA gen but were tested negative for PVL. Spa typing revealed one t032, with resistance to erythromycin, norfloxacin and gentamycin and one t067 with resistance to erythromycin, clindamycin and norfloxacin. The third MRSA with spa type t6613 was susceptible to all tested non beta-lactam antibiotics (Table 1).

3.3. VRE

VRE could be detected in four of eleven (36%) activated sludge samples represented by one Enterococcus isolate each. All four isolates were identified as *Enterococcus faecium* and harboured the *vanA* gene. All isolates showed highly similar resistance patterns. They were all resistant to ampicillin, teicoplanin, and vancomycin; three isolates showed additional resistance to trimethoprim/sulfamethoxazole (Table 1).

4. Discussion

The omnipresence of ESBL in environmental population of Enterobacteriaceae is widely demonstrated. The findings of this study go in full concordance with prior results. This includes also the isolated species (mostly *E. coli*) and the detected genes (CTX-M family) being dominant [8,11,29,30].

Other studies concerning *E. coli* from sewage sludge also reported tetracycline, ampicillin/clavulanic acid and trimethoprim/sulfamethoxazole as antibiotics with the highest non-susceptibility rate. These antibiotics showed the highest non-susceptibility in ESBL *E. coli* from Austrian sewage sludge as well [31,32]. Regarding co-resistance, the isolates did not show a reduced occurrence as can be observed in ESBL isolates from surface waters, without direct wastewater influence. Resistance to quinolones was very common and most of the isolates could be classified as multiresistant (resistance to three or more tested antibiotic classes). Environment and residence time in the WWTP seem not to favour a potential adaptation process in the ESBL population. The permanent entry of ARB from different sources in the activated sludge basin and the horizontal gene transfer are the dominant factors for the composition of resistant bacteria. Selection pressure due to different substances, does not seem to have enough time in this environment to contribute to resistance development [9,10,30,33–37]. Therefore, the isolates of this study reflect rather the situation of clinical ESBL isolates where this kind of co-resistance and multiresistance is dominant. Interestingly the majority of the ESBL Enterobacteriaceae isolates remained susceptible to the tested 4th generation cephalosporin (cefepim).

The isolation of a PER-1 producing *Aeromonas* spp. is more remarkable. There are reports of PER-1 based ESBL (also in *Aeromonas*) in European surface waters, nevertheless clinical isolates with this enzyme are reported rarely. In Austria, this is the first PER-1 producer documented so far [38,39].

The MRSA isolates from the sewage sludge can be linked to hospital settings. A multiresistant phenotype including the aminoglycoside gentamicin is a typical characteristic of hospital acquired (HA)-MRSA. T032 is a common *spa* type of the ST22-MRSA-IV (Barnim epidemic MRSA strain). It is the most prevalent HA-MRSA in Europe and has spread in Austria since the beginning of this decade. The second gentamicin resistant (t067) isolate can be linked to the so called paediatric clone. The resistance pattern of this MRSA isolate, with the exotic *spa* type t6613, showed similarity with CA-MRSA, but did not harbour the genes for the PVL toxin [40,41].

In general MRSA isolates from surface water are rather rare, with only low number of analysed isolates. Therefore an estimation which of the three MRSA types is more dominant in water environment is difficult to make [15,16,42,43].

VRE isolates showed nearly identical features in terms of species, gene and resistance pattern. Likewise MRSA, VRE isolates were only investigated and isolated in few studies compared to studies with ESBL isolates. This is remarkable because in contrast to Staphylococci, Enterococci have a much better ability to survive in surface water and they are indicator bacteria for water quality assessment [44–46]. Therefore, the exclusivity of *vanA* isolates is more likely to be based on the low number of sludge isolates. Furthermore other environmental VRE isolates from Austria revealed also *vanB* [44–49].

However, there is much evidence that confirms the presence of diverse and plentiful ARB in fertilizer produced from livestock animals [50,51]. There appears to be significant variability on wastewater management across different industrialized countries. In high income countries sewer connectivity is generally high, whereas in many middle and low income countries sewer connectivity is low and untreated sewage is discharged mainly to surface water bodies [52,53].

5. Conclusions

Wastewater treatment plants serve as a collection basin of multiresistant bacteria. In the investigated activated sludge samples all three screened multiresistant phenotypes were present, with ESBL harbouring Gram negative bacilli representing the most common ones. The study shows for the first time in Austria, the presence of VRE in WWTP and the first detection of a PER-1 mediated ESBL. All these multiresistant bacteria have the potential to spread in other ecological niches and therefore further monitoring and measures for reduction should be taken into consideration.

Supplementary Materials: The following tables are available online at www.mdpi.com/1660-4601/15/3/479/s1, Table S1: Antibiotics, disk content and breakpoints used for disk susceptibility testing according to the EUCAST guidelines (EUCAST V2.0, 2012).

Acknowledgments: This project was funded by "Hygienefonds der Medizinischen Universität Graz", Auenbruggerplatz 2, 8010 Graz, Austria.

Author Contributions: Gebhard Feierl, Franz F. Reinthaler and Gernot Zarfel conceived and designed the study; Herbert Galler and Doris Haas took samples and provided background information of the samples. Herbert Galler, Gernot Zarfel, Christian Petternel and Josefa Luxner performed the bacterial isolation, microbiological experiments, and analyses. Gernot Zarfel and Josefa Luxner performed molecular biology experiments. Herbert Galler, Gebhard Feierl and Gernot Zarfel analysed the data. Herbert Galler and Gernot Zarfel wrote the manuscript. Juliana Habib and Clemens Kittinger edited the manuscript.

Conflicts of Interest: The authors declare no conflict of interest.

References

1. Kummerer, K. Antibiotics in the Aquatic Environment—A Review—Part I. *Chemosphere* **2009**, *75*, 417–434. [CrossRef] [PubMed]
2. Sharma, V.K.; Johnson, N.; Cizmas, L.; McDonald, T.J.; Kim, H. A Review of the Influence of Treatment Strategies on Antibiotic Resistant Bacteria and Antibiotic Resistance Genes. *Chemosphere* **2016**, *150*, 702–714. [CrossRef] [PubMed]
3. Canton, R. Antibiotic Resistance Genes from the Environment: A Perspective through Newly Identified Antibiotic Resistance Mechanisms in the Clinical Setting. *Clin. Microbiol. Infect.* **2009**, *15*, 20–25. [CrossRef] [PubMed]
4. Rizzo, L.; Manaia, C.; Merlin, C.; Schwartz, T.; Dagot, C.; Ploy, M.C.; Michael, I.; Fatta-Kassinos, D. Urban Wastewater Treatment Plants as Hotspots for Antibiotic Resistant Bacteria and Genes Spread into the Environment: A Review. *Sci. Total Environ.* **2013**, *447*, 345–360. [CrossRef] [PubMed]
5. Baquero, F.; Martinez, J.L.; Canton, R. Antibiotics and Antibiotic Resistance in Water Environments. *Curr. Opin. Biotechnol.* **2008**, *19*, 260–265. [CrossRef] [PubMed]
6. Falagas, M.E.; Karageorgopoulos, D.E. Extended-Spectrum Beta-Lactamase-Producing Organisms. *J. Hosp. Infect.* **2009**, *73*, 345–354. [CrossRef] [PubMed]
7. Livermore, D.M.; Canton, R.; Gniadkowski, M.; Nordmann, P.; Rossolini, G.M.; Arlet, G.; Ayala, J.; Coque, T.M.; Kern-Zdanowicz, I.; Luzzaro, F.; et al. CTX-M: Changing the Face of ESBLs in Europe. *J. Antimicrob. Chemother.* **2007**, *59*, 165–174. [CrossRef] [PubMed]

8. Blaak, H.; Lynch, G.; Italiaander, R.; Hamidjaja, R.A.; Schets, F.M.; de Roda Husman, A.M. Multidrug-Resistant and Extended Spectrum Beta-Lactamase-Producing *Escherichia coli* in Dutch Surface Water and Wastewater. *PLoS ONE* **2015**, *10*, e0127752. [CrossRef] [PubMed]

9. Zarfel, G.; Lipp, M.; Gurtl, E.; Folli, B.; Baumert, R.; Kittinger, C. Troubled Water under the Bridge: Screening of River Mur Water Reveals Dominance of CTX-M Harboring *Escherichia coli* and for the First Time an Environmental VIM-1 Producer in Austria. *Sci. Total Environ.* **2017**, *593–594*, 399–405. [CrossRef] [PubMed]

10. Zurfluh, K.; Hachler, H.; Nuesch-Inderbinen, M.; Stephan, R. Characteristics of Extended-Spectrum Beta-Lactamase- and Carbapenemase-Producing Enterobacteriaceae Isolates from Rivers and Lakes in Switzerland. *Appl. Environ. Microbiol.* **2013**, *79*, 3021–3026. [CrossRef] [PubMed]

11. Zhang, H.; Gao, Y.; Chang, W. Comparison of Extended-Spectrum Beta-Lactamase-Producing *Escherichia coli* Isolates from Drinking Well Water and Pit Latrine Wastewater in a Rural Area of China. *BioMed Res. Int.* **2016**, *2016*, 4343564. [CrossRef] [PubMed]

12. Uhlemann, A.C.; Otto, M.; Lowy, F.D.; DeLeo, F.R. Evolution of Community- and Healthcare-Associated Methicillin-Resistant *Staphylococcus aureus. Infect. Genet. Evol.* **2014**, *21*, 563–574. [CrossRef] [PubMed]

13. Stryjewski, M.E.; Corey, G.R. Methicillin-Resistant *Staphylococcus aureus*: An Evolving Pathogen. *Clin. Infect. Dis.* **2014**, *58* (Suppl. 1), S9–S10. [CrossRef] [PubMed]

14. Cuny, C.; Wieler, L.H.; Witte, W. Livestock-Associated MRSA: The Impact on Humans. *Antibiotics* **2015**, *4*, 521–543. [CrossRef] [PubMed]

15. Skariyachan, S.; Garka, S.; Puttaswamy, S.; Shanbhogue, S.; Devaraju, R.; Narayanappa, R. Environmental Monitoring and Assessment of Antibacterial Metabolite Producing Actinobacteria Screened from Marine Sediments in South Coastal Regions of Karnataka, India. *Environ. Monit. Assess.* **2017**, *189*, 283. [CrossRef] [PubMed]

16. Lepuschitz, S.; Mach, R.; Springer, B.; Allerberger, F.; Ruppitsch, W. Draft Genome Sequence of a Community-Acquired Methicillin-Resistant *Staphylococcus aureus* USA300 Isolate from a River Sample. *Genome Announc.* **2017**, *5*, e01166. [CrossRef] [PubMed]

17. Concepcion Porrero, M.; Harrison, E.M.; Fernandez-Garayzabal, J.F.; Paterson, G.K.; Diez-Guerrier, A.; Holmes, M.A.; Dominguez, L. Detection of mecC-Methicillin-Resistant *Staphylococcus aureus* Isolates in River Water: A Potential Role for Water in the Environmental Dissemination. *Environ. Microbiol. Rep.* **2014**, *6*, 705–708. [CrossRef] [PubMed]

18. Shenoy, E.S.; Paras, M.L.; Noubary, F.; Walensky, R.P.; Hooper, D.C. Natural History of Colonization with Methicillin-Resistant *Staphylococcus aureus* (MRSA) and Vancomycin-Resistant Enterococcus (VRE): A Systematic Review. *BMC Infect. Dis.* **2014**, *14*, 177. [CrossRef] [PubMed]

19. Gastmeier, P.; Schroder, C.; Behnke, M.; Meyer, E.; Geffers, C. Dramatic Increase in Vancomycin-Resistant Enterococci in Germany. *J. Antimicrob. Chemother.* **2014**, *69*, 1660–1664. [CrossRef] [PubMed]

20. Mutters, N.T.; Mersch-Sundermann, V.; Mutters, R.; Brandt, C.; Schneider-Brachert, W.; Frank, U. Control of the Spread of Vancomycin-Resistant Enterococci in Hospitals: Epidemiology and Clinical Relevance. *Dtsch. Arztebl. Int.* **2013**, *110*, 725–731. [PubMed]

21. Reinthaler, F.F.; Galler, H.; Feierl, G.; Haas, D.; Leitner, E.; Mascher, F.; Melkes, A.; Posch, J.; Pertschy, B.; Winter, I.; et al. Resistance Patterns of *Escherichia coli* Isolated from Sewage Sludge in Comparison with those Isolated from Human Patients in 2000 and 2009. *J. Water Health* **2013**, *11*, 13–20. [CrossRef] [PubMed]

22. The European Committee on Antimicrobial Susceptibility Testing. *Breakpoint Tables for Interpretation of MICs and Zone Diameters, Version 2.0*; EUCAST: Växjö, Sweden, 2012.

23. Clinical and Laboratory Standards Institute (CLSI). *Performance Standards for Antimicrobial Susceptibility Testing: Twenty First International Supplement*; CLSI: Wayne, PA, USA, 2011.

24. Kiratisin, P.; Apisarnthanarak, A.; Laesripa, C.; Saifon, P. Molecular Characterization and Epidemiology of Extended-Spectrum-Beta-Lactamase-Producing *Escherichia coli* and Klebsiella Pneumoniae Isolates Causing Health Care-Associated Infection in Thailand, Where the CTX-M Family is Endemic. *Antimicrob. Agents Chemother.* **2008**, *52*, 2818–2824. [CrossRef] [PubMed]

25. Eckert, C.; Gautier, V.; Saladin-Allard, M.; Hidri, N.; Verdet, C.; Ould-Hocine, Z.; Barnaud, G.; Delisle, F.; Rossier, A.; Lambert, T.; et al. Dissemination of CTX-M-Type Beta-Lactamases among Clinical Isolates of Enterobacteriaceae in Paris, France. *Antimicrob. Agents Chemother.* **2004**, *48*, 1249–1255. [CrossRef] [PubMed]

26. Boye, K.; Bartels, M.D.; Andersen, I.S.; Moller, J.A.; Westh, H. A New Multiplex PCR for Easy Screening of Methicillin-Resistant *Staphylococcus aureus* SCCmec Types I–V. *Clin. Microbiol. Infect.* **2007**, *13*, 725–727. [CrossRef] [PubMed]

27. Lina, G.; Piemont, Y.; Godail-Gamot, F.; Bes, M.; Peter, M.O.; Gauduchon, V.; Vandenesch, F.; Etienne, J. Involvement of Panton-Valentine Leukocidin-Producing *Staphylococcus aureus* in Primary Skin Infections and Pneumonia. *Clin. Infect. Dis.* **1999**, *29*, 1128–1132. [CrossRef] [PubMed]

28. Ruppitsch, W.; Indra, A.; Stoger, A.; Mayer, B.; Stadlbauer, S.; Wewalka, G.; Allerberger, F. Classifying *Spa* Types in Complexes Improves Interpretation of Typing Results for Methicillin-Resistant *Staphylococcus aureus*. *J. Clin. Microbiol.* **2006**, *44*, 2442–2448. [CrossRef] [PubMed]

29. Literak, I.; Dolejska, M.; Radimersky, T.; Klimes, J.; Friedman, M.; Aarestrup, F.M.; Hasman, H.; Cizek, A. Antimicrobial-Resistant Faecal *Escherichia coli* in Wildmammals in Central Europe: Multiresistant *Escherichia coli* Producing Extended-Spectrum Beta-Lactamases in Wild Boars. *J. Appl. Microbiol.* **2010**, *108*, 1702–1711. [CrossRef] [PubMed]

30. Kittinger, C.; Lipp, M.; Folli, B.; Kirschner, A.; Baumert, R.; Galler, H.; Grisold, A.J.; Luxner, J.; Weissenbacher, M.; Farnleitner, A.H.; et al. Enterobacteriaceae Isolated from the River Danube: Antibiotic Resistances, with a Focus on the Presence of ESBL and Carbapenemases. *PLoS ONE* **2016**, *11*, e0165820. [CrossRef] [PubMed]

31. Holzel, C.S.; Schwaiger, K.; Harms, K.; Kuchenhoff, H.; Kunz, A.; Meyer, K.; Muller, C.; Bauer, J. Sewage Sludge and Liquid Pig Manure as Possible Sources of Antibiotic Resistant Bacteria. *Environ. Res.* **2010**, *110*, 318–326. [CrossRef] [PubMed]

32. Luczkiewicz, A.; Jankowska, K.; Fudala-Ksiazek, S.; Olanczuk-Neyman, K. Antimicrobial Resistance of Fecal Indicators in Municipal Wastewater Treatment Plant. *Water Res.* **2010**, *44*, 5089–5097. [CrossRef] [PubMed]

33. Korzeniewska, E.; Harnisz, M. Extended-Spectrum Beta-Lactamase (ESBL)-Positive Enterobacteriaceae in Municipal Sewage and their Emission to the Environment. *J. Environ. Manag.* **2013**, *128*, 904–911. [CrossRef] [PubMed]

34. Ojer-Usoz, E.; Gonzalez, D.; Garcia-Jalon, I.; Vitas, A.I. High Dissemination of Extended-Spectrum Beta-Lactamase-Producing Enterobacteriaceae in Effluents from Wastewater Treatment Plants. *Water Res.* **2014**, *56*, 37–47. [CrossRef] [PubMed]

35. Poeta, P.; Radhouani, H.; Pinto, L.; Martinho, A.; Rego, V.; Rodrigues, R.; Goncalves, A.; Rodrigues, J.; Estepa, V.; Torres, C.; et al. Wild Boars as Reservoirs of Extended-Spectrum Beta-Lactamase (ESBL) Producing *Escherichia coli* of Different Phylogenetic Groups. *J. Basic Microbiol.* **2009**, *49*, 584–588. [CrossRef] [PubMed]

36. Belmar Campos, C.; Fenner, I.; Wiese, N.; Lensing, C.; Christner, M.; Rohde, H.; Aepfelbacher, M.; Fenner, T.; Hentschke, M. Prevalence and Genotypes of Extended Spectrum Beta-Lactamases in Enterobacteriaceae Isolated from Human Stool and Chicken Meat in Hamburg, Germany. *Int. J. Med. Microbiol.* **2014**, *304*, 678–684. [CrossRef] [PubMed]

37. Miller, J.H.; Novak, J.T.; Knocke, W.R.; Pruden, A. Survival of Antibiotic Resistant Bacteria and Horizontal Gene Transfer Control Antibiotic Resistance Gene Content in Anaerobic Digesters. *Front. Microbiol.* **2016**, *7*, 263. [CrossRef] [PubMed]

38. Girlich, D.; Poirel, L.; Nordmann, P. Diversity of Clavulanic Acid-Inhibited Extended-Spectrum Beta-Lactamases in *Aeromonas* spp. from the Seine River, Paris, France. *Antimicrob. Agents Chemother.* **2011**, *55*, 1256–1261. [CrossRef] [PubMed]

39. Maravic, A.; Skocibusic, M.; Samanic, I.; Fredotovic, Z.; Cvjetan, S.; Jutronic, M.; Puizina, J. *Aeromonas* spp. Simultaneously Harbouring Bla(CTX-M-15), Bla(SHV-12), Bla(PER-1) and Bla(FOX-2), in Wild-Growing Mediterranean Mussel (*Mytilus Galloprovincialis*) from Adriatic Sea, Croatia. *Int. J. Food Microbiol.* **2013**, *166*, 301–308. [CrossRef] [PubMed]

40. Espadinha, D.; Faria, N.A.; Miragaia, M.; Lito, L.M.; Melo-Cristino, J.; de Lencastre, H.; Medicos Sentinela Network. Extensive Dissemination of Methicillin-Resistant *Staphylococcus aureus* (MRSA) between the Hospital and the Community in a Country with a High Prevalence of Nosocomial MRSA. *PLoS ONE* **2013**, *8*, e59960. [CrossRef] [PubMed]

41. Zarfel, G.; Luxner, J.; Folli, B.; Leitner, E.; Feierl, G.; Kittinger, C.; Grisold, A. Increase of Genetic Diversity and Clonal Replacement of Epidemic Methicillin-Resistant *Staphylococcus aureus* Strains in South-East Austria. *FEMS Microbiol. Lett.* **2016**, *363*, fnw137. [CrossRef] [PubMed]

42. Tolba, O.; Loughrey, A.; Goldsmith, C.E.; Millar, B.C.; Rooney, P.J.; Moore, J.E. Survival of Epidemic Strains of Healthcare (HA-MRSA) and Community-Associated (CA-MRSA) Meticillin-Resistant *Staphylococcus aureus* (MRSA) in River-, Sea- and Swimming Pool Water. *Int. J. Hyg. Environ. Health* **2008**, *211*, 398–402. [CrossRef] [PubMed]

43. Boopathy, R. Presence of Methicillin Resistant *Staphylococcus aureus* (MRSA) in Sewage Treatment Plant. *Bioresour. Technol.* **2017**, *240*, 144–148. [CrossRef] [PubMed]

44. Nakipoglu, M.; Yilmaz, F.; Icgen, B. *vanA* Gene Harboring Enterococcal and Non-Enterococcal Isolates Expressing High Level Vancomycin and Teicoplanin Resistance Reservoired in Surface Waters. *Bull. Environ. Contam. Toxicol.* **2017**, *98*, 712–719. [CrossRef] [PubMed]

45. Lata, P.; Ram, S.; Shanker, R. Multiplex PCR Based Genotypic Characterization of Pathogenic Vancomycin Resistant Enterococcus Faecalis Recovered from an Indian River along a City Landscape. *Springerplus* **2016**, *5*, 1199. [CrossRef] [PubMed]

46. Veljovic, K.; Popovic, N.; Vidojevic, A.T.; Tolinacki, M.; Mihajlovic, S.; Jovcic, B.; Kojic, M. Environmental Waters as a Source of Antibiotic-Resistant Enterococcus Species in Belgrade, Serbia. *Environ. Monit. Assess.* **2015**, *187*, 599. [CrossRef] [PubMed]

47. Zarfel, G.; Galler, H.; Luxner, J.; Petternel, C.; Reinthaler, F.F.; Haas, D.; Kittinger, C.; Grisold, A.J.; Pless, P.; Feierl, G. Multiresistant Bacteria Isolated from Chicken Meat in Austria. *Int. J. Environ. Res. Public Health* **2014**, *11*, 12582–12593. [CrossRef] [PubMed]

48. Nishiyama, M.; Iguchi, A.; Suzuki, Y. Identification of Enterococcus Faecium and Enterococcus Faecalis as vanC-Type Vancomycin-Resistant Enterococci (VRE) from Sewage and River Water in the Provincial City of Miyazaki, Japan. *J. Environ. Sci. Health A Tox. Hazard. Subst. Environ. Eng.* **2015**, *50*, 16–25. [CrossRef] [PubMed]

49. Larsen, B.; Essmann, M.K.; Geletta, S.; Duff, B. Enterococcus in Surface Waters from the Des Moines River (Iowa) Watershed: Location, Persistence and Vancomycin Resistance. *Int. J. Environ. Health Res.* **2012**, *22*, 305–316. [CrossRef] [PubMed]

50. Cheng, W.; Chen, H.; Su, C.; Yan, S. Abundance and Persistence of Antibiotic Resistance Genes in Livestock Farms: A Comprehensive Investigation in Eastern China. *Environ. Int.* **2013**, *61*, 1–7. [CrossRef] [PubMed]

51. Zhang, X.; Li, Y.; Liu, B.; Wang, J.; Feng, C.; Gao, M.; Wang, L. Prevalence of Veterinary Antibiotics and Antibiotic-Resistant *Escherichia coli* in the Surface Water of a Livestock Production Region in Northern China. *PLoS ONE* **2014**, *9*, e111026. [CrossRef] [PubMed]

52. Kookana, R.S.; Williams, M.; Boxall, A.B.A.; Larsson, D.G.; Gaw, S.; Choi, K.; Yamamoto, H.; Thatikonda, S.; Zhu, Y.G.; Carriquiriborde, P. Potential ecological footprints of active pharmaceutical ingredients: An examination of risk factors in low-, middle- and high-income countries. *Philos. Trans. R. Soc. B Biol. Sci.* **2014**, *369*, 20130586. [CrossRef] [PubMed]

53. WWAP (United Nations World Water Assessment Programme). *Wastewater: The Untapped Resource: The United Nations World Water Development Report 2017*; UNESCO: Paris, France, 2017; Available online: http://unesdoc.unesco.org/images/0024/002471/247153e.pdf (accessed on 4 March 2018).

International Journal of
Environmental Research and Public Health

Article

Disinfection of the Water Borne Pathogens *Escherichia coli* and *Staphylococcus aureus* by Solar Photocatalysis Using Sonochemically Synthesized Reusable Ag@ZnO Core-Shell Nanoparticles

Sourav Das [1], Neha Ranjana [1], Ananyo Jyoti Misra [1], Mrutyunjay Suar [1], Amrita Mishra [1], Ashok J. Tamhankar [1,2,*], Cecilia Stålsby Lundborg [2] and Suraj K. Tripathy [1,3,*]

[1] School of Biotechnology, Kalinga Institute of Industrial Technology (KIIT University), Bhubaneswar 751024, India; sddassourav52@gmail.com (S.D.); ranjana.nh@gmail.com (N.R.); ananyomisra@gmail.com (A.J.M.); msbiotek@yahoo.com (M.S.); amritamishrabio@gmail.com (A.M.)
[2] Department of Public Health Sciences, Karolinska Institutet, SE 17177 Stockholm, Sweden; cecilia.stalsby.lundborg@ki.se
[3] School of Chemical Technology, KIIT University, Bhubaneswar 751024, India
* Correspondence: ejetee@gmail.com (A.J.T.); tripathy.suraj@gmail.com (S.K.T.); Tel.: +91-943-832-1810 (S.K.T.)

Academic Editor: Paul B. Tchounwou
Received: 3 May 2017; Accepted: 5 July 2017; Published: 10 July 2017

Abstract: Water borne pathogens present a threat to human health and their disinfection from water poses a challenge, prompting the search for newer methods and newer materials. Disinfection of the Gram-negative bacterium *Escherichia coli* and the Gram-positive coccal bacterium *Staphylococcus aureus* in an aqueous matrix was achieved within 60 and 90 min, respectively, at 35 °C using solar-photocatalysis mediated by sonochemically synthesized Ag@ZnO core-shell nanoparticles. The efficiency of the process increased with the increase in temperature and at 55 °C the disinfection for the two bacteria could be achieved in 45 and 60 min, respectively. A new ultrasound-assisted chemical precipitation technique was used for the synthesis of Ag@ZnO core-shell nanoparticles. The characteristics of the synthesized material were established using physical techniques. The material remained stable even at 400 °C. Disinfection efficiency of the Ag@ZnO core-shell nanoparticles was confirmed in the case of real world samples of pond, river, municipal tap water and was found to be better than that of pure ZnO and TiO$_2$ (Degussa P25). When the nanoparticle- based catalyst was recycled and reused for subsequent disinfection experiments, its efficiency did not change remarkably, even after three cycles. The sonochemically synthesized Ag@ZnO core-shell nanoparticles thus have a good potential for application in solar photocatalytic disinfection of water borne pathogens.

Keywords: core-shell; disinfection; *Escherichia coli*; nanoparticles; pathogens; silver; solar-photocatalysis; *Staphylococcus aureus*; water; zinc oxide

---

1. Introduction

A large part of the population of developing countries is vulnerable to water borne diseases caused by pathogenic microbes present in the aquatic environment. Amongst the various enteric pathogens, *Escherichia coli* and *Staphylococcus aureus* are causal agents of various types of infections [1] and may lead to deterioration in the quality of drinking water in rural areas. The major causes behind this can be attributed to unawareness about personal hygiene practices and poor sanitation facilities. For disinfection of these microbes, composite nanoparticles-assisted photocatalysis has good potential for field application [2]. Currently, the simplicity and cost-effectiveness of sunlight- assisted photocatalysis using metal/metal oxide nanoparticles is gaining much attention for water treatment applications [3]. However, although the nanosized catalysts were successfully developed, certain issues such as short-shelf life of the nanoparticle systems due to catalyst poisoning, decreased active surface area of the nanoparticle systems by surface doping, and the possibility of leaching of reactive metal ions into the purified water have restricted their commercial exploitation. To deal with these limitations, core-shell structure nanoparticles were proposed. It is expected that nanoparticles with core-shell morphology will not only protect the metal catalysts, but also show promising results with regards to increased photocatalytic disinfection efficiency and extended shelf life of the material [4]. Metal@ZnO core-shell structure nanoparticles have been used for photocatalytic degradation of the organic dyes Rhodamine B and methyl orange in an aqueous solution [5,6]. *E. coli* has been extensively used as a good model micro-organism for studying photocatalytic disinfection but studies with *S. aureus* are mostly done with TiO_2 and its doped variants [7,8]. To the best of our knowledge, no such disinfection with core-shell nanoparticles has been carried out with the latter micro-organism. Ag/TiO_2 nanocomposites were previously explored for successful photocatalytic disinfection of *E. coli* [9]. Most of these reports have followed precipitation technique for coating metal oxide shell on noble metal (e.g., gold) nanoparticles. Although these materials have shown interesting photocatalytic properties, the high cost of gold is expected to hinder their practical application. Hence, Aguirre et al. tried to replace the gold core by a cheaper alternative, i.e., silver, and used this material for degradation of dyes [10]. Das and co-workers for the first time applied Ag@ZnO nanoparticles synthesized by a chemical precipitation technique for sunlight-assisted photocatalytic disinfection of the pathogenic bacterium *Vibreo cholerae* in synthetic as well as real water systems [11]. This could be a potential alternative to conventional disinfection techniques such as chlorination which are known to generate toxic byproducts [12]. However the conventional precipitation technique employed for the synthesis of metal@ZnO core-shell nanoparticles could not provide the well dispersed and porous materials required for catalytic applications [13]. Thus, it was necessary to investigate alternative synthetic protocols to obtain the monodispersed metal@ZnO core-shell nanoparticles and to check the potential of such materials for photocataytic applications. Recently, sonochemical techniques have been used extensively to obtain well dispersed and highly crystalline nanomaterials [4,14]. However, to the best of our knowledge, such techniques have never been exploited for the synthesis of metal@ZnO core-shell nanoparticles and examining their potential to disinfect bacterial pathogens. In the present paper, we report sunlight-assisted photocatalytic disinfection of two water borne pathogenic bacteria, *Escherichia coli* and *Staphylococcus aureus*, in saline solution (0.9%) and some real water systems using Ag@ZnO core-shell nanoparticles synthesized using an ultrasound assisted method.

2. Materials and Methods

2.1. Materials

All the reagents and chemicals used in the synthesis of photocatalyst as well as in the disinfection reaction were of research grade (99.99% pure) and were procured from MERCK (Mumbai, India). De-ionized water was used during all synthesis processes.

2.2. Synthesis and Characterization of Ag@ZnO Core-Shell Nanoparticles

Ag nanoparticles were synthesized by reduction of silver perchlorate monohydrate (Sigma-Aldrich, St. Louis, MO, USA) by $NaBH_4$ and trisodium citrate dihydrate [11]. Experimental procedures were as follows: distilled water (97 mL) was placed in a 250 mL glass beaker which was placed in an ice bath. Silver perchlorate monohydrate (1 mL, 1 mM) followed by 100 mM sodium borohydride (1 mL) and 3 mM of trisodium citrate (0.885 mL) were added to the beaker under vigorous stirring. A transparent bright yellow color was observed immediately due to the formation of the Ag nanoparticles. This colloid was aged for 12 h at room temperature. Zinc oxide nanoparticles were coated on the surface of Ag nanoparticles via an ultrasound assisted precipitation technique. To zinc nitrate hexahydrate aqueous solution of a known concentration (50 mL), sodium hydroxide solution (1 M) was added to obtain a white precipitate of zinc hydroxide, which was redissolved by adding excess of sodium hydroxide. This solution (20 mL) was added to the aqueous dispersion of Ag nanoparticles (10 mL) and exposed to ultrasound for 30 to 90 min. Then the solution was allowed to cool by natural process. The composite nanoparticles were then collected by centrifugation (at 12000 rpm) and dried at 80 °C for 12 h. Following this, the nanoparticles were sintered at 200 °C and 400 °C for 1 h. During centrifugation nanoparticles were washed with de-ionized water (three times) to remove the water soluble sodium chloride and other impurities.

Formation of Ag nanoparticles, and Ag@ZnO core-shell nanoparticles was investigated by UV-visible spectroscopy (Carry 100, Agilent, Santa Clara, CA, USA) respectively. Morphology and crystal structure of the nanoparticles was analyzed by transmission electron microscopy (JEM-2010, JEOL, Akishima, Tokyo, Japan) and X-ray diffraction (XRD, Rigaku, Tokyo, Japan) techniques respectively. The composition/functional property of nanomaterials were analyzed with FTIR spectroscopy at room temperature in an acquired range of 500–4000 cm^{-1}. Average surface area and porosity was measured by the Brunauer–Emmett–Teller (BET) technique.

2.3. Preparation of Bacterial Cultures

Bacterial strains of the Gram negative bacterium *E. coli* DH5-alpha and Gram-positive bacterium *S. aureus* were used as the target microorganisms in this study. The strains were purchased from the Microbial Type Culture Collection and Gene Bank (MTCC, Chandigarh, India). The strains were grown aerobically in a nutrient broth (HiMedia, Mumbai, India) at 37 °C in a shaking incubator (Daihan Labtech, New Delhi, India) at 200 rotations per minute (rpm). At optical density (OD_{600}) 0.6 for *E. coli* and 0.8 for *S. aureus*, corresponding to 10^8 CFU/ml (CFU = colony forming unit), the bacteria were harvested by centrifugation at 5000 rpm for 10 min. They were thereafter washed with 0.9% normal saline solution (NSS) to provide appropriate osmotic conditions [11]. All the glassware and plastic-ware used for media preparation, experimental purposes and analysis were sterilized by autoclaving at 121 °C, for 20 min before being used [15].

2.4. Photocatalytic Disinfection Experiments

Bacterial cells with a final cell concentration of 5×10^6 CFU/mL were put in 1 L of normal saline solution and multiple reactions were performed with varying concentrations of Ag@ZnO ranging from 1 to 5 mg/L. Photocatalytic disinfection reactions were carried out in 2 L reactor vessels under continuous and controlled agitation (500 rpm). The set up was kept under dark conditions for 30 min to attain equilibrium. After the dark phase, the system was exposed to sunlight for 120 min and samples were collected at 15 min intervals. To monitor and analyze the inactivation of microbes, 100 µL of collected samples were further diluted in 900 µL of sterile 0.9% NSS and a volume of 100 µL from the final diluted sample was spread on nutrient agar plates. The plates were left for overnight incubation at 37 °C. Following this, viable cell count was performed to obtain the results for the rate of disinfection [11,16]. The above steps were repeated using two commonly used catalysts ZnO and TiO_2 (Degussa P25) for comparative studies and with the optimum catalyst concentration for proper disinfection as obtained by Ag@ZnO. Additionally two experimental controls were performed. (1) In light control, under only photolytic condition the microbial population was exposed to sun-light in absence of Ag@ZnO. (2) In dark control, microbial population was reacted with Ag@ZnO in absence any light. The Intensity of sunlight was measured by a digital lux meter and found to be $90,000 \pm 5000$ lux. To evaluate whether the sun-light/Ag@ZnO assisted photocatalytic disinfection system is applicable to natural water systems, samples of tap (municipal supply, Bhubaneswar, India), river, and pond water were collected. Results were compared with de-ionized water. All water samples were collected and transported in clean and autoclaved sample bottles (Tarsons, Kolkata, India) at 4 °C and immediately were filtered by using Whatman filter paper and centrifuged at 5000 rpm for 15 min to remove insoluble materials followed by autoclaving to eliminate any microbial contamination. To demonstrate the efficiency of the synthesized catalyst, a calculated amount (as mentioned earlier) of targeted pathogens were spiked in the sterilized natural water samples and subjected to photocatalysis in presence of three different photocatalysts (Ag@ZnO, ZnO and TiO$_2$). The concentration of catalyst used was the one which was obtained as the optimum for the respective bacteria from experiments conducted in saline solution.

2.5. Determination of Lipid Peroxidation

Malondialdehyde (MDA) is an end product of lipid peroxidation. Therefore estimation of MDA through its reaction with thiobarbituric acid (TBA), forming a pink colored MDA-TBA complex, predicts the disintegration or, rather the damage of microbial cell membrane leading to death [16–18]. To establish this, analysis was performed by obtaining 1 mL samples from the reactor contents at regular time intervals (5 min) and the samples were mixed with 2 mL of 10% (wt./vol.) trichloroacetic acid. The mixture was subjected to centrifugation at 11,000 *g* for initial 35 min and then again for an additional 20 min to ensure the removal of precipitated proteins, catalyst, cells and other possible solid components from the system [16]. 3 mL of freshly prepared 0.67% (wt./vol.) TBA (Sigma Aldrich) was added to the supernatant. The samples were boiled in a water bath for 10 min and then the absorbance was measured at 532 nm using UV-visible spectrophotometer. The concentration of MDA in the system was calculated in nanomoles of MDA released per mg dry weight of bacteria [18].

2.6. Potassium Ion (K$^+$) Leakage Studies

To study the K$^+$ leakage from photocatalytically inactivated bacteria, 2 mL sample was collected at regular time intervals (2 min) from the reaction system and was subjected to centrifugation as per the details given in previous reports [18,19]. The supernatant was collected and analyzed using microwave plasma atomic emission spectrometer (4200 MP-AES, Agilent Technologies, Santa Clara, CA, USA).

2.7. Stability and Reusability of the Photocatalyst

The stability of the catalyst in post reaction condition was investigated using XRD. Additionally for further confirmation the post reaction water sample was analyzed using MP-AES to detect the leaching of Ag$^+$ and Zn^{2+} ions during the photocatalytic disinfection experiment [11]. Catalyst was recovered by centrifugation at 12,000 rpm for 30 min and dried at 80 °C and reused for the photocatalytic disinfection application. Unless otherwise mentioned all the experiments were conducted in triplicate.

3. Results and Discussion

3.1. Characterization of Nano-Photocatalyst

UV-visible spectra of the aqueous dispersion of Ag nanoparticles, and Ag@ZnO core-shell nanoparticles synthesized by ultrasonic hydrolysis of zinc nitrate hexahydrate are shown in Figure 1a. Aqueous dispersion of Ag nanoparticles showed a clear SPR band at 391 nm which showed a distinct red shift of about 22 nm immediately after addition of aqueous sodium zincate sol. This is attributed to an immediate change in the chemical environment around Ag nanoparticles. With increase in the ultrasonic irradiation time SPR band has shown a red shift to 397 nm with development of a shoulder peak at 487 nm. It is expected that during the formation of core-shell nanoparticles, Ag nanoparticles may have aggregated slightly to form large clusters. This may have caused dipole coupling between closely interacting metal nanoparticles. This hypothesis is supported by the electron microscopy images.

Results of our XRD study is shown in Figure 1b. For the synthesized nanoparticles, three distinct peaks at 2θ = 38.2, 44.9 and 64.8 corresponding to (111), (200) and (220) planes of metallic Ag with face-centered cubic structure (JCPDS Card No. 04-0783) is observed. Similarly three major peaks of ZnO at 2θ = 31.99, 34.63, and 46.51 corresponding to (100), (002), and (102) planes of synthetic ZnO with hexagonal wurtzite structure (JCPDS Card No. 36-1451) are obtained. Any peak corresponding to other Ag/Zn compounds was not obtained. This suggests that no alloy or solid solution is formed.

Mean crystallite diameter (MCD) was found to be ≈ 15 and 25 nm for Ag and ZnO nanoparticles respectively. It is also observed that the crystal structure and phase remained unchanged after heat treatment (at 200 and 400 °C). However the MCD and crystalinity have increased slightly after heat treatment. The results of FTIR spectroscopy are shown in Figure 1c. The broad band around 3400 cm^{-1} may correspond to O–H stretching mode of hydroxyl groups whereas the strong peak at 2345 cm^{-1} resembles to the stretching mode of acidic O–H group, which arises in the range of 2400–3300 cm^{-1}. The small vibration appearing at 1630 cm^{-1} may belong to the stretching peak of C=O group [11]. Vibration peaks at 1500 and 1280 cm^{-1} corresponds to C–H bending and C–O stretching mode respectively [11,14]. The peaks at 1630 and 637 cm^{-1} may correspond to Zn–O stretching and deformation vibration, respectively [14].

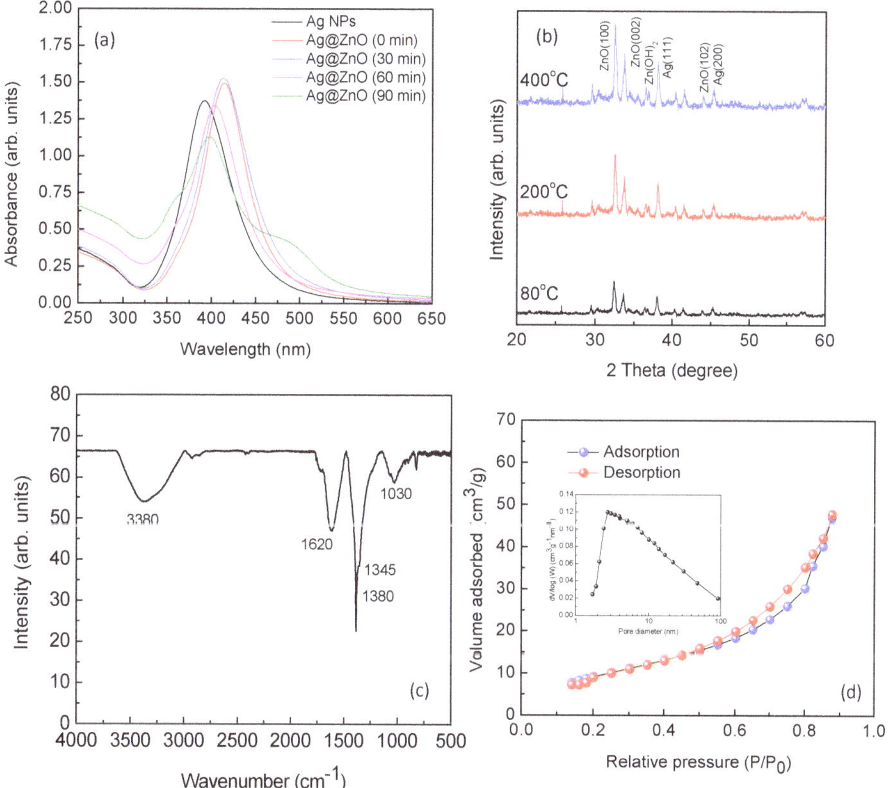

Figure 1. UV-Visible spectra of aqueous dispersion of Ag and Ag@ZnO core-shell nanoparticles (**a**), XRD pattern (**b**) and FTIR spectrum (**c**) of Ag@ZnO core-shell nanoparticles, (**d**) Nitrogen adsorption/desorption isotherms obtained at 77 K and inset shows the pore size distribution of the as-synthesized Ag@ZnO NCs synthesized by the sonochemical technique and dried at 80 °C for 2 h.

The adsorption-desorption isotherm plot for the nitrogen sorption (77 K) of the Ag@ZnO nanoparticles sample that was synthesized by sonochemical technique and dried at 80 °C for 2 h shows typical "type IV" isotherm in the Brunauer classification (Figure 1d). The sample exhibited average pore size in the range of 5–20 nm indicating the porous nature of the material. The specific surface area of Ag@ZnO core-shell nanoparticles was evaluated to be 65.5 m^2/g based on the BET result. This high surface area and porous nature are expected to be very beneficial for photocatalytic applications [12,14].

Morphology of the Ag@ZnO synthesized by the sonochemical technique were investigated by TEM. TEM samples were prepared by dipping the TEM grid in aqueous colloidal dispersion of NC followed by freeze drying for 12 h. Figure 2 shows TEM and HRTEM images of core-shell Ag@ZnO nanoparticles. Core-shell structure is observed for the materials. However, multiple silver nanoparticles were encapsulated within a single zinc oxide shell. Similar situation was also observed by Tripathy et al. [11]. A broad size distribution is observed for synthesized nano-Ag particles. The size of the Ag is found to be in the range of 10–30 nm and that of ZnO shell is about 5 to 10 nm. Metal core was found to have inter planar spacing of ~0.23 nm which corresponds to the (111) plane of the metallic silver with face-centered cubic structure. In ZnO shell, the spacing between adjacent lattice fringes is 0.16 nm, which is close to the *d*-spacing of the (110) plane of hexagonal ZnO (exact value is 0.168 nm).

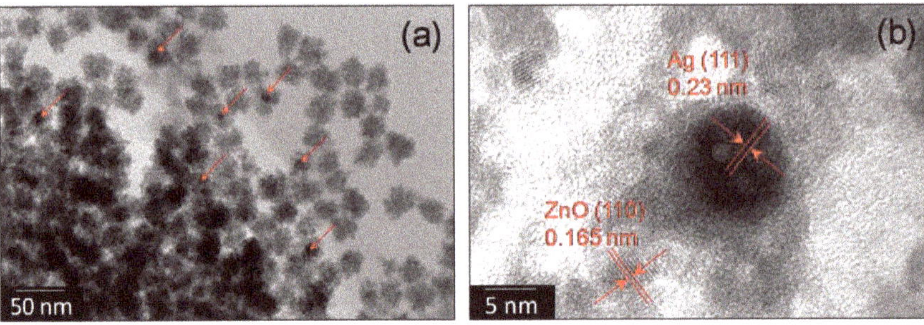

Figure 2. TEM (**a**) and HRTEM (**b**) images of Ag@ZnO core-shell nanoparticles synthesized by the sonochemical technique and dried at 80 °C for 2 h.

3.2. Photocatalytic Disinfection of Target Pathogens

Figure 3a,b show the photocatalytic disinfection achieved against the target pathogens at different catalyst concentrations. In Figure 3c,d bacterial disinfection is represented in its corresponding log reduction profile and the disinfection pattern is validated through comparison of the obtained profile with the standardized Chick-Watson model [11,20,21]. Figure 3a,b suggest that amongst the concentrations tested, 2 mg/L and 3 mg/L resulted in complete disinfection (6 log reductions) of *E. coli* and *S. aureus*, respectively, in 60 min and 90 min, respectively. It is observed that sunlight alone is not effective for the complete disinfection of the targeted pathogens as only 3 and 2.5 log reductions could be observed for *E. coli* and *S. aureus*, respectively, at 120 min. Experiments conducted under dark conditions did not show any remarkable change in the microbial colony counts, as less than 0.5-log reduction for both the microorganisms was achieved in 120 min (Figure 2c,d). Using the optimum concentration of Ag@ZnO nanoparticles for each of the bacteria for photocatalytic disinfection, comparative sunlight-assisted photocatalytic disinfection activity was evaluated with pure-ZnO and commercial TiO$_2$ (Degussa P25) and the results are shown in Figure 4a,b. Figure 4c,d shows the Chick-Watson disinfection kinetics of *E. coli* and *S. aureus* using different photocatalysts. These results suggest the superior disinfection efficiency of Ag@ZnO nano-photocatalyst compared to the conventional metal oxide systems. An increase in inactivation for both targeted bacteria was observed with the increase in catalyst concentration from 1 to 2 mg/L in *E. coli* and 1 to 3 mg/L in *S. aureus*. With further increase in the catalyst concentration beyond the mentioned range, a deterioration in disinfection rate was obtained for both the targeted microorganisms. With lower concentration of catalyst the amount of reactive oxygen species (ROS) generated is comparatively less. Thus complete disinfection required a longer irradiation time [22]. It is expected that as the rate of ROS production is slow at lower concentrations of catalyst, and under the initial conditions the microorganisms may activate their molecular resistance mechanisms. Therefore an extended disinfection time period is required for sufficient ROS generation and thus under the constant attack of ROS, bacteria may lose their capability of reactivation. With an increase in catalyst concentration the ROS generation rate increases, which is expected to improve the disinfection rate. Similarly, under the optimal conditions, the rate of ROS generation is maximum and therefore it may be expected that the interaction of the same with bacterial cells is more frequent. This may lead to an enhanced disinfection rate. It is further noticed that with increase in the catalyst concentration, disinfection gets delayed. This is mainly because with the increase in catalyst concentration the turbidity of the system increases, thereby blocking the sunlight irradiation from uniformly reaching the catalyst particles and cells, hence resulting in slower inactivation [23]. The current study involves *E. coli* and *S. aureus* bacteria. The photocatalytic performance of a photocatalyst depends both on its concentration and the irradiation time. *E. coli* was found more sensitive to sunlight-assisted photocatalytic disinfection process than *S. aureus*, as it requires comparatively less catalyst concentration and shorter sunlight irradiation time in comparison to *S. aureus* as evidenced from Figure 3a,b. The difference in susceptibility of both bacterial species to Ag@ZnO nanoparticles can be ascribed to the differences in their cell membrane/wall structures, chemical components, biological shape, and differences in robustness of Gram-positive and Gram-negative bacteria [24].

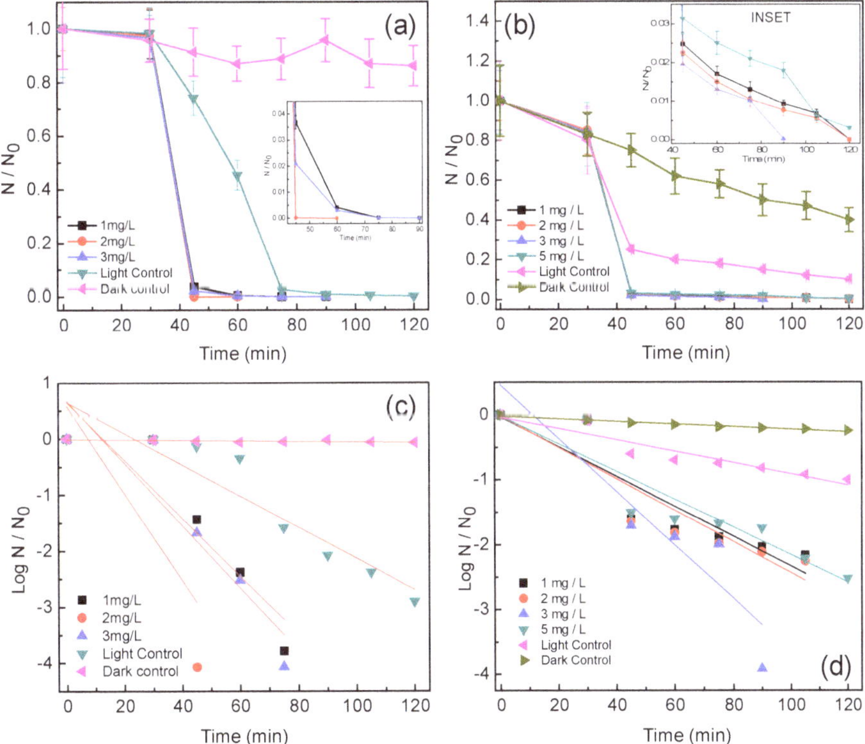

Figure 3. Effect of Ag@ZnO core-shell NPs loading on the solar-PCD kinetics of (**a**) *E. coli* and (**b**) *S. aureus*. Linear fitting plots of PCD kinetics of (**c**) *E. coli* and (**d**) *S. aureus* according to Chick-Watson model. Initial bacteria concentration = 5×10^6 CFU/mL, Temperature = 35 ± 2 °C. Error bars indicate the standard deviation of replicates (*n* = 3).

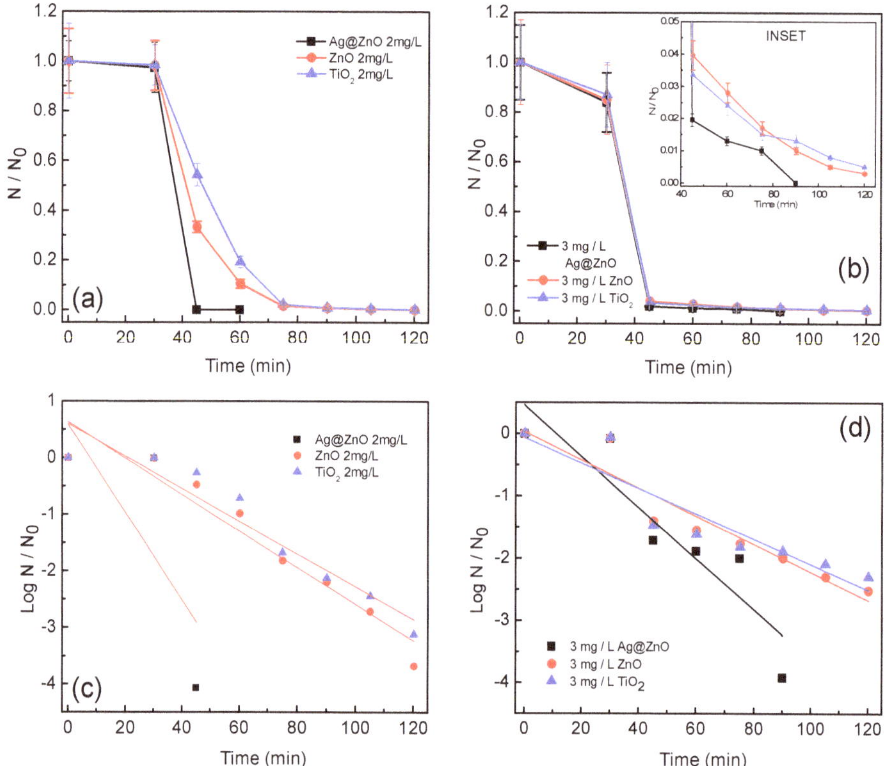

Figure 4. Effect of different catalysts on the solar-PCD kinetics of (**a**) *E. coli* and (**b**) *S. aureus* at a catalyst loading of 2 mg/L and 3 mg/L respectively. Linear fitting plots of PCD kinetics of different catalysts against (**c**) *E. coli* and (**d**) *S. aureus* according to Chick-Watson model at a catalyst loading. Initial bacteria concentration for each experiments = 5×10^6 CFU/mL, Temperature = 35 ± 2 °C. Error bars indicate the standard deviation of replicates ($n = 3$).

From Figure 4, it is observed that Ag@ZnO nanoparticles show enhanced disinfection efficiency for both targeted pathogens in comparison to the classical metal oxide systems (ZnO and TiO$_2$). The expected reason behind the enhanced efficiency may be the positioning of the noble metal (i.e., Ag) in the core and encapsulating it with a ZnO shell. Photocatalytic disinfection involves the excitation of the photocatalyst with light energy greater than or equal to that of the band gap [25]. On excitation the electrons forming the valence band of the metal oxide shuttle to the conduction band, where they are usually accepted by electron acceptors present in the reaction environment. This reduction pathway leads to the formation of ROS which results in killing of microbial cells by damaging their membrane integrity [24,26]. Therefore it leads to subsequent release of the intra-cellular components, which become vulnerable to the ROS attack [26,27]. Figure 5a–d show the effect of temperature on the photocatalytic disinfection of the targeted pathogens. These results show that as the temperature increased, a maximum process efficiency was observed at a reaction temperature of 55 °C. It is thus observed that, the rate of disinfection improved as the temperature of the reaction system increased. At 55 °C disinfection is achieved within 45 min and 60 min for *E. coli* and *S. aureus*, respectively. The post-disinfection reactivation of the target microbes was monitored for 24 h. None of the microbes showed an6y reactivation thus suggesting cell death due to damage caused by the ROS to both the target pathogens.

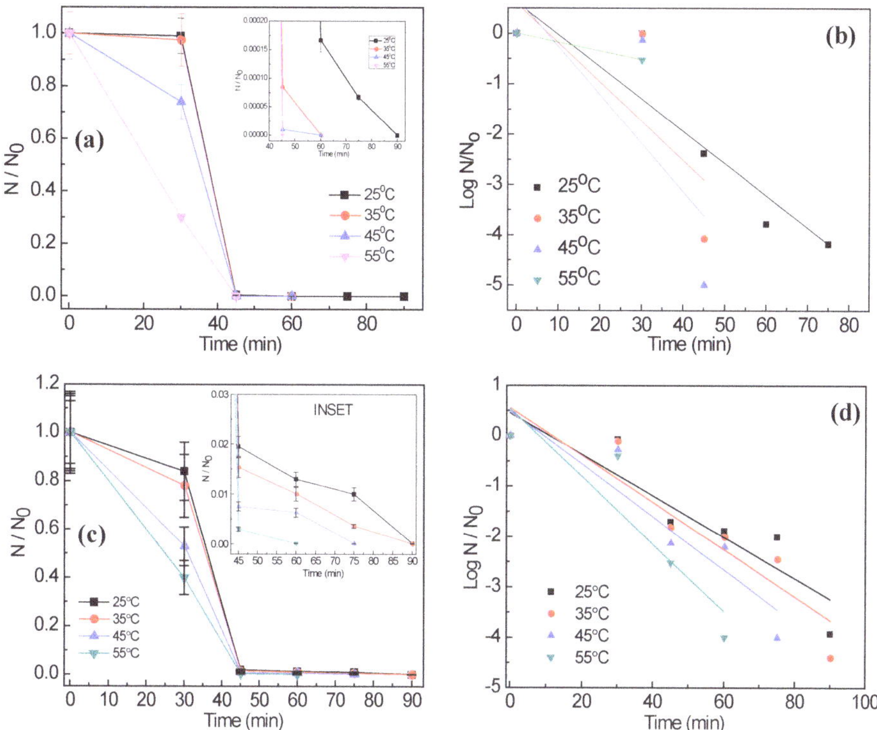

Figure 5. Effect of different reaction temperature on the solar-PCD kinetics of (**a**) *E. coli* and (**c**) *S. aureus* at a catalyst loading of 2 mg/L and 3 mg/L respectively. Linear fitting plots of PCD kinetics of different reaction temperature against (**b**) *E. coli* and (**d**) *S. aureus* according to Chick-Watson model. Initial bacteria concentration for each experiments ~ 5 × 10^6 CFU/mL, Error bars indicate the standard deviation of replicates (*n* = 3).

3.3. Determination of MDA to Study the Membrane Lipid Peroxidation

Time dependent generation of MDA (a key biomarker of membrane lipid peroxidation) for *E. coli* and *S. aureus* subjected to photocatalytic disinfection under their respective optimum catalyst concentration and temperature of 35 °C is shown in Figure 6a,b. Earlier experiments had shown complete disinfection at 60 and 90 min, respectively, for *E. coli* and *S. aureus* (Figure 3a,b). Hence, a similar correlative result can be inferred from the above mentioned figure. It is quite evident that maximum generation of MDA is observed after 75 min i.e., 0.03 nmol/mg cell dry weight and 90 min i.e., 0.0375 nmol/mg cell dry weight for *E. coli* and *S. aureus* respectively, which indicates cell membrane disintegration resulting in disinfection. Slight elevation in MDA production is seen within the first 30 min, which may be attributed to a loss of membrane integrity due to the action of shear stress produced on the microbial cells due to the continuous stirring conditions [28]. Additionally, the misbalance of ionic potential may also play a role in loss of membrane integrity leading to lipid peroxidation. It may also be noted that after the reported disinfection time, a decline in MDA concentration has been initiated. After a threshold level of MDA is generated in the photocatalytic system, it is expected to be mineralized being an organic compound itself [18,29]. When the microbial cells were exposed to sunlight without the presence of photocatalysts, less than even 0.01 nmol/mg cell dry weight generation was observed in both the microbes as shown in Figure 6a,b. However, the effect of Ag@ZnO on microbial cells without the presence of light is also found to be non-substantial, where the concentration was less, as 0.005 nmol/mg cell dry weight were quantified for both the test microbes. It is proposed that generation of ROS (such as OH• radical) in the photocatalytic process may lead to peroxidation of the cell membrane peptidoglycan layer and membrane proteins, followed by decomposition of cellular components and cellular disintegration [16–19,29], as ROS mainly (•OH) hit unsaturated membrane lipids to make lipid radicals. This, in the presence of oxygen is expected to give a lipid peroxyl radical capable of abstracting hydrogen from an adjacent unsaturated lipid and produce a lipid hydroperoxide and a lipid radical. This series of reactions continues until all the membrane unsaturated lipids are destroyed and malondialdehyde (a stable by-product of membrane lipid peroxidation) is subsequently produced. MDA generation patterns suggest that lipid peroxidation in *E. coli* maintains a uniform rate while a sporadic rate occurs for *S. aureus*, thus suggesting a higher robustness of the latter in comparison to the former [24].

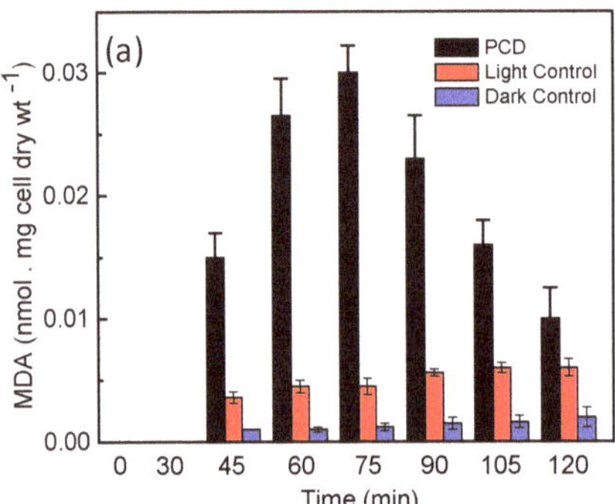
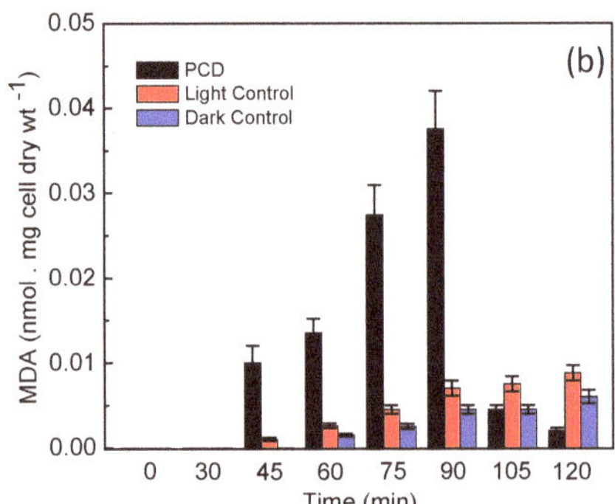

Figure 6. Lipid peroxidation kinetics of (**a**) *E. coli* and (**b**) *S. aureus* cells subjected to solar-photocatalysis in presence of 2 mg/L and 3 mg/L Ag@ZnO NPs respectively. Initial bacteria concentration = 5×10^6 CFU/mL, Temperature = 35 ± 2 °C. Error bars indicate the standard deviation of replicates ($n = 3$).

3.4. Analysis of Potassium Ion (K^+) to Study the Cell Membrane Damage

Leakage of K^+ ions is generally considered as a dominating evidence of compromised cellular integrity. The results obtained through K^+ leakage analysis are in agreement with many previous studies which mention the dysfunction of potassium channels of microorganisms on photocatalytic treatment [18,30–32]. It can be observed that the concentration of K^+ (in ppm) increases with the increase of the reaction time up till a particular time period beyond which the concentration in the reaction environment becomes constant. As shown in Figure 7a,b the maximum K^+ estimated after 120 min for *E. coli* and *S. aureus* after photocatalytic disinfection was found to be 575 ppb and 440 ppb, respectively. An interesting observation was made that the time required for complete disinfection for each of the target bacterium, as evaluated from the decreasing CFU count, does not correspond well with the K^+ leakage pattern. This must be because the primary target of photocatalytically produced ROS is membrane lipids. Once the entire membrane of the bacteria is compromised, an increase in K^+ ion is expected. This pattern of K^+ release suggests that the increase in the concentration of potassium ion in the reaction environment indicates a steady progress in the photocatalytic disinfection process. Once the entire bacterial death is achieved, it is expected that the total amount of K^+ will be maintained for the remaining reaction phase [33].

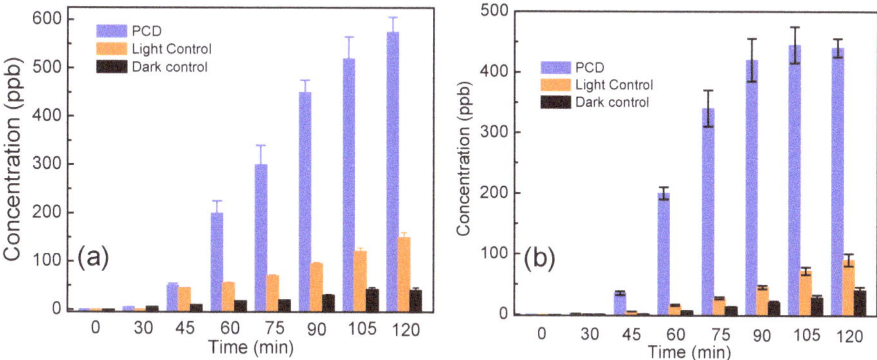

Figure 7. Leakage of K$^+$ ion from (**a**) *E. coli* and (**b**) *S. aureus* cells subjected to solar-photocatalysis in presence of 2 mg/L and 3 mg/L Ag@ZnO core-shell NPs, respectively. Initial bacteria concentration = 5×10^6 CFU/mL, Temperature = 35 ± 2 °C. Error bars indicate the standard deviation of replicates ($n = 3$).

3.5. Stability and Reusability of the Catalyst Post Disinfection

When the stability of the catalyst in post-reaction condition was investigated using XRD, no alteration in the crystal structure of Ag@ZnO was observed, suggesting its structural stability throughout the process [11,34]. It is known that leaching of material could re-toxify the system and it could also be argued that Ag$^+$ and Zn^{2+} ions which are reported to show antimicrobial properties may leach out of the system and hence, may be the actual cause of disinfection. However, the answer to this possibility is already communicated in our previous paper [11], there being no detectable amount of Ag$^+$ and Zn^{2+} ions in the system post-disinfection. If the catalyst could be recycled after photocatalytic disinfection then it may be suitable for commercial exploitation of the process. Ag@ZnO core-shell nanoparticles were recycled after the photocatalytic disinfection experiments and used for next batch of bacterial disinfection experiment (after heating at 80 °C). As shown in Figure 8, core-shell nanophotocatalyst exhibited insignificant reduction in *E. coli* and *S. aureus* disinfection efficiency, even after three consecutive cycles.

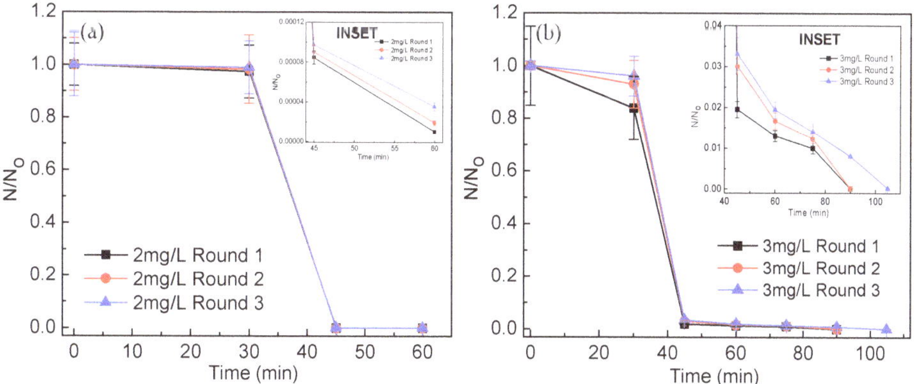

Figure 8. Effect of Ag@ZnO core-shell NPs reusability till three rounds of solar-PCD kinetics of (**a**) *E. coli* and (**b**) *S. aureus*. Initial bacteria concentration = 5×10^6 CFU/mL, Temperature = 35 ± 2 °C. Error bars indicate the standard deviation of replicates ($n = 3$).

3.6. Photocatalytic Disinfection Efficiency in Real Water Systems

As the results show (Figure 9), Ag@ZnO exhibits a better disinfection profile as compared to pure semiconductors in case of all the real water samples. The results correspond well with our previous results [11]. The superiority of the Ag@ZnO as compared to the traditional photocatalysts can be attributed to many causes. It is a well-known and established fact that the photocatalyst that is being used here has a core shell nanocomposite structure. The structure itself has many advantages over its traditional counterparts. It is a matter of general observation that the metal ions in the composite structure are protected by the shell in the composite structure. This has many advantages: firstly it solves the problem of leaching out of the silver metal ion. Silver is itself a very poisonous metal ion and detrimental and harmful to various organisms [4,5,11]. At the same time, the target pathogens *E. coli* and *S. aureus* are unable to survive and escape its effects. The core shell morphology also increases the surface area of the photocatalyst. As the surface area increases, so does the effectivity of the photocatalyst. Both the traditional photocatalysts used here, namely TiO_2 and ZnO lack in this property. The lack of a proper nanocomposite structure in the cases of TiO_2 and ZnO can also explain the lesser efficiency that these photocatalysts show in the photocatalytic degradation of real water samples.

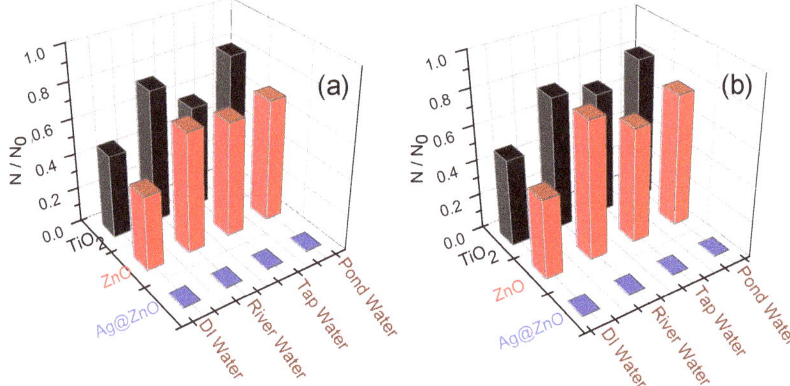

Figure 9. Effect of different photocatalysts on the relative reduction in the (**a**) *E. coli* and (**b**) *S. aureus* cell count (N/N_0) in real water samples after 120 min of solar irradiation at a catalyst loading of 2 mg/L and 3 mg/L catalyst concentration, respectively. In each case the initial bacteria concentration = 5×10^6 CFU/mL, Temperature = 35 ± 2 °C.

Various studies have already shown that at various concentrations both zinc and silver are detrimental to the growth of microorganisms [35]. The photocatalyst that we have used contains both these elements, so as a result, a better result can always be expected than that from the traditional ones, namely the likes of ZnO and TiO_2. The combined effect of toxicity of these two potent antimicrobial agents, combined with the lesser amount of leaching due to the unique structure is indeed a deciding factor in increasing the efficiency of the photocatalyst against the traditional players [11].

However, the issue of safety can be raised, regarding the compatibility of silver and zinc in various water streams and water bodies, as both of these metals are known to be toxic to organisms [36,37]. To attend these sensitive issues, we did an MP-AES assay, and it was observed that the concentration of zinc and silver was below the detectable levels. Thus it addresses most of the toxicity-related issues.

3.7. Proposed Mechanism of Photocatalytic Disinfection

The possible disinfection mechanism has been reported in the literature [9,11,15]. In the present case, we expected that the disinfection mechanism is contributed by the action of the photo-induced reactive oxygen species generated during the reaction (Figure 10). The initial site of attack is expected to be the lipopolysaccharide layer present in the external cell walls of the target bacteria [11]. It is assumed that the oxidative stress which is generated due to this process disintegrates the peptidoglycan layer and results in peroxidation of the lipid membrane, eventually causing oxidation of the membrane proteins [9]. This leads to rapid leakage of K^+ ions from the bacterial cells hence dysfunction of the potassium channels resulting in deregulation of cell signaling. Additionally the dwindling cell functionality and viability is also attributed by the peroxidation of polyunsaturated phospholipid components of the cell membrane, eventually leading to cell death [11,15].

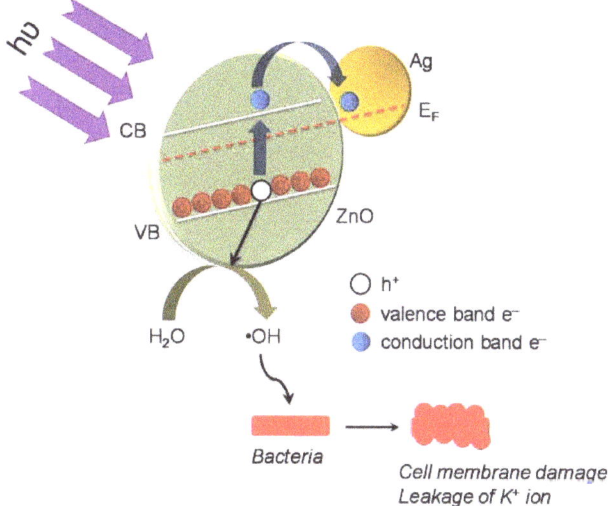

Figure 10. Proposed mechanism of sun-light assisted photocatalytic disinfection of bacteria using Ag@ZnO core-shell nanoparticles

There is an increasing demand regarding the issue of providing safe and potable drinking water to underdeveloped Third World countries. There is an urgent need to develop strategies that follow an alternate route to address this concern [18]. Based on the above statement, the concept of "Advanced Oxidation Process" can be proposed; based on the proven effectivity and superiority as compared to that of other traditional catalysts.

The catalyst that we have proposed, generally works well towards the basic range of pH values. All the real water samples, especially the likes of tap-water, and river water have basic pH. This can also be explain the better effectivity and working efficiency of the proposed catalyst, although further confirmation is required.

It can also be concluded from the MP-AES analysis that the proposed catalyst is completely non-toxic in nature and can be applied for a wide range of applications. It can also be concluded that, since there is no evidence for the proposed catalyst's toxicity to organisms, it can surely be used as a better, safer option than the traditional ones.

4. Conclusions

When DI water contaminated with *E. coli* and *S. aureus* was subjected to Ag@ZnO core-shell nanoparticles mediated photocatalytic disinfection under sun-light radiation, complete disinfection of *E. coli* and *S. aureus* was achieved within 60 and 90 min respectively at 35 °C and in 45 and 60 min at 55 °C. Quantitative analyses of K^+ ion release and MDA assay proposed the damage of bacterial cell wall by ROS generated during solar photocatalysis. The disinfection profile for both the bacteria was validated using the Crick-Watson disinfection model. Disinfection achieved using the Ag@ZnO system was also validated for real world samples of municipal tap, pond and river water. When the nanocatalyst was recycled and reused for subsequent photocatalytic disinfection experiments, its efficiency did not change remarkably, even after three cycles. The reportted photocatalytic system may find applications in designing a portable water decontamination system for pathogen infested geographical locations.

Acknowledgments: This work was supported by Department of Science & Technology, Government of India (IFA12-ENG-37) and Swedish Research Council, Government of Sweden (Grant Number K2013-70X-20514-07-5).

Author Contributions: Suraj K. Tripathy created the original study plan. Sourav Das, Neha Ranjana and Ananyo Jyoti Misra designed and executed the disinfection experiments under the guidance of Suraj K. Tripathy. Amrita Mishra and Mrutyunjay Suar helped in the molecular biology experiments. Cecilia Stålsby Lundborg and Ashok J. Tamhankar reviewed and edited the manuscript. All authors read and approved the manuscript.

Conflicts of Interest: The authors declare no conflict of interest.

References

1. Ashbolt, N.J. Microbial contamination of drinking water and disease outcomes in developing regions. *Toxicology* **2004**, *198*, 229–238. [CrossRef] [PubMed]

2. Adeleye, A.S.; Conway, J.R.; Garner, K.; Huang, Y.; Su, Y.; Keller, A.A. Engineered nanomaterials for water treatment and remediation: Costs, benefits, and applicability. *Chem. Eng. J.* **2016**, *286*, 640–662. [CrossRef]

3. Malato, S.; Fernández-Ibáñez, P.; Maldonado, M.I.; Blanco, J.; Gernjak, W. Decontamination and disinfection of water by solar photocatalysis: Recent overview and trends. *Catal. Today* **2009**, *147*, 1–59. [CrossRef]

4. Tripathy, S.K.; Mishra, A.; Jha, S.K.; Wahab, R.; Al-Khedhairy, A.A. Synthesis of thermally stable monodispersed Au@SnO$_2$ core–shell structure nanoparticles by a sonochemical technique for detection and degradation of acetaldehyde. *Anal. Methods* **2013**, *5*, 1456–1462. [CrossRef]

5. Haldar, K.K.; Sen, T.; Patra, A. Au@ZnO core–shell nanoparticles are efficient energy acceptors with organic dye donors. *J. Phys. Chem. C* **2008**, *112*, 11650–11656. [CrossRef]

6. Li, Z.; Zhang, F.; Meng, A.; Xie, C.; Xing, J. ZnO/Ag micro/nanospheres with enhanced photocatalytic and antibacterial properties synthesized by a novel continuous synthesis method. *RSC Adv.* **2015**, *5*, 612–620. [CrossRef]

7. Ribeiro, M.A.; Cruz, J.M.; Montagnolli, R.N.; Bidoia, E.D.; Lopes, P.R.M. Photocatalytic and photoelectrochemical inactivation of *Escherichia coli* and *Staphylococcus aureus*. *Water Sci. Technol. Water Supply* **2015**, *15*, 107–113. [CrossRef]

8. Petti, S.; Messano, G.A. Nano-TiO$_2$-based Photocatalytic disinfection of environmental surfaces contaminated by methicillin-resistant *Staphylococcus aureus*. *J. Hosp. Infect.* **2016**, *93*, 78–82. [CrossRef] [PubMed]

9. Reddy, M.P.; Venugopal, A.; Subrahmanyam, M. Hydroxyapatite-supported Ag–TiO$_2$ as *Escherichia coli* disinfection photocatalyst. *Water Res.* **2007**, *41*, 379–386. [CrossRef] [PubMed]

10. Aguirre, M.E.; Rodríguez, H.B.; San Román, E.; Feldhoff, A.; Grela, M.A. Ag@ZnO core–shell nanoparticles formed by the timely reduction of Ag+ Ions and Zinc acetate hydrolysis in *N,N*-dimethylformamide: mechanism of growth and photocatalytic properties. *J. Phys. Chem. C* **2011**, *115*, 24967–24974. [CrossRef]

11. Das, S.; Sinha, S.; Suar, M.; Yun, S.I.; Mishra, A.; Tripathy, S.K. Solar-photocatalytic disinfection of *Vibrio Cholerae* by using Ag@Zno core–shell structure nanocomposites. *J. Photochem. Photobiol. B* **2015**, *142*, 68–76. [CrossRef] [PubMed]

12. Chong, M.N.; Jin, B.; Chow, C.W.; Saint, C. Recent developments in photocatalytic water treatment technology: A review. *Water Res.* **2010**, *44*, 2997–3027. [CrossRef] [PubMed]

13. Zhang, N.; Liu, S.; Xu, Y.J. Recent Progress on metal core@ semiconductor shell nanocomposites as a promising type of photocatalyst. *Nanoscale* **2012**, *4*, 2227–2238. [CrossRef] [PubMed]

14. Xu, H.; Zeiger, B.W.; Suslick, K.S. Sonochemical synthesis of nanomaterials. *Chem. Soc. Rev.* **2013**, *42*, 2555–2567. [CrossRef] [PubMed]

15. Carré, G.; Hamon, E.; Ennahar, S.; Estner, M.; Lett, M.C.; Horvatovich, P.; Andre, P. TiO$_2$ photocatalysis damages lipids and proteins in *Escherichia coli*. *Appl. Environ. Microbiol.* **2014**, *80*, 2573–2581. [CrossRef] [PubMed]

16. Maness, P.C.; Smolinski, S.; Blake, D.M.; Huang, Z.; Wolfrum, E.J.; Jacoby, W.A. Bactericidal activity of photocatalytic TiO$_2$ reaction: Toward an understanding of Its killing mechanism. *Appl. Environ. Microbiol.* **1999**, *65*, 4094–4098. [PubMed]

17. Esterbauer, H.; Cheeseman, K.H. Determination of aldehydic lipid peroxidation products: Malonaldehyde and 4-hydroxynonenal. *Methods Enzymol.* **1990**, *186*, 407–421. [PubMed]

18. Das, S.; Sinha, S.; Das, B.; Jayabalan, R.; Suar, M.; Mishra, A.; Tripathy, S.K. Disinfection of multidrug resistant *Escherichia coli* by solar-photocatalysis using Fe-doped ZnO nanoparticles. *Sci. Rep.* **2017**, *7*, 104. [CrossRef] [PubMed]

19. Leung, T.Y.; Chan, C.Y.; Hu, C.; Yu, J.C.; Wong, P.K. Photocatalytic disinfection of marine bacteria using fluorescent light. *Water Res.* **2008**, *42*, 4827–4837. [CrossRef] [PubMed]

20. Chick, H. An investigation of the laws of disinfection. *J. Hyg.* **1908**, *8*, 92–158. [CrossRef] [PubMed]

21. Watson, H.E. A Note on the variation of the rate of disinfection with change in the concentration of the disinfectant. *J Hyg.* **1908**, *8*, 536–542. [CrossRef] [PubMed]

22. Choi, O.; Deng, K.K.; Kim, N.J.; Ross, L.; Surampalli, R.Y.; Hu, Z. The Inhibitory effects of silver nanoparticles, silver ions, and silver chloride colloids on microbial growth. *Water Res.* **2008**, *42*, 3066–3074. [CrossRef] [PubMed]

23. Rizzo, L.; Della Sala, A.; Fiorentino, A.; Puma, G.L. Disinfection of urban wastewater by solar driven and UV lamp–TiO$_2$ photocatalysis: Effect on a multi-drug resistant *Escherichia coli* strain. *Water Res.* **2014**, *53*, 145–152. [CrossRef] [PubMed]

24. Shang, K.; Ai, S.; Ma, Q.; Tang, T.; Yin, H.; Han, H. Effective photocatalytic disinfection of *E. coli* and *S. aureus* using polythiophene/MnO$_2$ nanocomposite photocatalyst under solar light irradiation. *Desalination* **2011**, *278*, 173–178.

25. Mills, A.; Le Hunte, S. An overview of semiconductor photocatalysis. *J. Photochem. Photobiol. A* **1997**, *108*, 1–35. [CrossRef]

26. Kiwi, J.; Nadtochenko, V. Evidence for the mechanism of photocatalytic degradation of the bacterial wall membrane at the TiO$_2$ interface by ATR-FTIR and laser kinetic spectroscopy. *Langmuir* **2005**, *21*, 4631–4641. [CrossRef] [PubMed]

27. Foster, H.A.; Ditta, I.B.; Varghese, S.; Steele, A. Photocatalytic disinfection using titanium dioxide: Spectrum and mechanism of antimicrobial activity. *Appl. Environ. Microbiol.* **2011**, *90*, 1847–1868. [CrossRef] [PubMed]

28. Doran, P.M. *Bioprocess Engineering Principles*; Academic Press: Cambridge, MA, USA, 1995.

29. Pulgarin, C.; Kiwi, J.; Nadtochenko, V. Mechanism of photocatalytic bacterial inactivation on TiO$_2$ films involving cell-wall damage and lysis. *Appl. Catal. B* **2012**, *128*, 179–183. [CrossRef]

30. Ennis, H.L. Role of potassium in the regulation of polysome content and protein synthesis in *Escherichia coli*. *Arch. Biochem. Biophys.* **1971**, *142*, 190–200. [CrossRef]

31. Helali, S.; Polo-López, M.I.; Fernández-Ibáñez, P.; Ohtani, B.; Amano, F.; Malato, S.; Guillard, C. Solar photocatalysis: A green technology for *E. coli* contaminated water disinfection. Effect of concentration and different types of suspended catalyst. *J. Photochem. Photobiol. A* **2014**, *276*, 31–40.

32. Zhang, L.S.; Wong, K.H.; Yip, H.Y.; Hu, C.; Yu, J.C.; Chan, C.Y.; Wong, P.K. Effective photocatalytic disinfection of *E. coli* **K-12** using AgBr−Ag−Bi$_2$WO$_6$ nanojunction system irradiated by visible light: The role of diffusing hydroxyl radicals. *Environ. Sci. Technol.* **2010**, *44*, 1392–1398. [PubMed]

33. Hu, C.; Guo, J.; Qu, J.; Hu, X. Photocatalytic degradation of pathogenic bacteria with AgI/TiO$_2$ under visible light irradiation. *Langmuir* **2007**, *23*, 4982–4987.

34. Paik, P. Recent advancement in functional core-shell nanoparticles of polymers: Synthesis, physical properties, and applications in medical biotechnology. *J. Nanopart. Res.* 2013. [CrossRef]

35. Chudobova, D.; Dostalova, S.; Ruttkay-Nedecky, B.; Guran, R.; Rodrigo, M.A.M.; Tmejova, K.; Kizek, R. The effect of metal ions on *Staphylococcus aureus* revealed by biochemical and mass spectrometric analyses. *Microbiol. Res.* **2015**, *170*, 147–156. [CrossRef] [PubMed]

36. Bondarenko, O.; Juganson, K.; Ivask, A.; Kasemets, K.; Mortimer, M.; Kahru, A. Toxicity of Ag, CuO and ZnO nanoparticles to selected environmentally relevant test organisms and mammalian cells in vitro: A critical review. *Arch. Toxicol.* **2013**, *87*, 1181–1200. [CrossRef] [PubMed]

37. Singh, R.; Gautam, N.; Mishra, A.; Gupta, R. Heavy metals and living systems: An overview. *Indian J. Pharmacol.* **2011**, *43*, 246. [CrossRef] [PubMed]

International Journal of
Environmental Research and Public Health

Article

Sunlight Assisted Photocatalytic Degradation of Ciprofloxacin in Water Using Fe Doped ZnO Nanoparticles for Potential Public Health Applications

Sourav Das [1,†], Soumen Ghosh [1,†], Ananyo Jyoti Misra [1], Ashok J. Tamhankar [1,2,‡], Amrita Mishra [1], Cecilia Stålsby Lundborg [2,‡] and Suraj K. Tripathy [1,3,*,‡]

[1] School of Biotechnology, Kalinga Institute of Industrial Technology (KIIT), Bhubaneswar 751024, India; sddassourav52@gmail.com (S.D.); soumen.cbel@gmail.com (S.G.); ananyomisra@gmail.com (A.J.M.); ejetee@gmail.com (A.J.T.); amritamishrabio@gmail.com (A.M.)

[2] Department of Public Health Sciences, Karolinska Institutet, SE 17177 Stockholm, Sweden; cecilia.stalsby.lundborg@ki.se

[3] School of Chemical Technology, Kalinga Institute of Industrial Technology (KIIT), Bhubaneswar 751024, India

[*] Correspondence: tripathy.suraj@gmail.com; Tel.: +91-943-832-1810

[†] S.D. and S.G. have equal contribution.

[‡] A.J.T., C.S.L. and S.K.T. have equal contribution.

Received: 25 August 2018; Accepted: 29 October 2018; Published: 1 November 2018

Abstract: Antibiotic residues in the aquatic environment have the potential to induce resistance in environmental bacteria, which ultimately might get transferred to pathogens making treatment of diseases difficult and poses a serious threat to public health. If antibiotic residues in the environment could be eliminated or reduced, it could contribute to minimizing antibiotic resistance. Towards this objective, water containing ciprofloxacin was treated by sunlight-assisted photocatalysis using Fe-doped ZnO nanoparticles for assessing the degradation potential of this system. Parameters like pH, temperature, catalytic dosage were assessed for the optimum performance of the system. To evaluate degradation of ciprofloxacin, both spectrophotometric as well as microbiological (loss of antibiotic activity) methods were employed. 100 mg/L Fe-doped ZnO nanoparticle catalyst and sunlight intensity of 120,000–135,000 lux system gave optimum performance at pH 9 at 30 °C and 40 °C. Under these conditions spectrophotometric analysis showed complete degradation of ciprofloxacin (10 mg/L) at 210 min. Microbiological studies showed loss of antibacterial activity of the photocatalytically treated ciprofloxacin-containing water against *Staphylococcus aureus* (10^8 CFU) in 60 min and for *Escherichia coli* (10^8 CFU) in 75 min. The developed system, thus possess a potential for treatment of antibiotic contaminated waters for eliminating/reducing antibiotic residues from environment.

Keywords: antibiotic residues; aquatic environment; ciprofloxacin; Fe-doped ZnO nanoparticles; photocatalysis; sunlight

---

1. Introduction

Antibiotic residues in the environment is pose a major public health challenge [1]. Fluoroquinolones (FQs) are a class of environmentally stable broad spectrum antibiotics, which inhibits the enzymes DNA topoisomerase II (Gyrase) and DNA topoisomerase IV in bacteria thus interfering with their DNA replication machinery [2,3]. FQs are effective against both Gram positive and Gram negative bacteria and are used both in humans and animals. Ciprofloxacin is the most commonly used FQ, Studies report the occurrence of FQs, including ciprofloxacin, in water bodies worldwide [4]. FQ reaches water bodies through excretion after incomplete metabolism within the human/animal gut [5]. Their presence at up to 87 microgram/L and 31 mg/L has been demonstrated in wastewater discharge [6]. Conventional wastewater treatment including biological oxidation and other chemical and physical process leads to only partial removal of these compounds [7]. As a consequence, the presence of broad spectrum antibiotics like FQs, even at very minute concentrations, poses a threat to the surrounding ecosystem and human health through the development of antibiotic resistance amongst environmental bacteria [8], which can potentially lead to further spread of resistance to other bacterial populations including human and animal pathogens through processes such as ingestion of untreated or partially purified water or horizontal gene transfer [9].

With the immediate necessity for substantive degradation of such organic environmental pollutants, semiconductor photocatalysis more appropriately, Advanced Oxidation Processes (AOPs) have proven quite useful [10]. They normally use a semiconductor metal oxide or one of its doped variants as a photo-oxidant which in presence of light charges up and leads to the generation of highly reactive oxidative species like hydroxyl radicals (OH·), superoxide anion ($O_2\cdot-$) and hydrogen peroxide (H_2O_2) for remediation of organic pollutants. The basic principle behind their action is shown in Figure 1. To date TiO_2 and ZnO has been reported to be the best catalysts for photocatalytic applications because of their optical properties, thus having a much better quantum efficiency under visible light [11]. Moreover, owing to their high chemical stability, high oxidation efficiency, low toxicity, less cost, easy availability and being abundant in Nature they are excellent photocatalysts for the mineralization of organic pollutants in both acidic and basic media [12]. ZnO absorbs a substantial amount in the UV range [12] and UV accounts for only 3–5% of the sunlight, thus there is insufficient usage of the total sunlight available, so efforts are needed to design catalysts which will show better photocatalytic efficiency in the visible region of sunlight [11]. In order to address such problems, modifying the metal oxide semiconductor with transition, alkaline and rare earth metals like Mn, Fe, Co, Ni, Ag, Mg, Pb, N, C, S, P, is done [11], which will shift the light absorption towards the visible range.

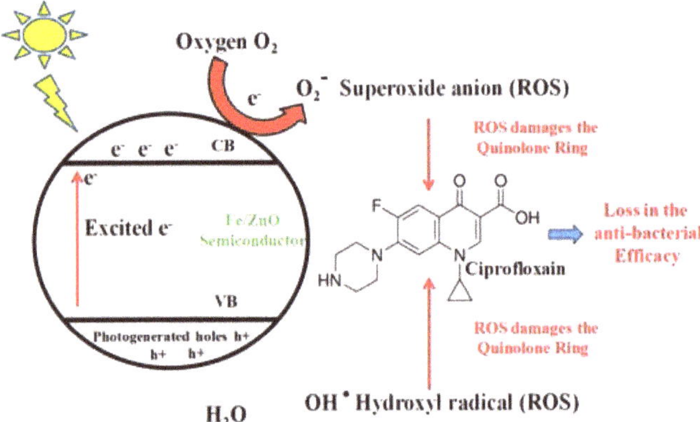

Figure 1. Schematic representation showing generation of reactive oxygen species (ROS) by Fe ZnO nanoparticles on activation with sunlight, and how these ROS attack active components of FQ to degrade them and reduce their anti-bacterial activity.

Photocatalysis with ZnO for the degradation of antibiotics like ciprofloxacin, amoxicillin, ampicillin, cloxacillin using different sources of light was performed earlier [13,14]. Nearly 50% degradation of antibiotics was achieved with high rate constant and maximium degradation was reported at pH 10–11. It has been previously reported in one of our studies that using Fe-doped ZnO for photocatalytic applications majorly contributes towards the generation of H_2O_2 in the system, which ultimately is detrimental for the photocatalytic oxidation. Moreover the presence of Fe in the system, serves as an added advantage for the photocatalytic oxidation, since it comes in contact with H_2O_2 in the system to generate more of hydroxyl radicals via the Fenton process [15]. This will ultimately magnify the oxidation of antibiotic-containing water. Thus doping the catalyst with iron has some added benefits as far as increasing the photocatalytic efficiency of ZnO are concerned. Earlier such Fe-doped ZnO has been used for the successful degradation of wastewater containing dye molecules [16]. The aim of this study was to evaluate sunlight-assisted photocatalytic degradation of ciprofloxacin using Fe-doped ZnO nanoparticles. Further, the residual antibacterial activity of the treated water was assessed against a Gram positive (*Staphylococcus aureus*) and a Gram negative (*Escherichia coli*) bacterium.

2. Materials and Methods

2.1. Materials

Chemicals used in this study include ciprofloxacin hydrochloride (MP Biomedicals, Santa Ana, CA, USA), zinc nitrate hexahydrate (98%, Sigma Aldrich, USA), trisodium citrate dihydrate (Sigma Aldrich, St. Louis, MO, USA), ferric chloride (Himedia, Mumbai, India), Luria agar and Luria broth (Himedia, Mumbai, India), sodium hydroxide (Merck, Kenilworth, NJ, USA), and hydrochloric acid (35.5%, Merck). All the chemicals were of molecular grade.

2.2. Preparation of Ciprofloxacin Stock Solution

Ciprofloxacin hydrochloride stock solution (100 mg L^{-1}) was prepared in deionized water (NaOH was used to solubilize the ciprofloxacin followed by 5 min of ultrasonication), 2 L at a time and stored in dark at 4 °C. Working solutions of 10 mg L^{-1}, (in 300 mL deionized water at a time) were prepared for each photocatalysis experiment, as required.

2.3. Synthesis of Nanocrystalline Fe-Doped ZnO

Fe-doped ZnO was prepared using a precipitation route as previously described [15]. Briefly, Zinc nitrate hexahydrate (5.948 g), ferric chloride (0.108 g) and trisodium citrate (5.882) were dissolved in 500 mL distilled water and stirred at 80 °C for 60 min. Then 250 mL of NaOH (250 mM) was slowly added dropwise into the solution using a burette until yellowish-white precipitate was formed. The precipitate was allowed to come to room temperature and was then centrifuged (10,000 rpm, which corresponds to 9391 g force, 10 min, Eppendorf 5424, USA), and rinsed with distilled water thrice. The precipitate was then dried at 70 °C overnight followed by calcination at 500 °C. The calcined Fe doped ZnO powder was characterized as mentioned in our previous paper and used for photocatalytic applications.

2.4. Photocatalytic Degradation of Ciprofloxacin

A 300 mL aqueous solution of ciprofloxacin with a concentration of 10 mg L^{-1} was placed in a 500 mL borosilicate beaker with the required amount (see below) of Fe-doped ZnO and mixed by a magnetic stirrer. The mixture was kept undisturbed in dark for 30 min to allow equilibrium. The experiments were performed with different catalyst concentrations 100, 150 and 200 mg L^{-1}, at pH 2, 3, 5.5, 7, 9, 10 and 11 (required pH was adjusted with 1 N HCl or 1 N NaOH), different reaction temperatures of 30 °C, 40 °C, 50 °C and 60 °C and different photocatalysts (TiO$_2$ and ZnO) at a light intensity of 80,000 ± 3000 lux, which corresponds to 650 W/m^2. At 15 min intervals, up to 210 min, collected samples were filtered through centrifugation (10,000 rpm, which corresponds to 9391 g force, 10 min, Eppendorf 5424) before spectrophotometric analysis (λ_{max}-280 and 320 nm using a Shimadzu UV-1800 instrument (Japan) and the microbiology experiments for assessment of residual antibacterial activity. The time dependent decrease in absorbance values at λ_{max}-280 and 320 nm suggests degradation of the antibiotic [14].

2.5. Residual Antibacterial Activity of the Treated Water

Qualitative assays were performed to assess the residual antibacterial activity of the treated water after photocatalytic degradation against the fully susceptible test organisms *Staphylococcus aureus* (MTCC code 3160) and *Escherichia coli* (MTCC code 7410) from the Microbial Type Culture Collection (MTCC, Chandigarh, India). The well diffusion method, according to the Clinical and Laboratory Standards Institute (CLSI) [17,18] was employed. All plates were prepared in 90 mm sterile Petri dishes (Tarsons, Mumbai, India) with 22 mL of Luria Bertani agar, yielding a depth of 4 mm. Test microorganism's 100 µL of inoculum suspensions (OD_{600}-0.5, corresponding to 1.0×10^8 CFU mL^{-1}) were poured into the agar plates when the temperature reached around 40–45 °C using a sterile micropipette, and homogenized thoroughly by mixing in a circular motion (pour-plate technique). After solidification, roundwells (6.0 mmin diameter) were punched into the seeded agar plates with a 6 mm cork borer. The wells were filled with 40 µL of the treated water samples (collected and filtered after regular time intervals) using a sterile micropipette. These plates were allowed to stand at 4 °C for 2 h and then incubated at 37 °C for 24 h. Three sets of simultaneous controls were used. One control was the organism control and consisted of a seeded Petri dish with no photocatalytically treated antibiotic sample. In the second control, samples were introduced in the holes of unseeded Petri dishes to check for sterility. Finally, to ensure the elimination of any solvent effect, wells filled with 40 µL of sterile double distilled water were run simultaneously as a third control. The diameters of the inhibition zones (zone of inhibition—ZOI) were measured in millimeters [19]. Each test was repeated six times and the mean values from the replicates along with standard error of mean (SEM) were calculated.

3. Results and Discussion

3.1. Photocataltytic Degradation of Ciprofloxacin and Process Optimization

Figure 2 shows the decrease in the C/Co absorption spectrum of ciprofloxacin (C = concentration at a particular time, Co = initial concentration of ciprofloxacin) at three catalyst concentrations (100, 150, 200 mg L^{-1}), during sun-assisted photocatalysis by Fe-doped ZnO nanoparticles. The values were calculated on the basis of intensity of the absorbance peaks at 280 and 320 nm. At both these λ_{max}, the absorbance showed a decreasing trend at all the three catalyst concentrations.

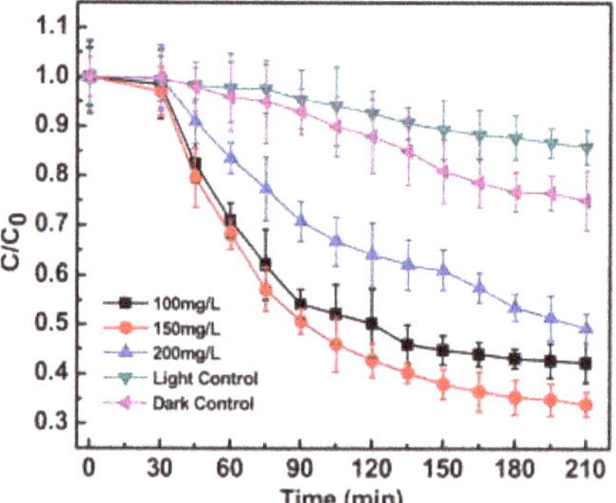

Figure 2. Photocatalytic degradation of antibiotic ciprofloxacin (10 mg/L) in water, in the presence of Fe-ZnO nanoparticles (at different concentrations of 100, 150 and 200 mg/L) irradiated with sunlight light intensity of 80,000 ± 3000 lux compared to photolysis (light control) and degradation in the absence of light (dark control). C_0 represents initial concentration of ciprofloxacin and C represents concentration of ciprofloxacin at a particular time point. C/C_0 denotes the time dependent change in ciprofloxacin concentration with respect to initial concentration.

A catalyst concentration of 150 mgL^{-1} caused a significant degradation of ciprofloxacin (10 mg L^{-1}) of up to 66% in 210 min and was found to be optimum. The other two concentrations were not as effective. The 100 mg L^{-1} catalyst may not have the capability for substantial generation of reactive oxygen species, while the 200 mg L^{-1} catalyst concentration may be high enough to create a catalyst shielding effect. Moreover the 200 mg L^{-1} may possess slow or improper degradation kinetics of only 51%. For further experiments, therefore all the degradation experiments were carried out with 150 mg/L of Fe-doped ZnO. There was no significant change in concentration of the ciprofloxacin due to the direct sunlight assisted photolysis (light control) which was found to be only 14% [14]. The decrease in C/Co value (up to 25%) of the antibiotic when subjected to dark control reaction (at the optimum photocatalyst concentration of 150 mg/L), may be attributed to direct adsorption of the antibiotic in the presence of doped ZnO nanoparticles [11].

The concentration of antibiotic in the wastewater system is a key parameter to optimize the photocatalytic degradation process. A study was performed with ciprofloxacin concentrations of 5, 10 and 15 mg L^{-1}. Figure 3 shows the photocatalytic degradation pattern of different concentrations of ciprofloxacin with the optimized concentration of Fe-doped ZnO nanoparticles. At 10 mg L^{-1} concentration no peaks were observed at 280 and 320 nm after 210 min of photocatalytic treatment, suggesting complete degradation of the quinolone ring. Five mg L^{-1} concentrations of ciprofloxacin were also completely degraded. Since studies with 10 mg L^{-1} concentrations were previously done and reported, the rest of the photocatalytic study were done with 10 mg L^{-1} concentration. With 15 mg L^{-1} ciprofloxacin concentration the degradation kinetics were a bit slower. Possible reasons could be a catalyst shielding effect and over-occupied catalyst active sites at 15 mg/L concentration [11,19].

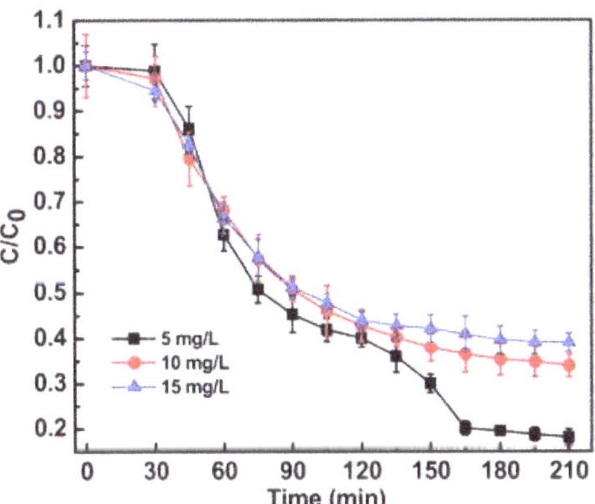

Figure 3. Photocatalytic degradation of antibiotic ciprofloxacin in water at different antibiotic concentration between 5, 10, 15 mg/L with optimum Fe-doped ZnO nanoparticles concentration of 150 mg/L and irradiated with sunlight intensity of 80,000 ± 3000 lux. C$_0$ represents initial concentration of ciprofloxacin and C represents concentration of ciprofloxacin at a particular time point. C/C$_0$ denotes, time dependent change in ciprofloxacin concentration with respect to initial concentration.

pH modifies the surface charge properties of Fe-doped ZnO and possibly the chemical structure of the antibiotic, therefore the influence of pH on the photocatalytic activity of Fe-doped ZnO nanoparticles was studied by altering the pH of the reaction mixture in both the acidic and basic range. Figure 4 shows the effect on the photocatalytic degradation on ciprofloxacin of different pHs in the presence of Fe-doped ZnO nanoparticles. The best degradation efficiency of ciprofloxacin with Fe-doped ZnO nanoparticles, nearly 65%, was seen at pH 9, while the lowest degradation, only 10%, was observed at pH 2 [14]. The maximum ciprofloxacin degradation was thus obtained at basic pH values between 9 and 11 under solar light, where the available hydroxyl ions in the system can react with the valence band holes (h+) to form reactive hydroxyl radicals (OH·), which possesses high oxidation capability under photocatalytic conditions, subsequently enhancing the rate of photocatalytic degradation of ciprofloxacin. Similar results for the degradation of aromatic compounds were reported earlier [20]. At an acidic pH value of 2, the solar photocatalytic degradation of ciprofloxacin was hindered due to the high proton concentration, which possesses higher attraction for the hydroxyl anions, quenching the formation of hydroxyl radicals. As free hydroxyl ions in the system are decreased, the formation of hydroxyl radicals becomes limiting. Thus photocatalytic degradation of ciprofloxacin decreased at lower pH. It may also be possibly due to dissolution of Fe-doped ZnO under acidic conditions. Similar observations were previously made in the photocatalytic degradation of azo dyes [16].

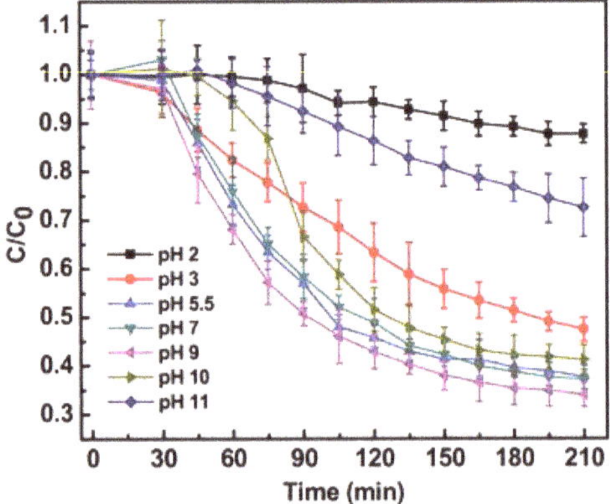

Figure 4. Photocatalytic degradation of antibiotic ciprofloxacin (10 mg/L) in water in the presence of Fe-ZnO nanoparticles (150 mg/L) irradiated with sunlight intensity of 80,000 ± 3000 lux at different reaction pH of 2, 3, 5.5, 7, 9, 10, 11. C_0 represents initial concentration of ciprofloxacin and C represents concentration of ciprofloxacin at a particular time point. C/C_0 denotes the time dependent change in ciprofloxacin concentration with respect to initial concentration.

Ciprofloxacin is an ampholytic compound with a pKa value of 6.09 for the carboxylic group and 8.74 for the nitrogen on the piperazinyl ring. The isoelectric or zwitterionic point is at pH 7.4. Thus ciprofloxacin seemed to be most sensitive to photocatalytic degradation at a pH closer to its zwitterionic form, i.e. at basic pH 9. It has earlier been reported that the maximum stability of the molecule was observed in reaction solution of pH 4.0 [21], where the carboxylic group is un-ionized and basic nitrogen is completely protonated. This adds an advantage to the ciprofloxacin pharmaceutically, because most of the pharmaceutical formulation possess pH between 3.5 and 5.5. This seems good from a pharmaceutical perspective but photocatalytic degradation at such low pH will be a challenge. Interestingly, it has been previously reported that, hospital wastewater flowing to drains has an pH in between 6.7 to 7.7 throughout the year, Moreover the pH of surface waters (mainly lakes and rivers) in India is between 6.5 to 8.5 [22]. The current study thus finds it application for degradation of antibiotics in hospital wastewater and surface water, since at this pH range the photocatalytic degradation was more than 60%, as shown in Figure 3.

From experimental observations and previous reports on the photocatalytic degradation of organic molecules like dyes [23] and antibiotics [24], we assumed that upon irradiation with solar light, within the Fe-doped ZnO nanoparticles, excitation of electrons takes place from the valence band into the conduction band. Photogenerated holes in the conduction band upon reacting with water molecules in the system generate hydroxyl radicals which possess oxidative nature and can get rid of antibiotics adsorbed on the Fe-doped ZnO surface. Moreover the high oxidative potential of valence band holes can also lead to the direct and indirect oxidation of antibiotics. The presence of Fe in the system possesses an added advantage to this photocatalytic degradation process. The presence of Fe delays the electron whole recombination, acting as one of the terminal acceptors of electrons, which eventually increases the generation of hydroxyl radicals and reactive species in the system. Also Fe as a Fenton agent is capable of producing reactive oxygen species like OH· radicals through the Fenton process, adding more ROS to the system for subsequent degradation of ciprofloxacin [11,25].

Temperature was found to modulate the degradation kinetics (Figure 5). Generally it has been reported that with an increase in temperature the degradation kinetics are enhanced [11], but in the current study, the opposite trend was observed. With increasing temperature, the degradation kinetics decreased up to 60 °C. A possible reason could be the increase in the stability of fluoroquinolones on exposure to heat stress. It has been reported by Roca et. al. [26], that FQs can be stable at temperatures up to 120 °C. In a country like India, where the atmospheric temperature can reach up to 50 °C, the technique presented in this paper can be employed for successful degradation of ciprofloxacin and maybe other fluoroquinolones also, in wastewater matrices. The technique presented in this paper may also find its application for the treatment of hospital, pharmaceutical or industrial wastewater for the degradation of many organic molecules.

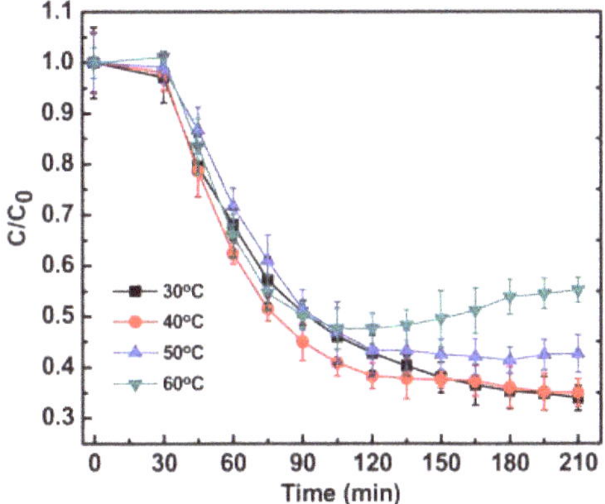

Figure 5. Photocatalytic degradation of antibiotic ciprofloxacin (10 mg/L) in water in the presence of Fe-ZnO nanoparticles (150 mg/L) irradiated with sunlight intensity of 80,000 ± 3000 lux and pH 9 with different reaction temperature. C_0 represents initial concentration of ciprofloxacin and C represents concentration of ciprofloxacin at a particular time point. C/C_0 denotes the time dependent change in ciprofloxacin concentration with respect to initial concentration.

3.2. Analysis of Residual Antibacterial Activity of Antibiotic after Photocatalytic Degradation

Ciprofloxacin, as already discussed, is an antibiotic that belongs to the FQ class of antibiotics. The antibiotics that belong to this group, generally inhibit the growth of several microorganisms via the inhibition of DNA Gyrase, which is a factor is responsible for the division of bacterial cells. ciprofloxacin is active against a wide spectrum of Gram positive and Gram negative bacteria and ciprofloxacin and antibiotics of the FQ group are widely present in wastewaters such as those from hospital, municipal, pharmaceutical industry sources, etc. [1,22,27]. The residues of these antibiotics in the wastewaters generate antibiotic resistant bacteria in the environment, which is a potential major threat to public health.

The current work aims to employ photocatalysis for the successful degradation of the antibiotic ciprofloxacin. After subjecting ciprofloxacin to photocatalytic treatment with Fe-doped ZnO nanoparticles, a confirmatory bacterial inhibition experiment was conducted to check whether the antibiotic was completely degraded in the experimental system using as test organisms *Staphylococcus aureus* and *Escherichia coli* [19]. The results of the experiments (Table 1 and Figures 6 and 7) showed that for both *Staphylococcus aureus* and *Escherichia coli*, ciprofloxacin lost its antibacterial activity after 60 minutes and 75 minutes post-irradiation, respectively. With increasing time, a decreasing zone of inhibition in both *Staphylococcus aureus* and *Escherichia coli* was evident. The zone of inhibition decreased from 12 mm to 5.5 mm and from 15 mm to 6 mm in the case of *Staphylococcus aureus* and *Escherichia coli* in 60 min and 75 min post-irradiation, respectively.

Table 1. Shows residual antibiotic activity of the antibiotic ciprofloxacin after photocatalytic degradation with Fe-doped ZnO nanoparticles against *Staphylococcus aureus* and *Escherichia coli*.

Test Bacteria	*Staphylococcus Aureus* (10^8 CFU)			*Escherichia Coli* (10^8 CFU)		
Time (min)	PCD ZOI in mm Mean $\pm$ SEM	DC ZOI in mm Mean $\pm$ SEM	PL ZOI in mm Mean $\pm$ SEM	PCD ZOI in mm Mean $\pm$ SEM	DC ZOI in mm Mean $\pm$ SEM	PL ZOI in mm Mean $\pm$ SEM
0	12 ± 0.3	12.5 ± 0.3	12.5 ± 0.2	15 ± 0.3	15 ± 0.2	14.5 ± 0.3
30	12.5 ± 0.3	12 ± 0.3	12.5 ± 0.2	14.5 ± 0.3	14.5 ± 0.2	14.5 ± 0.2
45	10 ± 0.3	8 ± 0.5	11 ± 0.2	11 ± 0.3	14 ± 0.2	$14 + 0.2$
60	7.5 ± 0.2	12 ± 0.5	10 ± 0.2	12 ± 0.4	14 ± 0.2	13.5 ± 0.3
75	5.5 ± 0.2	11 ± 0.3	10.5 ± 0.2	9.5 ± 0.3	14.5 ± 0.2	12 ± 0.5
90	0	11.5 ± 0.5	9 ± 0.3	6 ± 0.2	14.5 ± 0.2	11.5 ± 0.3
105	0	9 ± 0.3	8.5 ± 0.2	0	12 ± 0.3	12 ± 0.3
120	0	10 ± 0.2	7.5 ± 0.3	0	12.5 ± 0.4	10 ± 0.3
135	0	10 ± 0.3	8 ± 0.2	0	14 ± 0.3	9 ± 0.3
150	0	10.5 ± 0.2	7 ± 0.2	0	14 ± 0.2	8.5 ± 0.2
165	0	9 ± 0.3	6 ± 0.2	0	14.5 ± 0.3	7 ± 0.2
180	0	11 ± 0.3	5.5 ± 0.2	0	14 ± 0.2	0
195	0	11 ± 0.2	0	0	14 ± 0.3	0
210	0	12 ± 0.3	0	0	13 ± 0.5	0

SEM stands for standard error of mean, calculated from the standard deviation, PCD-photocatalytic degradation, DC-dark control, PL-photolysis, *n* (number of replicates) = 6. Zone of inhibition (ZOI) = total zone (including the disc)—diameter of the disc (6 mm). The well diffusion assays were performed in accordance with the Clinical & Laboratory Standards Institute (CLSI) Guidelines. No ZOI have been observed from the solvent controls i.e., with distilled water, no contaminating bacteria were found to grow around the treated samples when poured without the test bacteria. Experimental Conditions: catalyst concentration 150 mg/L, pH 9, antibiotic concentration 10 mg/L and temperature 30 °C.

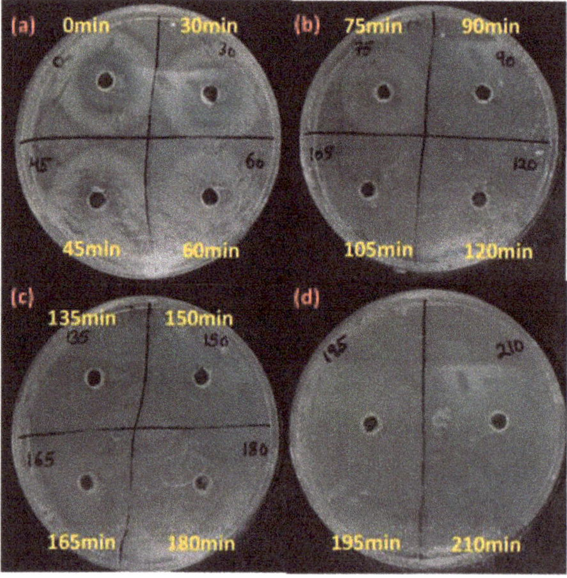

Figure 6. Residual antibiotic activity of the antibiotic ciprofloxacin after photocatalytic degradation with Fe-doped ZnO nanoparticles against *Staphylococcus aureus*. Yellow markings denote the time points at which sampling has been done. (**a**), (**b**), (**c**), (**d**) denotes the zone of inhibition shown by the antibiotic slurry collected at different time intervals.

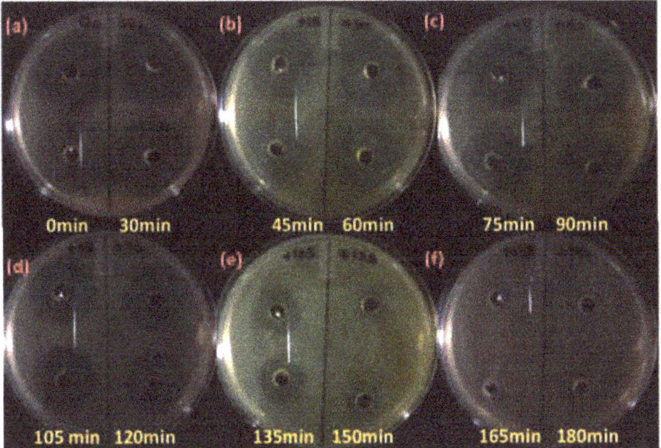

Figure 7. Residual antibiotic activity of the antibiotic ciprofloxacin after photocatalytic degradation with Fe Doped ZnO nanoparticles against *Escherichia coli*. Yellow marking denotes the time points at which sampling has been done. (**a**), (**b**), (**c**), (**d**), (**e**), (**f**) denotes the zone of inhibition shown by the antibiotic slurry collected at different time intervals.

It can be seen that *Escherichia coli*, a Gram negative organism, shows susceptibility to ciprofloxacin that has been collected 75 minutes post-irradiation, which is slightly less than that of *Staphylococcus aureus* (sample collected 60 minutes post-irradiation), before completely showing zero susceptibility in both cases. As a Gram negative microorganism *Escherichia coli* has a weak cell wall that is made up of lipopolysaccharides [28,29]. Therefore it is easy for a disinfecting agent to penetrate its cellular defenses. compared to *Staphylococcus aureus*, which is Gram positive. In the case of the light control and dark control, antibacterial activity was not lost even after 120 min for both *Escherichia coli* and *Staphylococcus aureus*. There was little decrease in the zone of inhibition parameters and it clearly signified that ciprofloxacin was still present in the case of experimental controls, suggesting that both the photocatalyst (Fe-doped ZnO) and sunlight are indispensable in the degradation process.

4. Conclusions

An Fe-doped ZnO nanoparticles-based sunlight-assisted photocatalytic system was developed for the degradation of the fluoroquinolone antibiotic ciprofloxacin in water, assessing its best performance parameters like pH, temperature, and catalyst dosage. The degradation of ciprofloxacin was proved both spectrophotometrically as well as microbiologically by the loss of antibiotic activity of the photocatalytically treated water. The developed Fe-doped ZnO nanoparticles-based photocatalytic system can potentially be used for the degradation of other fluoroquinolones and other antibiotics as well as other organic contaminants in water. Antibiotic residues in aquatic systems have the potential to induce resistance in bacteria, which has further the potential to infect humans and thereby become a serious threat to human health. The developed system has therefore potential to contribute to containing antibiotic resistance.

Author Contributions: S.K.T., A.J.T. and C.S.L. created the original study plan. S.D., S.G. and A.J.M. designed and executed the experiments under the guidance of S.K.T., A.M. helped in the microbiology experiments. S.D., S.G. and A.J.M. wrote the manuscript. S.K.T., A.J.T., and C.S.L. reviewed and edited the manuscript. All authors read and approved the manuscript.

Funding: This work was supported by Department of Science & Technology, Government of India (DST/TM/WTI/2K15/75C) and Swedish Research Council (Grant Number (Grant No. 2012-02889 and 2017-01327).

Conflicts of Interest: The authors declare no conflict of interest.

References

1. Lundborg, C.S.; Tamhankar, A. Antibiotic residues in the environment of South East Asia. *BMJ* **2017**, *358*, j2440. [CrossRef] [PubMed]
2. Walsh, C. Opinion—Anti-infectives: Where will new antibiotics come from? *Nat. Rev. Microbiol.* **2003**, *1*, 65–70. [CrossRef] [PubMed]
3. Paul, T.; Dodd, M.C.; Strathmann, T.J. Photolytic and photocatalytic decomposition of aqueous ciprofloxacin: Transformation products and residual antibacterial activity. *Water Res.* **2010**, *1*, 3121–3132. [CrossRef] [PubMed]
4. Yan, C.; Yang, Y.; Zhou, J.; Liu, M.; Nie, M.; Shi, H.; Gu, L. Antibiotics in the surface water of the Yangtze Estuary: Occurrence, distribution and risk assessment. *Environ. Pollut.* **2013**, *175*, 22–29. [CrossRef] [PubMed]
5. Cantón, R.; Horcajada, J.P.; Oliver, A.; Garbajosa, P.R.; Vila, J. Inappropriate use of antibiotics in hospitals: the complex relationship between antibiotic use and antimicrobial resistance. *Enferm. Infect. Microbiol. Clin.* **2013**, *31*, 3–11. [CrossRef]
6. Larsson, D.J.; de Pedro, C.; Paxeus, N. Effluent from drug manufactures contains extremely high levels of pharmaceuticals. *J. Hazard. Mater.* **2007**, *148*, 751–755. [CrossRef] [PubMed]
7. Selvam, K.; Swaminathan, M. Facile Synthesis of 2-Methylquinolines From Anilines on Mesoporous N-Doped TiO$_2$ Under UV and Visible Light. *Synth. React. Inorg. Met.-Org. Nano-Met. Chem.* **2013**, *43*, 500–508. [CrossRef]
8. Sturini, M.A.; Maraschi, F.; Pretali, L.; Ferri, E.N.; Profumo, A. Sunlight-induced degradation of fluoroquinolones in wastewater effluent: Photoproducts identification and toxicity. *Chemosphere* **2015**, *134*, 313–318. [CrossRef] [PubMed]
9. Rutgersson, C.; Fick, J.; Marathe, N.; Kristiansson, E.; Janzon, A.; Angelin, M.; Johansson, A.; Shouche, Y.; Flach, C.-F.; Larsson, D.J. Fluoroquinolones and qnr genes in sediment, water, soil, and human fecal flora in an environment polluted by manufacturing discharges. *Environ. Sci. Technol.* **2014**, *48*, 7825–7832. [CrossRef] [PubMed]
10. Ikehata, K.; Naghashkar, N.J.; El-Din, M.G. Degradation of aqueous pharmaceuticals by ozonation and advanced oxidation processes: A review. *Ozone-Sci. Eng.* **2006**, *28*, 353–414. [CrossRef]
11. Chong, M.N.; Jin, B.; Chow, C.W.; Saint, C. Recent developments in photocatalytic water treatment technology: A review. *Water Res.* **2010**, *44*, 2997–3027. [CrossRef] [PubMed]
12. Saleh, R.; Djaja, N.F. UV light photocatalytic degradation of organic dyes with Fe-doped ZnO nanoparticles. *Superlattices Microstruct.* **2014**, *74*, 217–233. [CrossRef]
13. Elmolla, E.S.; Chaudhuri, M. Degradation of amoxicillin, ampicillin and cloxacillin antibiotics in aqueous solution by the UV/ZnO photocatalytic process. *J. Hazard. Mater.* **2010**, *173*, 445–449. [CrossRef] [PubMed]
14. El-Kemary, M.; El-Shamy, H.; El-Mehasseb, I. Photocatalytic degradation of ciprofloxacin drug in water using ZnO nanoparticles. *J. Lumin.* **2010**, *130*, 2327–2331. [CrossRef]
15. Das, S.; Sinha, S.; Das, B.; Jayabalan, R.; Suar, M.; Mishra, A.; Tamhankar, A.J.; Lundborg, C.S.; Tripathy, S.K. Disinfection of multidrug resistant Escherichia coli by solar-photocatalysis using Fe-doped ZnO nanoparticles. *Sci Rep.* **2017**, *7*, 104. [CrossRef] [PubMed]
16. Srivastava, V.C. Photocatalytic oxidation of dye bearing wastewater by iron doped zinc oxide. *Ind. Eng. Chem. Res.* **2013**, *52*, 17790–17799.
17. Kerr, J. *Antibiotic Treatment and Susceptibility Testing*; BMJ Publishing Group: London, UK, 2005.
18. Bonev, B.; Hooper, J.; Parisot, J. Principles of assessing bacterial susceptibility to antibiotics using the agar diffusion method. *J. Antimicrob. Chemother.* **2008**, *61*, 1295–1301. [CrossRef] [PubMed]
19. Kusari, S.; Prabhakaran, D.; Lamshöft, M.; Spiteller, M. In vitro residual anti-bacterial activity of difloxacin, sarafloxacin and their photoproducts after photolysis in water. *Environ. Pollut.* **2009**, *157*, 2722–2730. [CrossRef] [PubMed]
20. Achouri, F.; Corbel, S.; Aboulaich, A.; Balan, L.; Ghrabi, A.; Said, M.B.; Schneider, R. Aqueous synthesis and enhanced photocatalytic activity of ZnO/Fe$_2$O$_3$ heterostructures. *J. Phys. Chem. Solids* **2014**, *75*, 1081–1087. [CrossRef]
21. Torniainen, K.; Tammilehto, S.; Ulvi, V. The effect of pH, buffer type and drug concentration on the photodegradation of ciprofloxacin. *Int. J. Pharm.* **1996**, *132*, 53–61. [CrossRef]

22. Gupta, N.; Pandey, P.; Hussain, J. Effect of physicochemical and biological parameters on the quality of river water of Narmada, Madhya Pradesh, India. *Water Sci.* **2017**, *31*, 11–23. [CrossRef]
23. Sturini, M.; Speltini, A.; Maraschi, F.; Pretali, L.; Profumo, A.; Fasani, E.; Albini, A.; Migliavacca, R.; Nucleo, E. Photodegradation of fluoroquinolones in surface water and antimicrobial activity of the photoproducts. *Water Res.* **2012**, *46*, 5575–5582. [CrossRef] [PubMed]
24. Chatzitakis, A.; Berberidou, C.; Paspaltsis, I.; Kyriakou, G.; Sklaviadis, T.; Poulios, I. Photocatalytic degradation and drug activity reduction of chloramphenicol. *Water Res.* **2008**, *42*, 386–394. [CrossRef] [PubMed]
25. Bousslama, W.; Elhouichet, H.; Férid, M. Enhanced photocatalytic activity of Fe doped ZnO nanocrystals under sunlight irradiation. *Optik-Int. J. Light Electron Opt.* **2017**, *134*, 88–98. [CrossRef]
26. Roca, M.; Villegas, L.; Kortabitarte, M.; Althaus, R.; Molina, M. Effect of heat treatments on stability of β-lactams in milk. *J. Dairy Sci.* **2011**, *94*, 1155–1164. [CrossRef] [PubMed]
27. Lamba, M.; Graham, D.W.; Ahammad, S.Z. Hospital wastewater releases of carbapenem-resistance pathogens and genes in urban India. *Environ. Sci. Technol.* **2017**, *51*, 13906–13912. [CrossRef] [PubMed]
28. Sonohara, R.; Muramatsu, N.; Ohshima, H.; Kondo, T. Difference in surface properties between Escherichia coli and Staphylococcus aureus as revealed by electrophoretic mobility measurements. *Biophys. Chem.* **1995**, *55*, 273–277. [CrossRef]
29. Li, J.; Ahn, J.; Liu, D.; Chen, S.; Ye, X.; Ding, T. Evaluation of ultrasound induced damage on Escherichia coli and Staphylococcus aureus by flow cytometry and transmission electron microscopy. *Appl. Environ. Microbiol.* **2016**, *82*, 1828–1837. [CrossRef] [PubMed]

MDPI

St. Alban-Anlage 66

4052 Basel

Switzerland

Tel. +41 61 683 77 34

Fax +41 61 302 89 18

www.mdpi.com

International Journal of Environmental Research and Public Health Editorial Office

E-mail: ijerph@mdpi.com

www.mdpi.com/journal/ijerph